The Significance of Word Lists

DISSERTATIONS IN LINGUISTICS

A series edited by
Joan Bresnan, Sharon Inkelas, and Peter Sells

The aim of this series is to make work of substantial empirical breadth and theoretical interest available to a wide audience.

The Significance of Word Lists

BRETT KESSLER

CSLI Publications
Center for the Study of Language and Information
Stanford, California

Library of Congress Cataloging-in-Publication Data

Kessler, Brett, 1956-
    The significance of word lists / Brett Kessler.
        p. cm.
    Based on the author's thesis (Ph.D.--Stanford University).
    Includes bibliographical references and index.
    ISBN 1-57586-299-9 (cloth : alk. paper) -- ISBN 1-57586-300-6
    (pbk. : alk. paper)
    1. Lexicostatistics. 2. Historical linguistics--Methodology. I. Title.

P143.3 .K47 2001
417'.7--dc21
                                        00-069912

∞ The acid-free paper used in this book meets the minimum requirements of the American National Standard for Information Sciences – Permanence of Paper for Printed Library Materials, ansi z39.48-1984.

Please visit our web site at
http://cslipublications.stanford.edu/
for comments on this and other titles, as well as for changes
and corrections by the authors, editors and publisher.

For my parents,
Lissia and Louis Kessler

Contents

Preface

Historical linguistics has no generally accepted methodology for calculating whether the connections it documents between languages are statistically significant. This lack has led to unusually strident controversies. One linguist spends years documenting evidence of special similarities between languages and calculates the odds as ten billion to one that they are related; another linguist dismisses the evidence as a tissue of coincidences. This polarization is particularly strong between those who accept and those who reject the technique of mass lexical comparison, which groups languages into families by collecting lists of words for the same concept, and noting exceptional similarities. Opponents (traditional comparativists) claim that the kind of evidence mass lexicalists gather is in principle incapable of revealing connections between languages, and may even look with disfavor on the whole idea of comparing lists of words. In this book I step back and look at the challenge from a statistical standpoint: What evidence is available for statistically proving that two languages are historically connected, and how can that evidence be treated in a mathematically convincing and unbiased manner? My conclusions are perhaps surprising. As currently practiced, the traditional comparativist method is much more reliable than mass lexical comparison. But when a rigorous statistical methodology is introduced, the most effective and reliable type of evidence turns out to be something that looks a lot like mass lexical comparison: collecting lists of words and comparing at a fairly superficial level those that name the same concept. The kind of evidence most favored by traditional comparativists turns out to be intractable or less powerful in the context of a statistical experiment. Hopefully this book will not simply antagonize both camps and lead to another schism. Rather, I hope that readers will be convinced that the difference between the two schools really hinges on a rather narrow issue of statistical significance, and that that difference can be bridged by an objective method of testing for such significance.

This is not one of those proposals that claim that traditional linguistic research can be thrown out the window and replaced by a simple numerical formula. Much of the book deals with the havoc that can be wreaked if one simply pulls words out of a dictionary and runs them through a computer without first conducting painstaking linguistic research into the structure and history of each individual word. It is for that reason that I refrain from directly investigating here the specific claims for large language families that are often put forth by mass lexicalists; it is much preferable for specialists in those areas to do so. Nevertheless, to avoid drifting off into the aether of the purely theoretical, I do give many examples of how my methodology would be applied, and what the results would be like, using a suite of eight languages. I also present explicit guidelines to experts who wish to apply my methodology to their own languages. They should be warned, however, that the statistics involved necessarily entail the use of specialized computer programs. Linguists who wish to explore these techniques before investing in a programing project are invited to contact me at bkessler@BrettKessler.com.

This book is based on my PhD dissertation at Stanford University. I would like to thank Stanford University and the Whiting Foundation for their financial support. I gratefully acknowledge the advice and encouragement of my reading committee: Edward Flemming; Thomas Wasow; my advisor, Paul Kiparsky, who inspired me to work on this topic; and Jarrett Rosenberg of Sun Microsystems, who taught me the elegance and power of permutation tests of significance and kept me enthused about statistical approaches to linguistic problems. Some of this material has been presented at talks and workshops at Stanford University and at a Linguistics colloquium at Wayne State University; I thank the sponsors for their support and the participants for their questions and comments. I also thank Rebecca Treiman for her contributions, and Christine Poulin and Willy Cromwell Kessler for innumerable instances of help and advice.

1

Introduction

When a linguist has finished compiling lists of correspondences between languages, there remains the problem of judging whether the amount of evidence is strong enough to prove that the languages are historically connected. One would expect that determining whether the amount of correspondence is statistically significant should be the easiest part of the task, and yet there is no commonly agreed upon method for doing so. Linguists looking at the same data can come up with widely differing appraisals as to whether the case for historical connection has been proved or not. Many debates over the sufficiency of the evidence for hypothesizing language families have raged since the beginning of comparative linguistics, and the furore is just as strong today. There have been for example various versions of the controversial Nostratic hypothesis, which seeks to demonstrate that the Indo-European languages are related to other nearby languages such as the Afro-Asiatic and Uralic families (Bomhard 1990, Collinder 1965, Illich-Svitych 1971–1984, Manaster Ramer 1993, Pedersen 1903). Many people feel there is convincing evidence for linking everything from Turkish to Japanese into one big Altaic family (Katzner 1986). Perhaps most famously, Greenberg claimed that most of the unclassified languages from the Andaman Islands to Tasmania, including New Guinea, are related (Greenberg 1971). He also argued for classifying most of the languages of North and South America in one family called Amerind (Greenberg 1987). With Ruhlen (Greenberg and Ruhlen 1992) he has even given the following data set: Proto-Afro-Asiatic *mlg* 'suck the breast' (Arabic *mlj* 'suck the breast' and Old Egyptian *mnd*^{y} 'woman's breast, udder'); Proto-Indo-European *melg-* 'to milk' (English *milk*, Latin *mulg-ēre*); Proto-Finno-Ugric *mälke* 'breast' (Saami *mielga*, Hungarian *mell*); in Dravidian, Tamil *melku*, Malayalam *melluka*, both 'chew', and Kurux *melkhā* 'throat'; Central Yupik (an Eskimo-Aleut language) *melug-* 'to suck'. The remaining forms are to

1

be referred to a Proto-Amerind *maliq'a* 'swallow, throat': in Almosan, Halkomelem *məlqw* 'throat', Kwakwala *m'lXw-'id* 'chew food for the baby', Kutenai *u'mqolh* 'to swallow'; in Penutian, Chinook *mlqw-tan* 'cheek', Takelma *mülk'* 'to swallow', Tfaltik *milq* 'to swallow', Mixe *amu'ul* 'to suck'; in Hokan, Mohave *malyaqé* 'throat', Walapei *malqi'* 'throat, neck', Akwa'ala *milqi* 'neck'; in Chibchan, Cuna *murki-* 'to swallow'; in Andean, Quechua *malq'a* 'throat', Aymara *malyq'a* 'throat'; in Macro-Tucanoan, Iranshe *moke'i* 'neck'; in the Equatorial family, Guamo *mirko* 'to drink'; in Macro-Carib, Surinam *e'mōkï* 'to swallow', Faai *mekeli* 'nape of the neck', Kaliana *imukulali* 'throat'.

The claim in Greenberg and Ruhlen (1992) is that there is a connection between an immense number of the world's language families whose connection was not formerly believed provable: first, all the language families of Amerind, plus Eskimo-Aleut, Indo-European, Uralic-Yukaghir, Turkic, Mongolian, Tungus, Korean, Japanese, Ainu, Gilyak, Chukchi-Kamchatkan, and apparently also Dravidian and Afro-Asiatic. (And this is by no means the broadest etymology in print; Ruhlen in particular has done a great deal of work proposing etymologies that he believes show that many words descend from a Proto-World, e.g., Bengtson and Ruhlen 1994, Ruhlen 1994:101–124.) In particular, Greenberg and Ruhlen claimed that there is some original entity, approximately *mälgi*, meaning 'suck the breast', that has evolved over time into the forms seen in the table. And one can certainly see the similarities between the words. Phonetically, all the words have /m/ as their first consonant, most also have an /l/ and a velar consonant. And certainly all of the meanings are only a step or two away from the core idea. When a baby *sucks* the *breast*, it *swallows milk* through the *throat*, which is inside the *neck*. If anything, similarities are perhaps more striking there than those one would encounter in a typical page from Pokorny (1959), a standard compendium of cognates in the Indo-European languages. No one doubts the interrelatedness of the Indo-European languages, but in Pokorny one constantly sees connections drawn between words as diverse as Albanian *pesë*, French *cinq*, and English *five*. If those connections are universally accepted, then why not Tamil *melku* and Kwakwala *m'lXw-'id*? And yet this and similar findings have been subject to intense criticism from linguists who believe such similarities are the product of mere chance (Campbell 1988, Hock 1993, Lass 1997, Matisoff 1990, Ringe 1996, Rosenfelder 1998a, Rosenfelder 1998b, Salmons 1992). How exactly does one know whether the words are similar enough to warrant positing a historical connection?

Statistical methodology gives guidance as to how one should go about gathering data, testing hypotheses, and evaluating whether the evidence is significantly different from what one could expect to get by chance. Such methodologies are embraced by most sciences, and are commonly practiced within certain subdisciplines of linguistics, such as certain schools of sociolinguistics and phonetics. However, statistical techniques have not been fully embraced by some other subfields, such as historical linguistics. To a large extent this failure is understandable. Historical sciences necessarily cannot share several important aspects of modern research methodology with synchronic disciplines such as phonetics. One obvious difference is that historical researchers have no opportunity to run experiments in the normal sense of the word; there can be no intervention in past events. Another difference is that historical researchers have much less data to work with. Not only are they unable to generate new data experimentally, but also the working of entropy necessarily implies that, in general, the farther back in time one goes, the less evidence one has. The problem is perhaps exacerbated for linguistics vis-à-vis other historical sciences. Speech is a particularly evanescent phenomenon. In the absence of intentional recording, such as by writing, speech is lost entirely. Furthermore, language is rather more finite than it may seem to a casual observer. I count only some 13,000 different words (including proper names) in the entire King James translation of the Bible, for example, and that number generally counts different inflected and derived forms as separate words (e.g., *walk, walketh, walked, walking; king, kingdom, kingly*). If one wants to take large samples of vocabulary, one must be aware that one is not drawing from an indefinitely large supply of morphemes.

I hold that, despite these limitations, there is a rôle for statistics in historical linguistics. In this study I focus on the question of how to decide whether the internal state of a particular set of languages is such as to justify concluding that the languages were ever historically connected. The focus on internal state means that I will be considering only such evidence as the vocabulary and grammar of the language, and not, in particular, information about the people that spoke that language. The focus on historical connection means that I am primarily interested in factoring out the workings of chance and coincidence; but, having ruled out chance, I do not specifically attempt to work out the details of the exact type of historical connection between the languages. Those are fascinating questions of sufficient complexity that one cannot hope to solve them by introducing a few statistical techniques. There will always be room for painstaking analysis and inference on the part of expert historical linguists.

What I do hope to accomplish is to make some progress toward developing a methodology that can reliably measure the strength of the evidence favoring the hypothesis that a set of languages is historically connected. Such a methodology should have a particular bearing on disputes such as the claim that factors other than chance unify languages as diverse as the purported Nostratic or Amerind languages, or indeed the claim that one can recover vocabulary patterns from an ancestor common to all known languages.

1.1 Types of Historical Connection

The basic linguistic terminology for discussing historical connections between languages can easily be conveyed by way of example. The Roman empire spread Latin throughout Europe. Over the years the common speech in different areas of the empire diverged; all languages continually undergo changes, but any given change is potentially limited in how widely it is adopted among speakers. Many languages, such as French, Spanish, Portuguese, Occitan, Catalan, Italian, and Rumanian, developed as a result of localized changes. They are said to be *related* to each other, because they are all descendants of Latin. Much of their grammar and vocabulary can be explained in terms of incremental change from the Latin ancestor. Elements of related languages whose history dates back to the same element in the ancestor language are called *cognates*. For example, the words for 'tongue' in most of these languages, such as French *langue*, Spanish *lengua*, Italian *lingua*, etc., are cognate, because they are all descended from the same Latin word, *linguam*, which underwent different changes in different speech communities.

More precisely, languages A and B are held to be *related* if for each of the languages there is a chain of speakers, such that the current speakers of the language learned it from other speakers, who learned it from other speakers, and so forth, until for each of those two languages there is a chain reaching back to a single common language C. And words and other elements are held to be *cognate* if they occur in related languages and similarly have a historical chain that links them via the ancestral tongue.

Languages can also be influenced by other languages. Japanese for example has thousands of vocabulary elements that have entered the language as a result of influence by Chinese and English words. It is not however considered to be related to either of those two languages; that is, it cannot be demonstrated that there is a chain of speakers leading back from Japanese and Chinese and English to a common ancestral language. The vocabulary of relatedness is not used for this type of situation; rather, one talks about *borrowing* elements, which then have the status of *loans*. (Some writers have preferred to call them *copies*, or to borrow the term *transfers* from biology, e.g., Lass 1997:121, because after all the receiving language is under no moral obligation to return them.)

To complicate matters, related languages can borrow elements from each other. The French word *virtuose* is a loan from Italian *virtuoso*. A moment's reflection makes it obvious that such loans can make it difficult to identify true cognates. In this case, the French cognate is *vertueux*, which has a somewhat different meaning ('virtuous'). Because the true cognate has been in French a longer length of time than the borrowed word, it has had more time to be subject to changes that took place only in French and not in Italian, and therefore has diverged more from its Italian counterpart. That is a very typical pattern; in general, the rule of thumb is that cognates are less similar to each other than are loans. Or, to state it even more broadly, if words are too similar, they are less likely to be cognates. The situation becomes more striking the more languages have diverged from each other. For example, English and French are related to each other, but virtually any pair of words that the casual observer would identify to be cognate are not. *Paternal* and *paternel* are loan words (both modeled ultimately on Latin); *father* and *père* are true cognates.

In any event, whether language elements share certain properties because they are inherited from a common ancestor language, or whether they share them through borrowing, the languages and the elements in question can be said to be *historically connected.* This study is concerned with distinguishing languages that are historically connected from those that are not. Or more precisely, my interest is in deciding whether patterns suggestive of historical connection are sufficiently strong to be considered proof of such connection. I am not proposing new methods for distinguishing the different types of historical connection from each other, although the reader needs to keep those distinctions firmly in mind, because I will from time to time take advantage of prior work in the field, suggesting that certain collections of evidence should or should not try to exclude loan elements.

This lack of emphasis on the status of loans may be considered a deficiency by some readers. Since the nineteenth century, the driving goal in investigating language relations has been to articulate the pattern of divergences, to draw trees whose branches indicate how languages have split up and developed independently. It is probably fair to say that for most researchers, borrowing has been considered noise in the system, a perturbation that makes it more difficult to discover a neat underlying tree. For them, loans are to be sought out and discarded before the real work can proceed. Nevertheless, there have always been researchers who have been at least as interested in focusing on how linguistic elements are borrowed across different languages and dialects (e.g., Ross 1996, Thomason and Kaufman 1991, Weinreich 1966). In linguistic historiography, this contrast is typified by the distinction between the focus of August Schleicher, whose work emphasized the diverging branches (the *Stammbaum* model), and the focus of Johannes Schmidt, whose work emphasized the borrowing of changes across dialects (the *Wellentheorie*); see for example Campbell (1999:chap. 7), Fox (1995:chap. 6), Hock (1991:sect. 15.3). Dialectologists in particular are often concerned with measuring synchronic distances between dialects (dialectometry), for which purpose the historical source of the similarities and differences may not be immediately relevant (e.g., Babitch and Lebrun 1989, Kessler 1995, Nerbonne and Heeringa 1998, Nerbonne et al. 1999, Séguy 1973). And when one is looking at long distance relationships, very often the status of loans is irrelevant. The main concern is often to figure out whether languages are connected at all, or whether observed patterns are simply a matter of chance.

1.2 Traditional Metrics

Conceptually, the most basic way to tell whether two speech forms are connected is to look at the historical record. There can be no question that the dominant languages of England, Australia, and the United States are closely related. Even if the speech forms had diverged so much as to now be mutually incomprehensible, the continuous historical documentation of British colonization would tell beyond question that the speech forms are not only historically connected, but that they are related to each other as descendants of (virtually) the same English ancestor. Unfortunately such excellent documentation is rarely available and one must fall back on internal evidence, that is, on the nature of the languages themselves.

The first linguistic necessity for exploring such questions is that there be variation in language. That goes without saying, for if everybody's language were identical, there would be no distinct speech forms to study. It is also vital that the variation be at least partially arbitrary. One might for example be able to partition the languages of the world into those that have a native word for 'penguin' and those that do not. But that fact alone would not tell much of interest about the connections between languages. Rather, it is reasonably true that cultures for which penguins are important will have a word for penguins, and those that do not know about penguins will not have a word for penguins. If a group whose language evolved in penguin territory moves far away it may eventually lose its lexical entry for penguins as the speakers stop talking about penguins; and when a culture moves into penguin territory, it will no doubt acquire a word for penguins. Therefore the presence or absence of a 'penguin' word will be a poor predictor of language connectedness; or rather, it will be no better a predictor than, say, geographic proximity.

A less obvious but even more important attribute that linguistic variation must have in order to be diagnostic of language connections is that it have a probability distribution that can be independently established. This is where linguistics hits a major roadblock. It is for all intents and purposes impossible to do experimentation in this area. Consider the simple question of determining the probability that a language will choose a particular predominant order in which to arrange subject (S), verb (V), and object (O) within a clause. It is impossible to imagine a definitive psycholinguistic experiment to answer this question, because it would be difficult if not impossible to factor out the degree to which subjects would be influenced by the predominant word order of their native language. It would be tremendously unethical to take a community of children and raise them away from all external language contact in order to examine the properties of the language they would invent, a fortiori to do such a trial dozens of times in order to get a good probability distribution. And it is not even obvious that a language would be developed and stabilize to mature probability distributions within a reasonable amount of time anyway. Conceivably a word order like OVS would be frequent in newly developed languages but would attenuate after a certain number of generations.

The practical impossibility of experimentation might drive one to try to infer probabilities from the frequency distributions within existing languages (Nichols 1996), but there is no definitive way to do that either. Existing distributions are a function not only of the probability that an isolated language would independently develop a word order like OVS, but also of the probability that that word order would spread by borrowing or by the spread of languages themselves. That is, the entire enterprise is circular: Both factors are unknown, one cannot predict both from each other. To be sure, there are some small corners of the world, mostly in Europe and Asia, where a certain amount of detail is known about language relationships and linguistic change over the course of recent centuries, and where the comparative method can hope to uncover the facts to a certain time depth. But such knowledge is so scanty and so much confined to certain language families that few linguists would be willing so much as to estimate the probability that an SOV word order would change to SVO in a course of a thousand years. And even that kind of detailed knowledge would not much help one to infer the likely probability distributions at time of language origin.

At the other extreme, one might imagine inferring the probabilities by tallying the various word orders in languages that are known to not be connected. However, that idea fails because it begs the question: Very often it is precisely the presence of long-distance relationships that is at issue. Indeed, it is objectively reasonable to hypothesize that *all* human languages are related. It is quite imaginable that language developed once, perhaps in tandem with the development of modern *Homo sapiens sapiens*, and spread around the globe with our species type. Or, even if several language families developed separately, it is also very likely that every family has had sufficient time to influence every other family, perhaps indirectly through other languages. Certainly polygenetic scenarios are also possible, but the point is that one cannot rule out the possibility of monogenesis or mutual influence. And even if one were so bold as to proclaim that language must have originated and developed completely independently in two particular areas—imagining for example that Australia was completely isolated from the rest of the world before languages were developed there—having just two data points is virtually useless. Would anyone be so reckless as to identify dozens of languages as all being culturally isolated from each other, both directly and indirectly, until historical times? For dozens of such languages would be necessary before one could get the sort of probability distributions that are needed.

A simple thought experiment should make this point clear. Imagine some catastrophe that resulted in English being the only language spoken on Earth, and all other languages completely forgotten. Linguists might then study the dialects of English analogously to the way they study separate languages today. They might find a few characteristics that are fairly rare among the dialects of English, perhaps voicing contrasts in glides, like /w/ versus voiceless /ʍ/, as in Scottish and Irish English *witch* vs. *which* (see International Phonetic Association 1996 or International Phonetic Association 1999 for explanation of phonemic symbols). These may be proportionately as rare in that future world as certain word orders are in ours. But there is no way that the rarity of such a property could possibly mean that it is inherently unlikely for a human language to have the property, nor, a fortiori, that it can in any wise be taken as evidence that Scottish and Irish English are completely unconnected with the /ʍ/-less dialects.

As if it were not problem enough that one does not know the probabilities that a language would develop properties like /ʍ/:/w/ contrasts or a particular word order in the absence of historical connections with other languages, there is another problem lurking around the corner. Probably most properties are not going to be so unlikely that they could be considered probative in themselves. If one knew somehow that the odds that a language would develop OVS word order on its own were one in a billion, and found two languages with OVS order, one would be able to state that it seems very likely that the languages were connected. But no doubt most probabilities would not be on that order of magnitude. It would be unreasonable to draw any conclusions about a pair of languages that share a property whose odds were only one in a hundred or even one in a thousand. The only recourse one would have in such cases would be to combine the probabilities of several properties. If languages shared many unlikely properties, then there may be more room to speculate that they are connected. Unfortunately, in order to do that, one would need an additional piece of information: not just the independent probabilities of those properties, but also their joint probabilities. If the probability of languages' having VSO word order were 1/100, and the probability of their placing prepositions before noun phrases were 1/4 (both numbers of course are drawn from thin air), would the probability of their having both be 1/400? No, because multiplication of probabilities crucially applies only when the properties are independent. In this case, Greenberg (1966) has concluded that those two properties are in fact strongly connected. One expects prepositions before noun phrases in VSO languages. The more independent properties are from each other, the more useful they would be, but in any event one would need to know the joint probabilities. And if the probability of individual properties is virtually unknowable, then it is all the more unlikely that one will acquire knowledge of those joint probabilities.

If one cannot get the needed probabilities from experiments or by observing existing languages, there is one remaining source to consider: a priori knowledge. If one understands well enough how language is put together, one can model the causal factors, and from the resulting model compute the probability of various language configurations. It is conceivable that someday our understanding of psychology, the sociology of communication, the biology of the speech apparatus, and other relevant fields of knowledge will be great enough that one could calculate and weigh all the factors that would work towards a language's structuring its word order so that the object comes before or after the verb in a clause. Unfortunately our current state of knowledge barely lets us conceive of what might go into such a calculation, much less plug in appropriate numbers.

There is however one important area in human language in which most linguists agree that the a priori probabilities are available, although the concept is rarely expressed in such terms. Even among those who believe that the human organism contains an innate language device that strongly constrains the possible forms that language can take, there is widespread acceptance of the thesis that the word used to denote a concept in any given language is arbitrary. This is codified as De Saussure's First Principle (Saussure 1989, 1:151–157), which is called *L'arbitraire du signe*. Of course, the expression for a certain concept will be constrained by the phonological pattern of a particular language. If one were told that the Mandarin Chinese word for 'head' is /qof/, one would wisely be skeptical, inasmuch as that language has no uvular consonants or word-final obstruents. Nor is language arbitrary in the sense that any speaker can freely make up any phonetic string to represent a particular concept. But vocabulary is arbitrary in the sense that words within the same language are theoretically exchangeable. Presented with the words for 'black' and 'white' in a language that is unrelated to any language one knows, one should not be able to determine which is which, with any accuracy better than guessing.

De Saussure's First Principle was well accepted long before De Saussure stated it, and forms the basis for all traditional metrics for determining whether there is a historical connection between languages. One checks whether there is more of a connection between words of similar meaning than between other pairs of words. If the word for 'sun' in Language A is more closely connected to the word for 'sun' in Language B, than, say, 'sun' in Language A and 'rock' in Language B; and so forth for several other concepts; then there would seem to be a historical connection. Otherwise, the arbitrariness of the sign should mean that the word for 'sun' in Language A should have no stronger a connection with the word for 'sun' in unrelated Language B than for the word for any other concept. Of course there will be exceptions, because even for related languages, words may be replaced over time and there may be sporadic shifts in pronunciation or even unrecognizably severe sound changes. But the relation should hold on the whole, across many different words. The exact nature of the metric is determined by how that vague word *connection* is defined in practice.

No doubt the oldest family of metrics for uncovering connections between languages is what I call *similarity*. Broadly speaking, languages are held to be connected if they are especially similar to each other. For example, researchers may group together languages that have a particularly high number of similarities between words or morphemes for the same or similar concepts. This appears to be the general approach followed by Jones, whose observations on the unity of Indo-European languages are credited as the great inspiration for modern work in historical linguistics. His demonstration that Latin, Greek and Sanskrit were historically related took the form of aligning verb roots and inflec-

TABLE 1   Words for the Numbers 1 Through 10 in Several Indo-European
Languages

A	B	C	D	E	F	G	H
een	eins	en	en	un	um	un	un
twee	zwei	två	to	deux	dois	dau	deu
drie	drei	tre	tre	trois	três	tri	try
vier	vier	fyra	fire	quatre	quatro	pedwar	peswar
vijf	fünf	fem	fem	cinq	cinco	pump	pymp
zes	sechs	sex	seks	six	seis	chwech	whegh
zeven	sieben	sju	syv	sept	sete	saith	seyth
acht	acht	åtta	otte	huit	oito	wyth	eth
negen	neun	nio	ni	neuf	nove	naw	naw
tien	zehn	tio	ti	dix	dez	deg	dek

Note: Languages are Dutch, German, Swedish, Danish, French, Portuguese,
Welsh and Cornish. Expected grouping: A–B (West Germanic), C–D (North
Germanic), A–D (Germanic), E–F (Romance), G–H (Brythonic Celtic).
Source: Hogben (1965:155).

tional endings of the same function, and pointing out how similar they
were (Nichols 1996). The metric still has strong support among some
researchers. Greenberg and Ruhlen, for example, work exclusively in
this metric. They claim that it is particularly appropriate for large-scale
investigations. After a researcher has gathered vocabulary items from
several languages and aligned them by meaning, it is particularly easy
for the human eye to pick out striking similarities between languages
(Ruhlen 1994). Examples such as that of Table 1 illustrate their contention. Even the untrained eye can discern connections such as those
between the Romance or West Germanic languages.

Similarity metrics are often strongly denounced by theoreticians of
linguistic methodology (e.g., Lass 1997:chap. 3, Salmons 1992), so it may
have surprised or even shocked the reader that I am taking them seriously. And yet the logical validity of the similarity metric is almost too
obvious to invite expatiation. It is based crucially on the observation that
not all linguistic elements necessarily change, or change very much, over
time. For example, it is universally agreed that Proto-Indo-European,
the theoretical ancestor of most of the European and many Southwest
Asian tongues, had a sound /m/, which has remained largely unchanged
in all descendant languages, after some 6,000 years of linguistic evolution. In those languages that have kept the Proto-Indo-European word
for 'mouse' at all (*mūs-*), the current form is very recognizable: Danish

mus, German *Maus*, Russian *myš*ʹ, Latin *mūs*, and so on, not to mention English *mouse*. And even where /m/ does change, the outcome is often quite similar to an /m/. For example, Greek changed word-final /m/ to /n/, which is still a nasal stop (Classical Greek *agrón* next to Latin *agrum*, 'field (accusative)'). The idea is that even after lengthy periods of time, historically connected languages should manifest more similarity between semantically matching lexical items than should unconnected languages.

Why then are similarity metrics frequently dismissed? One small reason has to do with the status of loans. As we saw earlier, loans are not germane to linguistic relatedness, yet they tend to be even more similar (*paternal, paternel*) than are true cognates (*father, père*). They can therefore be particularly confusing when similarity metrics are used. A larger part of the reason that similarity metrics tend to be dismissed out of hand is because the dismissers are thinking of entire methodologies that use similarity metrics, rather than of the similarity metric itself. It is true that linguistic arguments that consist primarily of remarking on subjectively noted similarities between word lists are in themselves not convincing. On the other hand, it is too hasty to say that nobody can ever develop a rigorous methodology that exploits the fact that languages that can be shown to be historically connected are more similar in some respects than are other languages. The problem lacking in most studies is that people are not working with precise definitions of similarity that are established in advance (Rosenfelder 1998b). That does not have to be the case. Bender (1969), for example, for the most part used explicit definitions of similarity in counting how many CVC correspondences occur by chance in a typical lexicostatistic experiment. Oswalt (1970) brought the idea to fruition in his similarity-based shift test. He determined in advance exactly what would constitute similarity and what would not, then applied the method uniformly across a preselected corpus.

The other important metric is *recurrent correspondences*, or *recurrence* for short. This metric reflects the basic procedure of the comparative method, which was introduced by Bopp and developed in the course of the nineteenth century (some good general textbooks on this methodology include Anttila 1972, Campbell 1999, and Hock 1991). It looks for unusually high numbers of matches between sounds in semantically matching morphemes. This is a little more subtle than similarity, but can be easily illustrated by an example (Table 2). By this metric, the fact that German *haben* sounds like the Latin *habere* and has the same meaning, 'have', does not count as evidence that German and Latin are related. What *does* count is that there are many German words beginning in /h/ that have the same meaning as Latin words in /k/.

TABLE 2 Recurrent Correspondences in German and Latin

Gloss	German	Latin
'heart'	*Herz*	*cord-*
'horn'	*Horn*	*cornū*
'dog'	*Hund*	*canis*
'hundred'	*hundert*	*centum*
'deer'	*Hirsch*	*cervus*

By the recurrence metric, it is irrelevant whether or not /h/ and /k/ share any properties. *Any* pair of sounds could constitute evidence of historical connection. The justification for this metric is the observation that changes in pronunciation tend to spread throughout the entire lexicon of a language, to affect all identical sounds. For example, some speakers of English may be aware of the fact that words spelled with *wh* used to be pronounced differently from words spelled with *w*. The former was unvoiced, the latter voiced. But for most speakers of the accents in England and overseas, all the former *wh* have now changed pronunciation and are pronounced just like *w*. The important point is that all words with *wh* changed their pronunciation, not just a few select words. The idea that all words are equally affected by sound changes is such an important doctrine in historical linguistics that it is often called simply the *regularity hypothesis* (Hock 1991:chap. 3). There is an important proviso to this hypothesis: The phonetic environment can condition the change. For example, in the southern United States, /ɛ/ has changed to /ɪ/ throughout the vocabulary, but only if the next sound is a nasal consonant, such as /m/ or /n/. Linguists since the nineteenth century have placed great emphasis on regular changes of this sort, whether conditioned or not. As a corollary, recurrence was held to be the most important metric for determining historical connections between languages, and it still is by the great majority of historical linguists. Again, the deceptively simple expression *comparative method* is a technical term in linguistics, and properly applies only when the recurrence metric is used.

Both of the metrics are substantially valid, but both have their drawbacks. The criterion of recurrence can fail to detect connections in certain cases. Some changes simply are not as regular as one might like. For example, it is impossible to say exactly what phonetic environment conditions the correspondence between General American /æ/ and English Received Pronunciation /ɑ/ in words like *dance, aunt, grass, plant*. Certainly a following cluster of /n/ plus consonant, or a fricative, is involved, but there are many exceptions, such as *lass, ass, ant.* Such exceptions are generally explained as the result of borrowing from other dialects after the completion of a sound change, although some linguists subscribe to a theory of lexical diffusion, which holds that sound changes often spread gradually and potentially irregularly through the vocabulary (see Labov 1994 for a discussion). Whatever the cause, such exceptions can be difficult. Another problem is that the conditioning environment can be rare. Many speakers pronounce *government* as /gəmənt/, at least when speaking fast and informally, but even if the changes involved in deriving this from earlier /gəvərnmənt/ are totally regular in some theoretical sense, there do not seem to be any other words that enter into any regular correspondence set with *government* (i.e., deletion of /vərn/). And at any rate, linguistic science has always recognized that there may be irregular or sporadic changes of pronunciation in language. These are simply not dignified by the technical term *sound change.* Correspondences such as /æsk/ versus /æks/ for *ask* (metathesis), /koˈvərt/ for earlier /ˈkəvərt/ (*covert,* by analogy with *overt*), /ˈbetə/ for earlier /ˈbitə/ (*beta,* a learned approximation of the Classical Greek vowel), all reflect well-recognized processes of linguistic change that are not expected to lead to recurrent correspondences. (See also Durie and Ross 1996 for discussions of several other problematic types of irregularity.) Nevertheless, the variants are obviously connected, and could potentially be recognized by a similarity metric.

The problem with a similarity metric is that its power would seem to decline faster than the recurrence metric, as the time distance between the languages increases. By *power* I mean its ability to detect a connection that really exists. The longer two or more languages have been separated, or rather, the more changes that have occurred in any of the languages involved, the less similar they will be. That almost goes without saying. In a system with sufficient room for movement, random changes in localized parts of the system will in general lead to increasing dissimilarity between the parts. A single example should suffice to illustrate this point. Proto-Italic and Proto-Hellenic, two daughters of Indo-European, apparently had exactly the same word for 'mother', approximately /maːteːr/. While in most of the descendants of Proto-Italic the first vowel has remained essentially the same in quality (Latin /maːter/, Italian /madre/), in Proto-Hellenic it has been gradually rising: Homeric Greek /mɛːtɛːr/, Classical Greek /meːteːr/, Modern Greek /mitera/. Nowadays the first vowels of /madre/ and /mitera/ have little similarity to each other. But the recurrence metric still has a chance of finding something, because there survive a significant number of other words containing the same vowel (e.g., both vowels in Italian /fama/,

Greek /fimi/ 'fame').

The difficulty is not the absolute amount of similarity that remains, but the amount of similarity relative to the prior variation in the system. The change of Proto-Hellenic /aː/ to Modern Greek /i/ is pretty trivial when considered as a gradual creeping up one side of the vowel triangle. But when one considers that the vowel changed from the lowest position to the highest, from a presumably central position to a front position, from a long vowel to a short (or at least, not distinctively long) one, one sees that virtually all distinguishing characteristics of a vowel have here been changed. Arguably the only distinctive property left is that it is not round. And of course vowels change in roundness all the time too. Perhaps the most important recent vowel change in North American English is that the low back vowels /ɒ/ as in *caught* and /ɑ/ as in *spa*, which previously differed only in rounding, are no longer distinguished for large numbers of speakers. If one wanted to propose a similarity metric that would be able to identify /aː/ and /i/ as connected and also /ɒ/ and /ɑ/, that metric would essentially have to say that *all* vowels are similar to each other. That is the same as saying that differences in vowels are irrelevant to historical linguistics. And that is what methodologies based on similarity metrics tend to say. For instance, Oswalt (1970) grouped all vowels into the same similarity set.

While that result may accord with Voltaire's apocryphal definition that etymology is a science where the vowels count for nothing, the same sort of considerations can quickly demonstrate that consonants count for very little. The Proto-Indo-European */p/ became /p/ or /β/ in Spanish (*padre* 'father', *cabra* 'goat'), /f/ or /v/ in English (*father, seven*). Already that just about covers the gamut of labial obstruents. Nor is it hard to find examples of obstruents becoming sonorants, or consonants changing place of articulation, or even vowels becoming consonants or vice versa. Any similarity class powerful enough to detect the important connections between even a handful of languages separated by several millennia will probably include *all* known sounds.

One way to avoid that paradoxical situation, where the identity of sounds is completely irrelevant, would be to have different similarity measures between different pairs of sounds. The sound /t/ could count as *very* similar to /d/, *somewhat* similar to /l/, and *barely* similar to /i/. Or one could just accept that some of the similarity sets will not be all-inclusive, and say for example that there is one consonant set for each place of articulation. On either approach, some possible connections simply will not be found or count as important, but one hopes that there will be enough sounds that have stayed the same or similar enough that the overall linguistic connection will be found. For example, even if one fails to catch that Tahitian /t/ has the same origin as Hawaiian /k/, one hopes there will be enough pairs like /p/:/p/ to allow one to uncover the fact that the languages are related.

In this study I use the recurrence metric. The main consideration is that the main use of a statistical methodology will be in exploring languages whose connection is not obvious. It simply seems likely that in such cases one will be dealing primarily with potentially large changes, so that much if not most of the information may be moderately obscure. And logically, if one does accept the regularity hypothesis, for which there is a tremendous amount of evidence (at the very least, all linguists would agree that a very large number of phonetic changes take effect regularly throughout the lexicon), recurrent correspondences of the most dissimilar sounds *should* count as much as recurrent correspondences of similar sounds. If all the words that start with /t/ in one language start with a sound as dissimilar as /i/ in another, there must be something going on. It is never particularly difficult to imagine a short sequence of small and natural steps that could account for such a correspondence. But while virtually all proponents of similarity metrics would recognize the validity of recurrent sound correspondences, there would probably be some difficulty in getting many linguists to accept the results of a study that used similarity alone. Imagine for example that it could be shown that for an unusually high number of meanings, the corresponding words in languages A and B both begin with labial consonants. If that consists of five words that match /p/:/b/ (i.e., /p/ in one language corresponds to /b/ in the other), that would probably satisfy everybody. But if that evidence consisted of a large number of unique matches, such as one /p/:/p/, one /p/:/b/, one /b/:/p/, one /p/:/m/, etc., then a large number of linguists would doubtless remain unconvinced of the connection, even if the results were equally significant statistically. That sort of reticence might not be justifiable—perhaps the evidence *should* be accepted as probative—but all things being equal (that is, provided other tests are not shown to be more powerful or accurate), it is wise to use a metric that has maximal intuitive appeal.

2

Statistical Methodology

2.1 Traditional Methodologies

The recurrence metric is not normally applied in the context of a rigorous methodology (Cowan 1962). It may be conceded that there is something in the metric itself that discourages unfounded claims. Recurring exact matches are much rarer than words that are merely similar, unless one has a very precise and restrictive definition of similarity. Even a handful of recurrent correspondences may prove to be significant, whereas a huge list of "similar" words, which may strike the imagination quite forcefully, could well be coincidental. Rosenfelder (1998a, 1998b) offered long lists of words that have similar meanings and pronunciations in languages as remote (or unrelated) as English and Chinese, just to show how easily it can be done. Hock (1993) cited other experiments in this vein. Nevertheless, there is more to rigor than picking a good metric. No matter what the metric is, there is the question of whether its value as observed in any particular study is significantly different from what one might get with any random collection of data. Is Table 3, with six /s/:/h/ correspondences, sufficient to prove that English is related to Hawaiian? Would a dozen correspondences suffice? What if they were joined by a half dozen recurrences for several other pairs of sounds? One sometimes reads claims that one must look for pairs of words that have two (or three) consonants that participate in recurrent matches (Campbell 1973, Hock 1991:558, Meillet 1925, Nichols 1996). Certainly such evidence would be more probative, but it does not really answer the question. Would one such pair of words be sufficient? Is it actually impossible to prove anything based on correspondences involving a single consonant?

Although one occasionally encounters informal rules of thumb, and, more rarely, actual research to support them (e.g., Bender 1969), for the most part the traditional methodology has been silent on this is-

17

TABLE 3 Multiple /s/:/h/ Correspondences Between English and Hawaiian

English	Hawaiian
sew	*humu*
smell	*honi*
snow	*hau kea*
stab	*hou*
star	*hōkū*
swell	*hoʻopehu*

sue. In one sense this silence is wise. Certainly there can be no general rule that will hold in all cases. It is intuitively obvious that three recurrences among very frequent phones in the two languages must count for less than three recurrences involving fairly rare sounds. Consequently recurrences involving languages like Hawaiian, which have small sound inventories (and therefore many words containing each phone), would in general be less probative than recurrences for languages like Georgian, which have much larger inventories (and therefore fewer words per phone). No single number will work for all pairs of languages.

More importantly, though, traditional historical work does not proceed along lines that invite statistical analysis. In a canonical statistical study, it is important that one select in advance (i.e., before even looking at the languages) the attributes that one is going to study and the metric for measuring those attributes. Furthermore, one needs an unbiased method of selecting the units whose attributes are going to be studied. But historical linguists are used to looking *everywhere* for evidence of language relatedness. It can be fairly said that a distinguishing characteristic of the historical linguist is a particular fascination with details and the surprises that are to be found in the dark corners of a language. How could they bring themselves to overlook the evidence of the peculiar genitive plural in the word for 'sister-in-law', just because it was not randomly selected as a unit to be considered in somebody's statistical study?

In the best cases, it is difficult to come up with a good list of comparanda for a statistical study of languages. What does it mean to compare genitive plural endings when one language may have a genitive plural ending, another has a genitive plural prefix, another has separate morphemes for plural and genitive, another always conflates genitive and accusative cases, another has postpositions for indicating case, another uses word order? What does it mean to ask if 'be' has a suppletive paradigm, when one language may have two words that translate 'be', another has none at all, a third only has two inflections for verbs and a fourth has 164? If even these simple (and typical) questions defy a truly satisfactory solution, the task of developing a complete a-priori statistical model that would precisely accommodate any observation a linguist might come up with is clearly far beyond hope.

One might argue that even if it is difficult to decide on all these comparanda before the fact, one can at least come up with some elegant comparanda that fit the data after one has looked at it. Perhaps for a particular pair of languages one finds that the ablative and dative cases have the same form in the plural. What would be the harm of throwing in that observation a posteriori? It turns out that the harm would be considerable. Indeed, as a rule of thumb, one might even generalize that the use of such a posteriori observations is almost guaranteed to result in the conclusion that two languages are connected. In any system where there is a lot of variation, such as language, coincidences are to be expected. As long as one has the freedom to sift through large amounts of information, ignoring comparanda that "do not work" and selecting comparanda that do, one can easily prove any assertion. Perhaps the only way to work around such problems is to know what the possible universe of comparanda is. One could put some bounds on the predictions by assuming first that all the ignored comparanda point in the direction of language connectedness, and see how those figures work out, and then by assuming that they all point in the direction of unconnectedness, and see how the figures work out then. Unfortunately, the very notion that one can count all possible comparanda is equivalent to the notion that one can mathematically model all of human language in the first place: at the very least, a practical impossibility.

So the traditional lack of any actual guidance on how much evidence is "enough" in comparative linguistics is appropriate, given the actual methodology that is applied in practice. The only reasonable way to add some degree of mathematical precision to the comparative method is to achieve a rapprochement with the typical methodologies used in more experiment-oriented disciplines: unbiased preselection of attributes, metrics, and units of study, even though that may mean the discarding of interesting observations. Doing so with the best available reliability will be the goal of this book. However, I in no way wish to imply that other practices have no place in historical linguistics. For one thing, not all studies have to be mathematically precise. There is plenty of room for presenting observations and argumentation that cannot be shoehorned into a rigorous statistical study. More importantly, the mere fact of demonstrating historical connectedness is just a small part of historical linguistics. By far the richer part of the field is concerned with studying the types of relations that can be found between languages already determined to be related. My ambition is to develop methods that can make that determination more reliable, so that linguists may more confidently pursue the larger tasks of historical and comparative linguistics as a whole.

2.2 Tests for Association

We wish to find out whether there is a true connection between two languages, or whether the sound recurrences we find should be chalked up to chance. In statistical terms, this amounts to looking for a correlation, or an association, between variables. There are so many excellent books about statistical methodology in general (e.g., Hays 1981, Woods et al. 1986) and association statistics in particular (e.g., Agresti 1990, Wickens 1989) that it seems a shame to state here what has been better stated elsewhere. But it would also be a shame to ask readers unfamiliar with statistics to put down this book until they have reviewed the literature of that field, especially since the core concepts required in this book are surprisingly few and simple. I'll give here a very basic example drawn from sociology. Sometimes our own discipline can be understood a little more clearly by stepping back and seeing how problems analogous to ours would be handled in other disciplines.

2.2.1 Contingency Tables

Suppose we were interested in finding out whether, among women, there is any association between having children at home and working outside the home. That is, we would like to see whether women who work outside the home are more or less likely to have children at home, or whether those two conditions are totally independent. The way to start would be to select a sample of women to get information about. Perhaps it should go without saying, but in fact it bears frequent repeating, that the sample should be selected in a way that would not bias the sample toward answering the question one way or the other. Obviously it would make no sense to conduct our interviews at the annual convention of the fictitious National Association of Working Mothers. But other potential dangers may not be as obvious. Interviewing every adult woman in one's neighborhood may fail if the neighborhood happened to have an especially high number of people from a class or religion that had particular views on these issues. An excursus on how samples are selected in sociology research would take us a bit far afield, but suffice it to say that for many types of questions it is commonly considered appropriate to randomly generate a list of telephone numbers and interview the people who answer at those numbers, provided they fit the basic assumptions of

TABLE 4 Number of Women Categorized by Whether they Work Outside
the Home and Have Children at Home

		Children		
		Yes	No	Sum
Work	Yes	349	325	674
	No	169	362	531
Sum		518	687	1205

Source: Newport (2000).

the research: in this case, that they are adult women. When the Gallup
Organization did just that across the United States, they came up with
the numbers in Table 4, approximately (I have converted back from reported percentages).

Table 4 is an example of a contingency table. It simply provides the
tally of people or objects that are cross-categorized by certain characteristics, or variables. These raw numbers, or observations, can be hard to
interpret at a glance. Is there an association or not? This question can
be answered by comparing what would happen if there were no association between having children at home and working outside the home,
and we had interviewed 1,205 women whose answers gave us the same
sums as in our table. What should the numbers in the four Yes × No
cells be, if 518 of the 1,205 women have children in the home and 687
do not, and if 674 of the same 1,205 women work outside the home,
and the other 531 do not? We would expect the same proportions to
hold up within any subcategory as hold up across all categories. Thus if
the sums tell us that 674 out of 1,205 (55.9%) women work outside the
home, we would expect 55.9% of women with children to work outside,
and 55.9% of women without children to work outside. So if 518 of the
women have children at home, 55.9% of those 518, or 290, would work
outside, if there were no association between children and working. Of
course one could have just as easily arrived at that number by reasoning
from the other category. 518 out of the 1,205 women have children, or
43.0%. So 43.0% of the 674 women who work outside the home should
have children, or 290. This reasoning can be reduced to a mechanical
rule. To find out what count a cell of a table would be expected to have
under the assumption of no association, multiply the sum of its row R by
the sum of its column C, and divide that product by the grand total of
the table, N, i.e., RC/N. In Table 5 I have added these expected values
to the observed values. Now it is obvious that there is an association between those variables in the sample of women. Among working women,

TABLE 5 Expectations for the observations in Table 4

| | | | Children | | |
			Yes	No	Sum
Work	Yes	Observed	349	325	674
		Expected	290	384	
	No	Observed	169	362	531
		Expected	228	303	
Sum			518	687	1205

more have children at home than one would expect from the sums: 349 were observed instead of the expected 290.

It is also useful to be able to express just how unexpected the observed data is. Intuitively, the unexpectedness of 349 vis-á-vis 290 is a function both of the difference (59) and the size of the expected value; a value 59 greater than expectation would be a lot more important if the expectation were 10 rather than 290. One standard way of expressing this unexpectedness is the formula for deviance: One squares the difference and divides by the expectation, $(E - O)^2/E$, where E is expectation and O is observation. For example, the deviance for the cell representing the number of women who work outside the home and have children at home is $(290 - 349)^2/290 = 12$. Summing up these deviances for all four cells gives a measure of the unexpectedness of the table, namely, approximately 48. This measure of deviance is called Pearson's χ^2 statistic.

The number 48 might not mean much in itself, but it is useful when one is comparing across different tables. As an overall measure of deviance, it is useful in comparing across contingency tables of the same shape (2×2), even if the exact expectations or even the grand total N are different. If one did a similar study on Swedish women and came up, hypothetically, with a χ^2 of 20, then, regardless of the N or the individual expectations or observations, we could say that the American study presents stronger evidence for an association between working and having children than does the Swedish study.

2.2.2 Significance

But there is a wrinkle in all this that may not be at all obvious at first glance, but which is paramount to researchers. Statisticians ask first and foremost how significant the observations are. *Significance* is a technical term in statistics. It is the probability that the deviances from expectation could be due to chance. To get a better intuition of this, let us back off and consider an extremely simple case. We know that if we take a fair coin and give it a good flip and let it land, it will come up heads half the time and tails half the time. It follows therefore that if we toss a coin two times, our expectation is that we will get one head and one tail. But what if we got two heads? Do we immediately conclude that the coin must be loaded, or that the laws of probability must be amended? Not at all. Each possible outcome of a series of coin tosses has a particular probability: We had simply predicted the most likely outcome (one head), but others are possible. In particular, the probability of getting two heads is .25: The probability of the first head is .5, and the probability of the second head is .5, and the probability of two independent events occurring is the product, .5 × .5. If you did the experiment (two coin tosses) four times, you would probably get two heads in one of the experiments. So it is not a particularly shocking outcome. In the same way, the expected frequencies we worked out for the survey are simply the most probable values that one would expect, given the sums, if the two variables are independent. But other outcomes are certainly possible. If one got the observations 291, 383, 227, 304 instead of the expected 290, 384, 228, 303, one might well intuitively understand that one should not make too much of the fact that the observations differed from expectation.

The sad fact is that all outcomes are possible. If one threw a coin even a dozen times, it could come up all heads. Even if there were no association between working and having children, one could, just by accident, come up with a large sample where all of the women who work outside the home have no children—as far from expectation as is possible. There is never any way to absolutely prove anything from a sample beyond a shadow of a doubt. So researchers have to accept that they are reasoning under uncertainty. In any experiment drawing from a sample, there are two possible explanations that could account for any difference from expectation. Either the expectation is wrong, and there is an association between the variables (and women who work outside the home really are more likely to have children at home); or, the expectation is right, there is no association, and the difference from expectation is due to chance. This sounds tautological, and it is, except for one detail. If we can quantify just what the probability of such a chance event is, then we are saying something interesting. If one gets heads twelve times in twelve coin tosses and is told that the probability of that happening is one chance in five thousand, then one begins to get a feel for how much one should be impressed by that feat. On the one hand, it could be a chance event that would happen about once every five thousand experiments. On the other hand, the lower the probability of a chance event, the higher the probability that something else is going on. One might begin to take more seriously the idea that the coin tosser is telekinetic, or cheating.

The goal of statistical science is to work out ways of figuring out exactly what the probability of chance is for various kinds of experiments. This probability is usually abbreviated p in most papers. Obviously our goal is to design experiments where we have a good chance of getting a low p, and the lower the p is, the more confident we will be about reporting our results as valid. Of course it is up to human judgment to decide just how convincing a particular value of p is. At what point do we decide that the probability that our results are chance events is low enough that we are willing to publish our results? Different fields have different conventions. In the social and behavioral sciences, one is traditionally content with a p of .05; the results are accepted if there is no more than one chance out of twenty that the results are accidental. Such a number reflects both the fact that there is much variation in human behavior, making it hard sometimes to get firm numbers, and that if one guesses wrong about the significance, people are not likely to die. If in the field of medicine one is testing the safety of a recreational drug, one would probably want to look for much lower p values, such as .001, accepting perhaps no more than one chance in a thousand that the experimental results are wrong.

There are, by the way, two ways of reporting p. The more traditional approach is for the researcher to decide in advance that any p no higher than, say, .05, will be considered significant. The researcher then checks to make sure that the probability is indeed within that range, and reports the fact: $p \leq .05$. Although there may be theoretical reasons for wanting to report things this way, the practice was certainly abetted by the fact that for many types of experiments, it has traditionally been much easier to determine whether one's results fall on one side of a particular p level than to work out the exact p value. Nowadays, the use of computers makes it just as easy to give exact results, e.g., $p = .032$. I prefer this type of reporting myself and will use it in this book, because it is more informative, and gives the reader the option of applying his or her own criteria. If the author likes cutoffs of .05 but the reader likes .01, both can be satisfied if the exact value is reported. Of course it is still useful for the author to state whether individual tests are significant by the author's own explicitly stated criterion. Incidentally, p is often called the significance level of a test. This is perhaps an obvious name, considering its use. But it is a little unfortunate in that the values go the wrong way: The lower the significance level, the more significant the experimental results are. This often confuses people new to statistics.

This leaves only one question: How does one compute p? Statisticians and mathematicians have worked out ways of computing p for different types of experimental designs. Sometimes that computation is straight-forward and exact. An experiment of coin tosses is one such case. If one has a series of N trials (coin tosses) where each trial has two possible events r (heads) or s (tails) and the probability p of getting an r on each individual trial is known (.5), then one can compute the probability that the series will have a specific number of r events (heads) by plugging the values into the binomial equation: $(N!/r!(N-r)!)p^r(1-p)^{N-r}$ (Hays 1981:119). If for example one tossed a coin 10 times and got 9 heads, one would plug in the numbers like this: $(10!/9!(10-9)!).5^9(1-.5)^{10-9}$, and end up with .009766. That is, there is slightly less than 1 chance out of a hundred that this experiment would have given exactly 9 heads.

That is close to the answer we are looking for, but there is one additional step. One is not really interested in the probability of getting a particular outcome such as exactly 9 heads. In most experiments, the odds of a particular outcome are going to be pretty small. Even the most likely outcome, getting 5 heads, has a probability of only .246, less than one chance in four. If we had just a few hundred trials (coin tosses) in an experiment, every possible outcome, even the one with 50% heads, would have a probability of less than .05. Obviously this number is not exactly what we are looking for. Rather, we want to find the probability of getting results that are *at least* as far from expectation as the observations we got. For the coin toss experiment, if our goal is to find out what the probability is of having gotten so many more heads than the expected 5, what that means is that we must calculate the probability of getting 9 or more heads. In this case it is easy. The only two possible cases are 9 heads and 10 heads, so we apply the binomial formula for each, then add the results together: .009766 + .000977 = .010743. And *that* is the p value we would report for the experiment. If we do the same thing for the case where we get 50% heads in any experiment of coin tosses, no many how many, we will always get a p greater than .5, which clearly has no significance whatsoever.

So an experiment based on coin tosses, or any analogous process, can be exactly evaluated for p. What about contingency tables such as the one we started out with above? They too can be handled exactly. The binomial equation does not apply because there are more than two possible events. More subtly, it is also invalidated by the fact that the probabilities do not remain constant throughout the experiment. Once we have assigned a particular woman to a particular cell of the table, then the odds that the next woman will go into that cell are slightly decreased (because we know the sums for the row and column that cell is in), and the odds that she will go into some other cell are slightly increased. Both of these problems can be handled by means of the hypergeometric formula (Hays 1981:136). In practice, however, applying this formula (which requires many factorials) is very difficult and may in some cases even tax computers. Consequently, most researchers use an approximation called the chi-squared distribution. The insight is that if a table is large, the χ^2 statistic discussed above is a pretty good predictor of p. Instead of grinding away at the hypergeometric formula, one can simply look up the χ^2 value in the tables printed at the back of almost all statistic handbooks. The only complication is that the p values depend not only on the χ^2 value, but also on the degrees of freedom of the table. In a nutshell, the degrees of freedom are the number of cells one could vary freely and still have enough cells left over so that one could adjust them so as to make sure the sums turn out right. For example, in our 2 × 2 table above, after we change the value in one cell, we would have no freedom to tinker with the values in the other three cells if we want the sums to stay the same; there is only one value they could all take. So there is just one degree of freedom. The general formula for the degrees of freedom for a table with R rows and C columns is $(R-1)(C-1)$. If we look for the χ^2 value 48 in a chi-squared distribution table under one degree of freedom, we will probably find it is off the chart. Even a χ^2 of 10.828 corresponds to a p of .001, and our χ^2 value is much higher than that. Consequently, by virtually anyone's criterion, the association we found between working outside the home and having children at home is significant.

Although the chi-squared distribution is an approximation to what in principle could be computed exactly, people use it quite freely and without any hesitation, provided that the assumptions hold. One key assumption is that of independence of the data points. That can be a difficult concept to grasp because there are many ways in which independence could be violated. But here is just one example. What if the people doing the poll decided they had not sampled enough women and went back and did some more sampling a month later to get more numbers to add to the table? And furthermore they had forgotten who they had asked before, and ended up polling many of the same women twice? That could seriously skew the results, because the answers a woman gives a month later are likely to be very similar to the ones she gave originally. That could end up making the results look much firmer than they are, because the variation would not rise as fast as the number of data points. In general, any kind of relation where one observation could help predict another invalidates the use of the chi-squared distribution.

The other major thing to watch for is that the chi-squared distribution works best for large numbers. As the size of the table gets smaller, the approximation gets less and less accurate. This is true both for the size of the table in terms of its dimensions (numbers of rows and columns) and in the numbers that go into each cell. In fact, 2×2 contingency tables themselves can give rather inaccurate results; some researchers try to tinker with the way they calculate χ^2, in an attempt to lower it in compensation; others ignore the problem. But everybody is concerned by the problem of low cell counts. Once expected frequencies start to get below about 5, inaccuracies start mounting up; expectations below 1 can be very serious. Tables with a high percentage of cells with low expected frequencies test as more significant (have a lower p) than they would if calculated, say, by the exact hypergeometric formula. In our sample table, we have no difficulties with the size of the expected frequencies. It is to be sure a 2×2 table, which tends to increase the significance (i.e., lower the reported p). But our p value is so low that we would be foolish to worry about the small effect that that problem adds. Clearly there is an association.

So far I have discussed two main approaches to significance testing. One is using formulas, which are exact, sometimes difficult to compute, and fairly mysterious to people who do not like to spend a lot of time thinking about mathematics. The other is using standard distributions, specifically, the chi-squared distribution, which are approximate, especially so when small numbers are involved; easy to compute, because one just looks them up in tables; and even more mysterious to the lay person. A third possibility is to explicitly model what the very meaning of chance is. In the poll we have been discussing, the raw observed data are in the form of Table 6. One way of thinking about chance is this: If there were no association between working and having children, then the entries in the last column could have been arranged in any order with respect to the data in the middle column. One possible chance arrangement would be just like the one we observed: The "Children" column would have the entries corresponding to observed 1, 2, 3, 4, 5, 6, 7, 8. Another chance arrangement might be 2, 1, 3, 4, 5, 6, 7, 8, which would of course look identical. Another might be 5, 2, 3, 4, 1, 6, 7, 8, and that would make a difference. In that arrangement, we would have more "Work Yes" paired with "Children Yes", and more "Work No" paired with "Children No" than we actually observed. If we continued this exercise until all possible arrangements were tried, we would fully exhaust the possibilities that chance could throw at us. Therefore, we could exactly compute the probability that chance would lead to a χ^2 that is at least as great as the χ^2 we observed. That number would simply be the fraction of those rearrangements that have a χ^2 at least as great as the observed one. There is no complicated math involved at all. One rearranges a col-

TABLE 6 Hypothetical Data Points for Table 4

Respondent	Work	Children
1	Yes	No
2	Yes	No
3	Yes	Yes
4	No	No
5	No	Yes
6	Yes	Yes
7	No	No
8	Yes	Yes
...		

umn of the raw data every possible way. For each of those arrangements, one fills in a contingency table and computes the χ^2. If the χ^2 of that arrangement is greater than or equal to the observed χ^2, one makes a tally mark. At the end of the exercise, one divides the number of tallies by the number of arrangements, and one has the probability p.

Such a procedure is called a permutation test. It is not often applied in everyday research, but it is statistically unexceptional (see Good 1995 for a much fuller explanation). In fact, it is the theoretical goal to which things like the chi-square distribution are striving. The only problem with it is that most experiments use so many data points that finding all the possible arrangements would require the fastest computer in the world to compute for the entire lifespan of several universes. That is seldom practical. One way around this is to employ a Monte Carlo technique. Instead of working with every possible rearrangement, one works with a very large number of them. If those rearrangements are randomly selected, then the answer will very closely approach the true one. Of course we are now back in the realm of approximations, but this technique has several advantages over approximations like the chi-squared distribution. One is that the nature of the approximation is easier to understand. Its variance is $p(1-p)/N$, that is, inversely proportional to the number of rearrangements one tries (Good, p. 155). So the more arrangements one tries, the closer one is likely to get to the true answer. Another advantage is that some of the requirements of the chi-squared distribution, especially that of large numbers, no longer apply. And finally, the idea of the permutation test extends easily to other tests of association, and even to types of tests for which no one has ever worked out any other method for determining significance. We will look more into those possibilities later in this book. For now, the important thing is that enumerating all chance combinations is a very direct and indeed a common-sense way of determining just how unlikely is the association that one has observed.

I would like to close this discussion of statistical generalia with a few general warnings. First, it is important to be very careful in drawing conclusions from the fact of association. One is tempted to conclude, "Women who have children living at home need to go out and earn extra money to care for them", or "Women who work outside the home feel a particularly strong need to have children". These conclusions may or may not be true, but the fact that variables are associated does not mean there is a causal effect, and it certainly does not mean that any explanation that pops into one's head is correct. Often the cause, if indeed there is one, is external to the data. For example, one possible explanation for the association is that women tend to have jobs and have children at about the same age. College age women, and elderly women, often have neither jobs nor at-home children.

Another misunderstanding that sometimes arises is to confuse p with strength. It is perhaps natural to assume that a very low p means that the association between the two variables must be very strong. This is natural because it is often true. Strength of the association can be one factor that causes one to get a large χ^2 and therefore a small p. But on reflection, one can see that another cause of a large χ^2 is simply large numbers. From the formula $(E - O)^2/E$, it follows that the deviances, and hence the total χ^2, that are reported are proportional to E, the expectation. Thus if two tables have observations about 10% different from observation, but one has low Es and the other has high Es, the second one will have the greater χ^2. One can easily verify that assumption with a few tests. Pretend for example that the Gallup organization had only polled a tenth of the number of women, and had gotten exactly the same results, proportionately (i.e., counts of 35, 33, 17, and 36 instead of 349, 325, 169 and 362). The χ^2 in this case would be much lower, 4.57 instead of 48. That would not even satisfy the .01 significance level. But the strength would have logically been the same. One popular measure of strength of association for 2 × 2 tables, Pearson's index of mean square contingency, $\sqrt{\chi^2/N}$, divides the χ^2 by the grand total N in an attempt to correct for that effect of sample size (Hays 1981:556). In this book I will not however report strength statistics very often, because my main concern will be reporting how certain we are that two languages are connected, not how closely connected they are.

Perhaps a more important lesson to take from that thought experiment is that taking larger samples does increase the significance of experiment results. All things being equal, it pays to look at more data. Of course not all things are always equal. It may be too expensive to get more data, or (especially in historical research) there might not be any more data, or additional data may be less reliable. For example, if the only way we could increase the poll sample size would be by including information that people gave us about their next-door neighbors, we might prefer to demur.

In the course of this book we will have occasion to see examples of how association statistics have been attempted in historical linguistics. In many cases, it will be startling to see just how far many of these attempts are from the standard methodology as exemplified by the Gallup poll. These variations may be alternately amusing and confusing. Confusing because it always behooves us to entertain the possibility that new ways might be correct, if we can just figure them out. Amusing because very often the aberration is as basic as getting one's polling data at the National Association of Working Mothers. Methods often seem absurd once one sees the parallel with mistakes that one could make in fields that people have sharper intuitions about. But it is useful to remember that linguists are not exploring these variations out of sheer perversity. As I have mentioned before and will mention again ad nauseam, it is unusually difficult to do standard statistical analysis in historical linguistics, where data is sparse, notoriously interdependent, and not amenable to experimental manipulation. We cannot go back and add a few dozen well chosen words to the Proto-Indo-European vocabulary.

2.3 Statistical Approaches to Historical Linguistics

Most past work in using statistical or mathematical methods in historical linguistics has been in pursuit of a rather different question from the one posed here. It starts out with the assumption that languages are known to be related, then asks just how closely they are related to each other. Usually, the interest in that question stems from a desire to construct a genetic tree showing how the current set of languages diverged from each other. If Languages A and B are structurally more similar to each other than either is to Language C, then it is inferred that Languages A and B shared a common history longer than either did with C. The first major Western treatment of such matters (the authors credit Czekanowski 1928 as providing inspiration) was Kroeber and Chrétien (1937), who attempted to show the subgrouping of Indo-European languages by computing a tetrachoric correlation on phonetic and morphological features known to be different in different sets of languages. Ross (1950) also approached the same problem in a similar way, but looking at the lexicon. One problem with both studies was that their sources had serious biases. For example, Kroeber and Chrétien drew their data from

Meillet (1922/1950), who presented data specifically selected to prove the existence of such groupings as Italo-Celtic and Northwest European (the National Association of Working Mothers fallacy). Swadesh (1950) introduced a bit more rigor into the endeavor by formulating a standard list of comparanda that most subsequent researchers followed, though with certain adjustments over the years (by Swadesh 1955 the list had been trimmed from 225 to 100 words). This innovation is analogous to Gallup's use of randomly generated telephone numbers, in that it guards against selecting evidence that is biased for (or, as happens much more rarely, against) one's hypothesis. In addition, he adduced a hypothesis that words on his list were, in general, replaced (primarily via semantic shifts) at a constant rate, so that knowing how many cognates two languages had in common among concepts on his list would enable one to conclude when the languages separated from each other. Thus was born the very prolific field of glottochronology, which, according to its adherents, made it possible to date the branchings of a genetic tree, or even generate the tree itself, simply by counting cognates (see Embleton 1986 for a broader review of such research). The method assumed that the researcher already knew which words were cognates, information that is often hard to come by; in practice, many researchers simply decided that similar words were cognates, and would even cite high percentages of similar words as proof that languages were related. Such abuses of glottochronology gave it a bad reputation, over and beyond the ill repute attributable to the shakiness of its key assumption that all basic words are replaced at a constant rate in all languages. Glottochronology is also referred to as lexicostatistics, although in careful usage the latter refers more broadly to any statistical manipulation of word lists (Campbell 1999:177). The offshoot known as phonostatistics computed similarities between languages by measuring the phonetic distance between the sounds in semantically matching words, as Grimes and Agard (1959) did for the Romance languages. More recently, researchers knowledgeable in cladistic methods used in biology have taken very sophisticated computerized approaches to reconstructing genetic trees based on features of related languages (e.g., Warnow 1997, Warnow et al. 1996).

Such studies have almost all been based on measures of similarity between words, or even on explicit lists of differences between languages. A notable exception is Guy (1980), who used counts of recurrent phoneme matches, computing a χ^2 statistic on a contingency table matching phonemes from one language with phonemes from another, in words of the same meaning. He took many statistical liberties which one would not wish to emulate in a study emphasizing statistical significance, but his heuristics are practical and free the researcher from the need to know in advance which forms are cognates.

More relevant to this study is research that addresses whether languages are related in the first place, but that literature is not as well developed. Collinder (1947) typifies the probabilistic family of approaches. He took thirteen similarities he had found between the Uralic and Altaic languages, showed that the probability of that particular constellation of details was vanishingly low, and concluded that the languages were all related. Hymes (1956) argued that Tlingit must be related to the Athapaskan languages because the odds against their sharing a certain sequence of verb affixes in the same order was 1,216,189,440,000 to 1. Greenberg and Ruhlen (1992) offered the following demonstration that data provided at the beginning of this book prove that all of the American Indian languages involved were related: Assuming that the consonants in the words cited were drawn from 14 equally frequent candidates, and assuming that the first consonant is /m/, the second /l/ or /r/, and the third /k/, /k'/, /q/, or /q'/, the odds that two words would match are $1/13 \times 2/13 \times 4/13 = .004$. And for that to hold up across six language families, the odds are $.004^5$ or one in ten billion. Despite the presence of several arithmetic errors in that demonstration which are surely typographical, one can discern that the basic strategy is to demonstrate that their tableau of words could have been completely different (in each of billions of possible scenarios), and so must be special. Nichols (1996) also contains many examples of such reasoning. For example, she offered the fact that the reconstructed Proto-Indo-European stem *widhew-* 'widow' contains four consonants (/w/, /j/, /dʰ/, /w/) in a certain order. The probability of a specific word having those four consonants in a particular order is very low (.00000625). Nichols multiplied this by the number of languages in the world to get a number less than .05. From this she concluded that any language in the world that has a word meaning 'widow' with those four consonants must be Indo-European. In all these examples, the use of probabilities and even significance levels suggests statistical reasoning, but in their overall structure, the examples are really not very much like the Gallup poll. The main difference is that instead of collecting large amounts of data in an unbiased manner, these approaches zero in on the most compelling cases or striking coincidences, as if one had indeed chosen to poll at the National Association of Working Mothers. In the same vein, the very test metric is typically invented after the evidence is uncovered. Furthermore, these kinds of calculations at best give the probability of the current state of the world. But almost any state of the world will be extremely rare when looking at things as complicated a language, in the same way that getting 50% heads would be rare in an experiment involving a few hundred coin tosses. Other differences from statistical procedure are found in various individual cases, such as reasoning from the number of languages in the world (the significance of the Gallup poll does not depend on the size of the population of the United States) and determining the probability of cooccurrence of features in unrelated languages by counting their frequency among languages not provably unrelated. While some of this research has its merits, with Nichols' work in particular serving to show what kinds of comparative evidence are more compelling than others, such computations are best classified as probabilistic rather than statistical, and it would be a mistake to treat the probabilities offered as being commensurable with significance levels in a statistical experiment.

More explicitly statistical attempts appeared in some of Swadesh's work (e.g., 1956), which were extended by Justeson and Stephens (1980). But to my knowledge the first full-scale methodology was that of Oswalt (1970). He provided a fixed (and therefore relatively unbiased) set of words (a Swadesh list) for researchers to find in the two languages; had them compare the words based on phonetic criteria well defined in advance; and offered a reasonably accurate way to test for significance of the findings (the shift test). It is unclear to me why his methodology has not gotten more attention than it has; a rare exception is Villemin (1983), who used it in one of his tests for connections among Japanese, Korean, and Ainu. Perhaps one reason is that the shift test is very tedious without the aid of a computer, which was not a common desktop item in 1970. The situation may change in this era of the personal computer, especially after the favorable mention in the text and exercises of Trasks's textbook (1996).

2.4   Tests Using Sound Recurrence

Ross (1950) was the first person to propose a solid statistical technique for proving historical connections between languages using sound recurrence. Ross's key idea was that De Saussure's First Principle implies that no attribute of a morpheme will help one predict what its meaning is. So if one knows that an English word begins with a /d/, that does not help one figure out what the word means. And the other way around, if one knows a concept, the probability that the word for that concept will begin with a /d/ is the same as for any other word; namely, it is just the proportion of all words that begin with a /d/ (about .045). Similarly, the probability that a German word will begin with, say, /t/, would be just the proportion of all words that begin with /t/ in German (about .025). From there it is a very small step to conclude that for any specific concept, the probability that the English word for it begins with a /d/

and the German word for it begins with a /t/ should be the product of those two probabilities, or .0011, provided the languages have selected their vocabulary independently, that is, provided that the languages are not historically connected. If in an unbiased sample of concepts, it turns out that the observed counts of /d/:/t/ pairs is very different from .0011 of all words, then one may suspect that English and German are historically connected. And of course the same reasoning should apply to all other pairs of sounds. One can build up a large contingency table for each sound that can occur in a particular position, one language being the rows of the table, the other being the columns. The cells in the table, for example, row *d* and column *t*, tell how many concepts are expressed by a word that begins with /d/ in English and by a word that begins with /t/ in German. In case a particular concept could reasonably be translated several different ways in a given language, some explicit criterion is selected for choosing between them. Almost any criterion would work, such as picking the most commonly used word, as long as the criterion is not biased with respect to the research hypothesis. For example, it would be a serious error to pick an English word because it is the one that sounds most like the German word, or because it maximizes the number of sound correspondences.

By way of concrete example, Table 7 gives the counts that one would get by applying Ross's technique to 200 word-pairs of English and German, languages that even casual observers perceive as being very closely related. (In order to make the size of the table more manageable, I borrow an idea from Ringe 1992 and group initial vowels together under the symbol ∅, i.e., the absence of a consonant.) One can see by inspection that the distribution is very different from expectation, that is, from the numbers one would expect from working out the probabilities in the manner just discussed. To continue the /d/:/t/ example, one would expect .0011 of the 200 words to have that pair, that is, that it would be surprising to get even one such match. Instead, 3 word pairs matched /d/:/t/. By way of comparison, Table 8 gives counts for a comparison of English and Hawaiian, languages that are much more distantly connected. (They are not known to be related, but they have borrowed a few words from each other.) There it is harder to see any patterns by inspection. The biggest numbers tend to be in the rows or columns for the most frequent initial sounds, such as English /s/, which is what one would expect by chance. To take a sample calculation: From the marginal totals, one would have expected about 5.9 words to begin with English /s/ and Hawaiian /ʔ/ ($33/200 * 36/200 * 200$), and in fact 6 such word pairs were observed. For English /s/ and Hawaiian /h/, one would expect 4.45 pairings, and 6 were observed.

TABLE 7 Initial Consonants in English and German

Eng	Ger																		Sum
	ʃ	f	Ø	v	h	g	z	k	b	n	d	r	m	l	ts	t	j	pf	
s	12	0	1	0	2	2	10	2	0	1	1	2	0	0	0	0	0	0	33
f	0	14	0	1	0	0	0	1	1	1	0	0	0	0	0	0	0	0	18
h	2	0	1	1	8	1	0	1	0	0	0	0	1	0	0	0	1	0	16
w	0	2	1	10	0	1	0	0	0	1	0	0	1	0	0	0	0	0	16
Ø	0	0	14	1	0	0	0	0	0	0	0	0	0	0	0	1	0	0	16
b	2	1	1	1	0	1	0	1	6	0	0	2	0	0	0	0	0	0	15
n	0	0	1	1	1	0	0	1	0	7	0	0	1	0	0	0	0	0	12
d	4	0	0	0	1	1	0	0	0	0	0	0	0	0	0	3	0	0	9
r	0	2	0	1	0	0	0	0	0	0	0	5	0	0	0	0	0	0	8
l	0	0	0	0	0	0	1	0	1	0	0	0	0	6	0	0	0	0	8
m	0	2	0	0	0	0	0	0	1	0	0	0	4	0	0	0	0	0	7
t	1	0	0	1	0	0	0	0	2	0	0	0	0	0	3	0	0	0	7
k	1	0	0	1	0	0	0	3	0	0	0	0	0	0	1	1	0	0	7
g	0	1	0	0	0	5	0	0	0	0	1	0	0	0	0	0	0	0	7
θ	0	0	0	1	0	0	0	0	0	0	4	0	0	0	0	0	0	0	5
j	0	0	1	0	0	1	0	0	0	0	1	0	0	0	0	0	1	0	4
p	2	0	0	0	0	0	0	0	0	0	0	0	0	0	1	0	0	1	4
ð	0	0	0	0	0	0	1	0	0	0	3	0	0	0	0	0	0	0	4
ʃ	1	0	0	0	0	0	0	1	0	0	0	0	0	0	0	0	0	0	2
tʃ	0	0	0	0	0	0	1	0	0	0	0	0	0	0	0	0	0	0	1
v	0	0	1	0	0	0	0	0	0	0	0	0	0	0	0	0	0	0	1
Sum	25	22	21	19	12	12	12	11	11	10	10	9	7	6	5	5	2	1	200

Note: Frequency of initial consonant pairs of words matching in the Swadesh 200 list.

What Ross's procedure yields, of course, is a contingency table, just like in the Gallup poll example, although with somewhat more rows and columns. From this familiar station one might assume that the next steps—calculating a deviance measure and its significance—would be straightforward. But it is not, for a couple of reasons. For one thing, one can tell at a glance that in any of these tables, so many of the expected values would be so low that not even the most devil-may-care researcher would dare to apply the chi-squared distribution. And in Ross's day, without computers, exact tests or even Monte Carlo approaches to permutation tests were too laborious to consider. Ross himself had an additional concern. For comparative linguists, the presence of exactly one correspondence (such as the sole /f/:/ʔ/ correspondence in Table 8) counts just as little as the presence of zero correspondences (such as the complete lack of /h/:/ʔ/ correspondences). Chi-squared statistics certainly do not work that way. To address that concern, Ross developed an

TABLE 8 Initial Consonants in English and Hawaiian

Eng	Haw									Sum
	ʔ	∅	h	k	m	l	p	w	n	
s	6	4	6	4	3	3	3	2	2	33
f	1	2	6	2	2	2	2	1	0	18
h	0	1	1	5	0	3	3	2	1	16
w	2	3	1	3	3	0	1	3	0	16
∅	2	5	3	0	2	3	1	0	0	16
b	5	3	1	2	1	0	1	0	2	15
n	4	2	1	2	0	1	2	0	0	12
d	3	1	0	1	2	2	0	0	0	9
r	3	2	0	1	0	0	2	0	0	8
l	2	1	1	0	1	3	0	0	0	8
m	1	0	0	1	3	0	0	1	1	7
t	1	1	2	0	0	2	0	0	1	7
k	1	2	2	0	1	0	1	0	0	7
g	2	0	2	0	2	0	0	0	1	7
θ	0	0	0	1	2	0	0	1	1	5
j	2	0	0	0	2	0	0	0	0	4
p	0	1	1	0	0	0	2	0	0	4
ð	0	0	0	2	0	2	0	0	0	4
ʃ	1	0	0	0	0	0	1	0	0	2
tʃ	0	0	0	1	0	0	0	0	0	1
v	0	0	0	0	0	1	0	0	0	1
Sum	36	28	27	25	24	22	19	10	9	200

Note: Frequency of initial consonant pairs of words matching in the Swadesh 200 list.

immensely complicated alternative equation, presented it without proof, and apparently found it too cumbersome to apply himself. Without a workable way of determining statistical significance, the data of the sort that Ross accumulated was useless, and so his proposal languished.

Despite that failure, Ross must be credited with the first attempt to apply serious statistical techniques to evaluating the significance of recurring sound correspondences. He established in advance the sample he worked with, rigorously defining the metric he was going to apply, and did so blindly, without getting distracted by compelling coincidences along the way. Then he took that results of the study and discussed how likely it was that they happened by chance. Crucially, that calculation was to be done within the context of that entire sample, which was selected in a way that would not bias the results in favor of (or against)

TABLE 9 Contingency Table for /d/:/t/ matches

English	German		Sum
	/t/	Not /t/	
/d/	3	6	9
Not /d/	2	189	191
Sum	5	195	200

the metric that he was looking at.

One of the few researchers to attempt working within Ross's general framework was Villemin (1983), who compared Japanese, Korean, and Ainu. Villemin improved on Ross's method by using a predefined word list (200 words from a list used in Swadesh's early work) to build up a large contingency table with the general appearance of Table 7. Villemin ignored Ross's work on the statistics of significance, and instead substituted a series of χ^2 tests, one for each pair of matched sounds. For example, if Villemin had been analyzing the data in Table 7, then for the pair /d/:/t/, the 2×2 contingency table in Table 9 would have been set up. Villemin would then compute Pearson's χ^2 statistic for the table, then compare that to the standard χ^2 distribution for 1 degree of freedom. If the statistic was found to be significant at a certain significance level, then Villemin would conclude that the languages were related. The same test was repeated for all possible pairs of sounds.

Ringe (1992, 1993, 1995) used the same basic strategy as Ross (1950) and Villemin (1983), but introduced yet a third measure of significance. Ringe proposed that one look at a cell in the table as if it obeys the probability laws that apply to 100 coin tosses (he used the Swadesh 100 word list). That is, for each of the 100 words, count a type r event if the pair of words has an /d/:/t/ match, and a type s event if it does not, then apply the binomial formula to see if the phoneme pair has significantly more words than expected. The two events are not equiprobable as they are for coins, but the binomial distribution parameterizes for unequal probabilities, as shown in the equation (p. 25). Ringe's book-length treatment finally brought the issue of statistical analysis to the attention of the linguistic community, and it contained many thoughtful insights and suggestions concerning what variations of the basic procedure may be useful and valid. Therefore I will have many occasions to refer to Ringe's work in these pages. If at times it appears that I am taking aim at his work rather often, it is because it is by far the worthiest target.

In the following chapters I deal with some problems inherent in the

Ross-Villemin-Ringe procedure, and offer some suggested corrections. Often I will illustrate the points with real data. I based this data on that used by Ringe (1992, 1993). The first reason for doing this was to ensure that any empirical difference between our results could be attributed to the methodology and not to incidental differences in word selection. But I also admired the quality of Ringe's data and the generosity with which he made it available. In the same spirit I list my version of the data set in the appendix. For reasons that will be discussed later, instead of Ringe's 100-word list (based on Swadesh 1955) I started from a longer version of the Swadesh word list (Swadesh 1952), which has 200 words. Ringe (1992) gave Swadesh-200 words for English and Latin, and I filled in the rest from other sources. For French, I relied on the native knowledge of Christine Poulin. For the other languages, I relied on bilingual dictionaries with English listings, attempting to find the closest semantic match, but otherwise preferring the first-listed of multiple alternatives. For Albanian, I used primarily Kiçi and Aliko (1969), reinforced by Drizari (1957), Kiçi (1976), and the grammar of Kacori (1979); for German, Roy (1970); for Hawaiian, Pukui and Elbert (1986); for Navajo, primarily Young and Morgan (1992), with help from Haile (1926), Hoijer (1945), Kari (1976), Young and Morgan (1980), and Young and Morgan (1987); for Turkish, İz and Hony (1955), with help from Hony and İz (1984). As will shortly become clear, the exact inflectional form in which words are cited is not very important, but to ensure some consistency, I tried to cite words in a common form chosen to be fairly unmarked across languages. Nouns are given in a nominative singular; in Navajo, words that require possessor inflection were given in the form appropriate for 'someone's'. Adjectives are given in a masculine singular nominative. Verbs are given in an indicative present tense third person singular (a few Navajo verbs are cited in first person when I could not readily verify the third person form). Within an inflectional category, I preferred longer phonological forms to shorter ones; in practice, this meant choosing liaison forms for French. In choosing words, strenuous attempts were made to avoid being influenced by known relationships with the vocabulary of other languages, and for this reason ordinary working dictionaries were used instead of historical or comparative ones. But of course subtle psychological experimenter bias is hard to totally disavow in situations where the selection process is not absolutely objective. And it should be noted that there is a certain force toward systematic bias in the fact that all the words are produced as translations of the English word. When synonyms occur in the target language, there is the real possibility that the lexicographer (or informant) will have tended to give a word that sounds like the English word, or perhaps will have

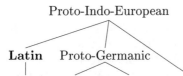

FIGURE 1 Genetic relationships between the languages in the sample.

done just the opposite and avoided words that sound similar.

The eight languages represent a variety of commonly-encountered language relationships (Figure 1). As has been mentioned, French is a descendant of Latin, and has been perceived as a separate language at least since the ninth century. In the etymological discussions I occasionally distinguish the Classical Latin of literature (as represented in the Latin word list; I follow Ringe in citing Latin words in approximately the form in which they would have been used by Cicero in the first century BC) from the Popular Latin which was the direct ancestor of spoken French. German and English are also very closely related to each other; their common ancestor, Proto-Germanic, was probably spoken in the early centuries of the Common Era. Latin and Proto-Germanic, in turn, are descendants of Proto-Indo-European, a language which was spoken perhaps six thousand years ago. Albanian is a descendant of the same Proto-Indo-European language. The connection between those three branches of the Indo-European family is much less apparent than that between Latin and French or between German and English. The remaining languages, Turkish (a Turkic language), Hawaiian (a Polynesian language in the Austronesian family), and Navajo (an Athapaskan language), are not commonly thought to have any particular known relationship to each other or to the Indo-European languages, although various claims have been advanced over the years that there are very ancient connections. On the other hand, there have been connections between these languages other than those devolving from their relatedness. The five Indo-European languages have influenced each other in several directions. But, at least in the word lists being considered here, there is negligible influence between the Indo-European and the three non-Indo-European languages, nor among the latter.

Table 10 gives a rough measure of how closely these various languages are connected to each other. No precise quantification is widely accepted. For present purposes, I adapted a technique used by glottochronologists—I simply counted what fraction of the words in the Swadesh 200 list have the same origin as their counterpart in the other language.

Swadesh (1952), for example, reported that English and German have 59% to 65% of their vocabulary in common, depending on which lists were used. One difference in my procedure, however, was that I counted relatively secure etymologies twice as heavily as more tentative ones. The table gives the overall closeness judgment as a number scaled between 0 (no pairs of words have the same source) and 1 (all pairs securely have the same source). It is important to note that even though the measure of closeness is here defined mathematically, it essentially distills human judgments about which words are cognates or loans, as gleaned from many etymological sources (for Albanian, my primary sources were Huld 1984, Janson 1986, Mann 1977, and Çabej 1982; for English, Barnhart 1988, Simpson and Weiner 1989, and Watkins 1985; for French, Bloch and von Wartburg 1964; for German, Kluge 1995; for Hawaiian, Blust 1993 and Pukui and Elbert 1986; for Latin, Ernout and Meillet 1979 and Sihler 1994; for Navajo, Young and Morgan 1992; for Turkish, Dankoff 1995, Räsänen 1969, Stachowski 1986, Steingass 1930 for Persian loans and Wehr 1979 for Arabic loans; for the Indo-European languages, Pokorny 1959.

TABLE 10 Cognate Levels for Language Connections

Languages		Cognates
Albanian	English	.130
Albanian	French	.203
Albanian	German	.152
Albanian	Hawaiian	.000
Albanian	Latin	.230
Albanian	Navajo	.000
Albanian	Turkish	.005
English	French	.285
English	German	.593
English	Hawaiian	.005
English	Latin	.292
English	Navajo	.000
English	Turkish	.000
French	German	.253
French	Hawaiian	.000
French	Latin	.565
French	Navajo	.000
French	Turkish	.000
German	Hawaiian	.000
German	Latin	.290
German	Navajo	.000
German	Turkish	.000
Hawaiian	Latin	.000
Hawaiian	Navajo	.000
Hawaiian	Turkish	.000
Latin	Navajo	.000
Latin	Turkish	.000
Navajo	Turkish	.000

Note: Human judgments as to how closely connected languages are, expressed as the ratio of the Swadesh word pairs that are believed cognate; numbers closer to 1 represent closer connections.

3

Significance Testing

Reviewers have already pointed out that Ringe's approach to significance testing is not correct because the trials (individual consonant pairs) do not have a constant probability. If one concept matches English /s/ with German /z/, then there is one less /s/ and /z/ respectively for the next pair of words to use. Instead of the binomial distribution, one needs to use the rather more complicated hypergeometric distribution. Ringe (1996) acknowledged this problem but declined to correct it because he considered the difference to be minor. And indeed, statisticians not infrequently use the binomial function as a computationally more tractable approximation to two-category hypergeometric computations, when the sample size is large (Winkler and Hays 1975:226). But Baxter and Manaster Ramer (1996) showed how disastrous the difference can be by looking at a boundary case: What if each of the 100 words in each of the two languages began with a different consonant? In this case, the only possible arrangement in any chance combination would be for each pair of consonants to occur only zero or one times, that is, each of the 100 consonants in language A will co-occur with exactly one of the consonants in language B, but with none of the other 99 consonants. So, logically, no outcome based on relative frequencies could be statistically different from any other, so they all must be nonsignificant. But by Ringe's test, the probability of each of those unique pairings is $1/100 \times 1/100 = .0001$, and the probability of such a rare event occurring once in 100 independent trials by the binomial theorem is just less than 1%. Yet in this case that must happen 100 different times, so each of the 100 word pairs will independently testify that the relationship between the languages is statistically significant, a remarkably strong but completely spurious showing.

Another obvious issue is how to coordinate different findings in different cells. Both Villemin and Ringe essentially advocated running separate statistic tests on each cell (phoneme pairing) in the contingency table. Although it is clear what to do if none of the cells gives a significant result—the languages are then not provably connected—it is less clear what to do if only some of them give a significant result. (The third possibility, that they all are significant, is impossible, because both researchers advocated only looking at cells with higher than expected frequencies. If some cells are higher than expected, others must be lower.) Villemin took the position that if any cell proved significant, then we can conclude that the languages are related.

Unfortunately, if one is searching through a large table looking for cells that have counts that are statistically significant, one is in effect doing dozens or even hundreds of experiments. The very definition of significance testing at, say, the .01 level, is that one is finding that the frequencies are so far from the expected value that one would expect them to turn up by chance only one time in a hundred. So if one is doing 100 trials, approximately one of them is expected to show significance, even if in fact the results are truly random.

To cope with that problem, Ringe himself invoked the binomial theorem again. If there are 100 different tests on pairs of phonemes at the .01 significance level, and, say, k of them give a positive result, he calculated the probability that k events of probability .01 would occur in 100 independent trials. Or, looking at it the other way, he used the binomial theorem to compute that 5 positives at $p = .01$ would be necessary before claiming significance in a 100-cell table. Languages on the average have more than $\sqrt{100} = 10$ consonants, and the threshold rises with table size (e.g, a 12×12 table requires 6 such cells). Unfortunately the binomial theorem becomes very unwieldy when numbers become very high, and Ringe did not provide tables to help out with this problem of multiple tests.

In practice, Ringe mentioned this second application of the binomial theorem only to explain why English and Turkish should not be considered provably related even though two consonant pairs are observed to have significantly high frequencies. He was less insistent on invoking the correction when known relatives are found to have only a small number of statistically significant correspondences. Thus he concluded (Ringe 1993) that English and French are shown to be connected on the basis of three consonant pairs, even though the test compared 16×17 consonants, or 272 different tests. If in fact the binomial correction is well founded, then three pairs is well within the range of expectation for such a large table. In the same paper, he also claimed that the relationship between French and Albanian was proved when he found three significant pairs in an even larger, 400-cell, table. It appears that Ringe only intended this second binomial test as a theoretical demonstration that a few cells will turn up with a significantly high frequency. Perhaps it is appropriate that he does not apply the test in practice, because the binomial distribution does not directly apply in this case, any more than it does for individual cells. Again, the problem is that the binomial distribution requires that the events in question have equal probability. But if one cell has an unusually high frequency, then it is going to be much less likely that some other cell will have an unusually high frequency. There are after all only 100 tally marks to distribute among all the cells.

An empirical reason for doubting the appropriateness of the binomial tests is presented in Table 11, which shows that desirable results were attained in very few of Ringe's tests when his binomial correction is taken into account. The *Needed* column in that table shows the tail of the distribution that would have probability of .01. For example, the first row states that 8 or more cells would have to be significant at .01 for $17 \times 17 = 289$ trials to be significant at .01. For smaller table sizes (less than 100 cells) the binomial distribution was computed directly. For the others, the distribution was approximated by the Poisson distribution for the same mean, which is probability (.01) times the number of cells. It can be seen that Ringe's correction almost never leads to the conclusion that languages are connected. The test works for the extremely closely related pairs English and German (which had a closeness judgment of .593 in Table 10), but not for more moderate connections like English and Latin (.292). Although it is not clear whether Ringe eventually decided that computing a second binomial was theoretically invalid or simply too cumbersome, by the time of his 1995 study, Ringe simply declared as an empirical rule of thumb that three or more cells are required to demonstrate a connection.

A better corrective would be the Bonferroni correction: lowering the probability threshold for each individual pair. If one wants to do 100 separate tests and accept an overall result if it satisfies the .01 significance level, one could require significance at $.01/100 = .0001$ for each test (Hays 1981:299). That is, significance for any single pair at .0001 would be enough to justify accepting overall significance at .01. To take the example of English /s/ = German /z/ again, each language has 17 consonants, so one is testing 289 pairs, therefore one would have to reset the significance level to around .000035. Because the probability of getting the 6 attested matches is .000481, the /s/:/z/ match does not prove to be significant. In the case of French vs. English or Albanian, one could see whether any of the three phoneme pairs is significant when the .01 threshold is lowered to take into account 272 or 400 tests, respectively

Table 11 Ringe's Examples with Binomial Corrections

Table	Size	Found	Needed	Connected?
English–German				
1992:8	289	16	8	yes
English–Latin				
1992:18	272	7	8	no!
1992:29 (200 words)	336	7	9	no!
English–French				
1993:2	272	3	8	no!
Albanian–French				
1993:4	400	3	10	no!
English–Turkish				
1992:24	221	2	7	no

Note: Table shows how many significantly frequent cells ($p \leq .01$) were found for each of Ringe's tests on initial consonants, and how many would be needed to show overall table significance, if Ringe's binomial correction is applied. Last column tells whether overall test concludes that the languages are connected; ! marks undesirable results. References in the Table column are to table numbers in Ringe (1992) and Ringe (1993). Most tests were performed with the Swadesh 100-word list.

(i.e., .000037 or .000025). By my calculations, none of them comes close.

Although such corrections would be an incremental improvement to the methodology, they would not address the problem that the statistics for each of the pairs of sounds are not the result of independent tests. Wrongly assuming independence can cause one to systematically overestimate the significance of the individual results (as Villemin did when he decided that a single significant cell was enough), or, conversely, cause one to overcompensate for the effects of multiple comparisons (as when applying the second binomial in Table 11 caused us to falsely reject many language relationships). Moreover, treating each cell separately makes it difficult to present a single number estimating the overall certainty that chance is involved in the table as a whole. Ringe's procedure is geared toward making the binary decision of whether or not one should consider correspondences to be due to chance, at a particular significance level. Although there is certainly nothing wrong with that approach, one might wish to turn the situation around and report a particular significance level: that there is a .01 chance that the resemblances between language A and B are the result of mere coincidence, but only a .0001 chance that the relation between A and C is accidental. There is a tendency in Ringe (1992) to treat in this fashion the number of phoneme pairs whose count is significantly higher than chance. The fact that he found 7 such pairs of initial consonants for English vs. Latin but 16 for English vs. German was taken to show that German is more closely related to English than Latin is (p. 42), which is indeed the case. But such an assumption is problematic, because the procedure discards information about the relative significance of each of those pairs. If the significance level of all seven of the English–Latin pairs were below .0001, but the significance levels of all sixteen English–German pairs were near .01, would it still be as obvious that German is more closely related? Furthermore, such a statistic would not be comparable across language pairs with varying numbers of consonant types, or across tests on different size vocabularies. Such factors might in themselves influence the significance of the test results without saying anything about the strength of the correlations. For example, increasing the number of words one looks at will increase the significance of the results (provided the additional words have the same properties as the old ones), but should not be taken to mean that those languages thereby become more closely connected historically.

Fortunately there is an obvious model for treating the figures in a contingency table as one big interdependent test: the χ^2 model. It will be recalled that Villemin interpreted contingency tables by performing a separate χ^2 test on each phoneme pair, which is inappropriate because that results in a hundred or more separate tests, just as in Ringe's procedure. What is here under consideration is doing one big χ^2 test over the whole table. Such a procedure gives a summary statistic telling how much deviation from expected values there is in the table overall. But as we have seen, computing the significance of the χ^2 statistic can be complicated. The standard approach of consulting the chi-squared distribution is not even thinkable, because the expected frequencies are always far too low in these tables. A 100-word list might be okay for populating a 5×5 table, but not for the 18×18 table one gets by comparing all the initial consonants of English and German. One surmises that it is this difficulty which led Ringe to propose his rather more complicated method for determining significance, and it certainly was a factor in leading Ross to reject the χ^2 statistic (1950:26).

There are a couple of solutions to the problem of low frequency counts in contingency tables. One approach would be to increase the number of words rather radically. A very rough estimate would be that with about 10,000 pairs of words, the table would fill out nicely. I do not in fact think that that is feasible, for reasons that will become clear later in this paper. In brief, not nearly so many words in a language are good candidates for inclusion in this kind of study.

Perhaps the best solution available to the linguist who chooses to do

TABLE 12 Contingency Table Test of English–German Correspondences by
Feature

English		German						
		Labial		Coronal		Other		Sum
Labial	20	*(7.68)*	2	*(9.12)*	2	*(7.20)*		24
Coronal	4	*(13.76)*	34	*(16.34)*	5	*(12.90)*		43
Other	8	*(10.56)*	2	*(12.54)*	23	*(9.90)*		33
Sum	32		38		30			100

Notes: Compares initial consonants of Swadesh-100 list. Upright numbers give observed counts; slanted numbers in parentheses are expected frequencies. Labial consonants are /b, f, m, p, pf, v/; Coronal are /d, ð, l, n, r, s, ʃ, t, ts, z/; Other are /g, h, k, w, y, ∅/.

the computations by hand is to collapse rows and columns along some dimension that is determined in advance, such as by grouping together into one row or column all consonants that have a particular feature in common. The place of articulation would be a good choice, because that seems to be reasonably stable through time, compared to other features. Oswalt (1970) recognized that fact when he required sounds to have the same place of articulation before they could be counted as similar at all. By using such grouping techniques one could ensure that the expected frequencies in each cell are high enough for the χ^2 distribution to be a reasonable approximation. For example, in Ringe's English–German comparison, none of the expected values for any of the consonant pairs was as high as 2 (Ringe 1992:20–21). But if one collapsed them into categories such as "Labial/Coronal/Other", one would end up with expected frequencies ranging from 7.2 to 16.7, a quite comfortable range for the χ^2 distribution (Table 12). Of course, such a solution would reduce the amount of information one is working with. In particular, if different members of a group in one language correspond to consonants found in different groups in the other language (e.g., in Hawaiian coronal stops have systematically become velar while coronal nasals remain coronal) then the overall measure of correspondence will come out lower than it really should be. In the case of English and German, to be sure, the χ^2 statistic is very high: 86.74 at 4 degrees of freedom is significant at a threshold much lower than .001.

Fortunately there is an even simpler solution to the problem: Do not rely on the standard χ^2 distribution. The permutation test for significance does not impose the requirement of large expected values. This test was discussed earlier in the Gallup poll example, but it might be helpful to spell out how it would be applied for these phoneme contingency tables. When one asks how likely it is that one will get so large a χ^2 representing so great a difference from expectation, what one really means is, if there is no association between these sounds, if there is no principle guiding English–German phoneme pairings other than pure chance, then any arrangement matching up the phonemes between English and German would be equally likely. Instead of the arrangement seen in the data fragment in Table 13, where concept 'skin' pairs an English word starting with /s/ with a German word starting with /h/, concepts 'sleep', 'swims', 'stand', 'star', and 'stone' pair English /s/ with German /ʃ/, etc., any of the English phonemes could have been paired up with any of the German phonemes, if it were a matter of pure chance. That is, concept 'sleep', expressed in English by an /s/-initial word, could just have easily been expressed in German by the /k/-initial word that in fact represents concept 'small' instead of by the /ʃ/-initial word that in fact represents concept 'sleep'. There are a very large number of possible worlds, each corresponding to one way of matching English phoneme tokens and German phoneme tokens one-to-one, and if there is no principle other than chance guiding that mapping, then all of those worlds are equally likely. What one wants to see is what percentage of those worlds has a χ^2 statistic at least as extreme as the one found in reality. That percentage directly corresponds to the likelihood that the actual arrangement that was found is due to chance; it is just a matter of mathematically defining an intuitive concept of chance. For example, the χ^2 statistic for the full English–German table is 895. The question is, for all possible arrangements of the phonemes, what percentage of them has a χ^2 of 895 or more?

There is a very simple algorithm for answering that question: just do all the possible arrangements, compute the χ^2 statistic for each of them, and determine what percentage is at least as great as the actual statistic. To be precise, a permutation test of χ^2 checks how often a χ^2 deviation statistic is expected to occur by chance, by explicitly enumerating the possibilities. If one considers all possible ways in which the words in the two languages could be paired up, and sees how often those combinations would give a χ^2 deviation statistic that is at least as high as that obtained when the words are sense-aligned, then one knows the likelihood that the original statistic reflects chance rather than some correlation among words of the same meaning. If 99% of the orderings give a χ^2 smaller than the observed statistic, then there is only a .01 probability that the statistic is due to chance. Of course, for large tables it is practically impossible to do all the permutations, even by computer. A list of 100 words would require 100! (almost 10^{158}) arrangements. One

TABLE 13 Initial Consonant Correspondences for a Subset of the English-German List

Concept	English	German
skin	/skɪn/	/haut/
sleep	/slip/	/ʃlɑːfən/
swim	/swɪm/	/ʃvɪmən/
stand	/stænd/	/ʃteːən/
star	/stɑr/	/ʃtɛrn/
stone	/ston/	/ʃtain/
small	/smɔl/	/klain/
bird	/bərd/	/foːgəl/
blood	/bləd/	/bluːt/
belly	/bɛlɪ/	/baux/
breast	/brɛst/	/brʊst/
bite	/bait/	/baisən/
burn	/bərn/	/brɛnən/
black	/blæk/	/ʃvarts/

can get around this problem by using Monte Carlo techniques, that is, by taking a random sample of the possible orderings. One can make the resultant estimate as precise as necessary simply by doing more reorderings. For example, at the critical p of .01, the variance after 1,000 orderings is just 0.0000099. This is where the computer has the upper hand. It is fairly trivial to make thousands of arrangements on a computer, while even a dozen arrangements would try the patience of the linguist working by hand. In this paper my computations are all based on 10,000 arrangements.

Table 14 shows the results for the cases where Ringe (1992, 1993) gave complete tables so that his computations can be redone with exactly the same data he based his conclusions on. The p column reports the probability that the correspondences found between each of the language pairs is due to chance. Any number below .01 meets the threshold that Ringe accepted as proof that languages are historically connected. The number .05 is the threshold most commonly used in the social sciences, but of course researchers are free to choose the value they feel most comfortable with. I myself would feel that a p less than .001 is close to a dead certainty (and in fact the numbers reported as such in the table are actually closer to one in 10,000, but have been rounded off to three digits), and a p less than .05 means the relationship is probably true, but perhaps not worth staking one's reputation on. In this set, among the tests that look at word-initial consonants, that would mean that the

TABLE 14   Application of Permutation Tests to Ringe's Examples

Table	Permutation Method		Ringe's Method
	p	Cramér	Connected?
English–German:			
1992:8	.001	.748	yes
English–Latin:			
1992:18	.001	.533	no!
1992:29 (200 words)	.001	.416	no!
English–French:			
1993:2	.001	.496	no!
Albanian–French:			
1993:4	.040	.521	no!
English–Turkish:			
1992:24	.090	.443	no

Notes: Table compares Monte Carlo estimate (at least 10,000 samplings) of χ^2 significance (Column p) and strength (Cramér) to Ringe's results on initial consonants when his binomial correction is applied. References in the Table column are to table numbers in Ringe (1992) and Ringe (1993). Most tests were performed with the Swadesh 100-word list.

connections between English on the one hand and German, Latin, and French on the other are certain; with Albanian, probable; with Turkish, not proven but perhaps worth a second look. These are precisely the results that Indo-European scholarship would lead one to expect. The connections with German, Latin, and French were known from the cradle of modern linguistic investigation; that with Albanian has required a good deal more scholarship to ascertain; and any relationship with Turkish is regarded by the great majority of scholars as not proven. In contrast, Ringe's statistical approach appears to be less powerful. If his binomial correction is applied, only English and German can be taken to be connected at $p \leq .01$. Note that the testing must be reapplied if one wishes to change Ringe's preferred cutoff from $p \leq .01$ to, say, .05.

Table 14 also includes the Cramér statistic, which is a measure of the strength of the correspondences that were found. It is an extension of the Pearson index of mean square contingency, generalized for tables larger than 2×2 (Hays 1981:556). This would be the safest number to use when, after deciding that two languages are probably connected historically, one begins to ask just how close that connection is. One can see that among the consonant-initial tests, which are the only ones that are truly comparable simply because Ringe reported complete results most often for those, the Cramér statistic is much higher for the relation that English has with German than for the one it has for Latin or French, and the former is somewhat larger than the latter. This is exactly what one expects. Although French is cladistically equivalent to Latin (being descended from it), the greater chronological separation from the common Proto-Indo-European starting point means that it has had more time to diverge from English's branch than Latin had.

The Cramér statistic also demonstrates that using a permutation test over the individual consonant pairs is better than the lumping into categories like "labial" that was demonstrated above. Although in both cases the p value was very low, the Cramér statistic for the lumping case was .645, which can be compared to the .748 seen in the table above. In more borderline cases, the diminished lack of specificity in the lumping case could result not just in reduced Cramér statistics, but also in increased p values, leading one to reject the historical connection unnecessarily. In comparison, the strength statistic offered by Ringe, namely, the number of consonant pairs that occur significantly more often than expected, is rather more difficult to interpret. The values reported (Table 11) for the certain cases (my $p < .001$) range from 3 to 16; and 3 itself can correlate with anything from $p < .001$ (English–French) to $p = .04$ (Albanian–French).

I also computed the p statistics from the raw word data, as listed in the appendix. The results are shown in Table 15. Note that these results may not be identical to those shown in the previous table. To a certain extent divergences are due to the fact that there may be slight differences in how the words were analyzed. But the main source of any divergence is the fact that the Monte Carlo technique is stochastic. Because the randomizations are random, results will not come out exactly the same way twice.

Like many other tables that will be presented in this book, Table 15 is sorted by descending order of cognate judgments, as discussed in Chapter 2. Thus the languages at the top of the table are closely connected to each other, and a desirable result would be for them to have a very low p value, near 0. The languages in the bottom half have no known connection at all in the Swadesh word lists, and a desirable result would be to have higher p, near 1. In the case of this experiment, it can be seen that the test successfully identified language connections with a cognate index above 15%, but was less successful in identifying the more remote connections between Albanian and the Germanic languages (German and English). There were also a few false positives. At a .05 cutoff, one would conclude that Albanian is related to Turkish and Hawaiian, and Latin to Navajo. These false positives, while undesirable, are not alarm-

TABLE 15 Chi Squared Technique, All Language Pairs

Languages		p	Cognates[1]
English	German	.000	.593
French	Latin	.000	.565
English	Latin	.000	.292
German	Latin	.000	.290
English	French	.000	.285
French	German	.001	.253
Albanian	Latin	.002	.230
Albanian	French	.005	.203
Albanian	German	.300	.152
Albanian	English	.062	.130
Albanian	Turkish	.015	.005
English	Hawaiian	.441	.005
Albanian	Hawaiian	.012	.000
Albanian	Navajo	.896	.000
English	Navajo	.630	.000
English	Turkish	.117	.000
French	Hawaiian	.247	.000
French	Navajo	.111	.000
French	Turkish	.209	.000
German	Hawaiian	.534	.000
German	Navajo	.526	.000
German	Turkish	.710	.000
Hawaiian	Latin	.188	.000
Hawaiian	Navajo	.052	.000
Hawaiian	Turkish	.335	.000
Latin	Navajo	.002	.000
Latin	Turkish	.076	.000
Navajo	Turkish	.417	.000

Note: p gives significance of the finding that the listed languages are connected, using all the words in the Swadesh 100 list. Calculations were performed over the initial consonant of word stems.

[1] Closeness judgments of Table 10.

ing. It must be kept in mind that the very definition of $p \leq .05$ means that the metric—here the χ^2 value—would occur by chance about one time out of twenty. Because I am presenting more than twenty separate tests, one or two false positives at that significance level are to be expected. In fact, if these tests were independent, one could use the binomial formula to show that one false positive is actually more likely than not ($p = .60$), and one would have to get four false positives ($p = .011$) before one begins to suspect the number of positives is not due to chance at a .05 significance level. If this level of false positives is unsettling, one can always set the significance level lower, such as to the .01 level that Ringe preferred. With that cutoff, the tests do no worse with the related Indo-European languages, and have only one false positive, Latin–Navajo. Of course, one does not tinker with significance levels in order to make sure one gets the expected results. But it is legitimate to choose significance levels to reflect one's own philosophy about the relative dangers of missing a legitimate linguistic connection, which danger can be alleviated by using higher significance levels, and of wrongly stating that languages are connected, which danger can be alleviated by using lower significance levels.

In the course of this book, I will have occasion to display several tables like Table 15. The reader should take these tables as illustrative rather than probative. It would be rash to assume that one statistical approach is better than another simply because it classifies one or two language relationships more accurately. For one thing, there are actual coincidences. Unrelated languages may coincidentally share properties that are so unlikely that they falsely suggest historical connection; if a test picks up on these shared properties, it is only doing its job. For another, the language pairs are not independent of each other. Reusing the languages by taking all available pairings of the eight available means that some coincidences could propagate. To take a hypothetical example, if English and Turkish had some unusually strong coincidental recurrences, then the pairing of Latin and Turkish might as well, simply because English has many (noncoincidental) recurrences with Latin; English could well correspond with Latin in the same consonants where it coincidentally corresponds with Turkish, meaning that Latin and Turkish have a high likelihood of corresponding as well. To do a controlled experiment to rigorously evaluate the effectiveness of the metric, one would randomly pick pairs of languages and not reuse them. I do not do this because seeing the 28 tests is more interesting and potentially more revealing than seeing just the 4 tests that would be possible if the eight languages were paired only once. The most important reason for not taking these as definitive tests is that there is no objective metric against which to judge whether the tests are correct when it comes to testing languages not known to be related. After all, one goal of this research is to investigate whether one can reasonably expect to use statistics to demonstrate relations between languages whose connections are not otherwise proven. It would be begging the question if I were to declare a test a failure because of a positive in one of those tests. If a test shows a connection between English and Turkish, it could be a mistake in the test; it could mean that the two languages coincidentally share some unexpected number of recurrent phonemes; or it could be that the Nostratic theories linking those languages together are correct (e.g., Bomhard 1990, Collinder 1965, Pedersen 1903). These tables must therefore be considered illustrative, rather than as experiments conclusively evaluating the relative merits of various metrics. The tables are meant simply to reassure the reader that the metrics are essentially on the right track. One cannot conclude that a metric is invalid if it finds a connection between supposedly unrelated languages, but one can suspect something is wrong if it finds that such connections are more certain than those between English and German.

To summarize this chapter, Ringe's methodology can be improved by replacing his method of computing significance with one based on the χ^2 statistic and a Monte Carlo estimation of the exact p value for the contingency table as a whole. Not only is this methodology statistically valid, but also it appears empirically to be stronger than Ringe's statistical analysis, which reliably shows language connections only in the most compelling cases. In short, the exact methodology is as follows.

1. For each of the two languages under consideration, find the most appropriate translations for each of the items on the Swadesh 100 list. In cases where there is no single obvious translation, choose one by any reasonable method (such as corpus frequency, or precision of semantic correspondence) that does not take into account any knowledge of the other language.

2. Make a table arranging the word-initial consonants of one language along the top and the word-initial consonants for the other language along the side, leaving space for "∅", representing the absence of a consonant. For each concept, consider which consonant begins the word in each of the two languages, and tally the count in the table. In the end, one will have for each consonant pair a number O telling how many concepts were represented by words beginning with that pair of consonants.

3. Calculate for each cell i in the table its expectation E under the assumption of no association, by multiplying its row sum by its

    column sum and dividing by the grand total for the table: $R_i C_i / N$.

4. Calculate the χ^2 statistic for the table, summing $(E-O)^2/E$ across all the cells of the table.

5. Rearrange the words of one language randomly. Repeat steps 2 through 4 to get a χ^2 statistic for the rearrangement.

6. Repeat the previous step 10,000 times, tallying how many times these random rearrangements have a χ^2 statistic at least as high as that observed for the original table. Divide by the number of iterations (10,000) to get p, the significance measure to be reported.

I believe the statistical method just described is substantially correct from a purely mathematical point of view, and it will be used as the point of departure in the following discussion. However, for various linguistic reasons, blindly following this recipe can lead to seriously wrong results. Subsequent chapters will address various problems and suggest improvements where possible.

4

Tests in Different Environments

Although Ross (1950) only looked at the first phoneme of a word, Ringe (1992) experimented with tests on sounds in many different environments. In his first case study, English and German, Ringe compared the efficacy of the test when he used various other elements defined according to the word's consonant–vowel (CV) template. For most of these, he presented complete tables and statistics (summarized here in Table 16). For this pair of languages at least, which are known to be closely related (they got the highest rating, .593, by the cognate metric, Table 10), it would appear that the most efficacious tests are on consonants that are not preceded by other consonants (only $C_{1.1}$ and $C_{2.1}$ give expected results). And the word-initial consonant appears the most efficacious of all.

I confirmed this finding with the other pairings of the languages in the data set, using the χ^2 metric developed in the preceding chapter. Table 17 is arranged in descending order of judged closeness, as was Table 15. One immediately obvious result is that tests on environments $C_{1.2}$ and $C_{2.2}$ are not even valid for many language pairs, because there are few if any contrasts in that position. Hawaiian, for example, does not have any consonant clusters at all. In such cases, the significance level is necessarily 1.00; there is no evidence at all that the languages are connected. But even when applicable, tests in positions other than $C_{1.1}$ appear to be very weak for this set of languages. $C_{2.2}$ never finds a connection, and $C_{1.2}$, V_1 and $C_{2.1}$ do so only in the easiest cases, where the cognate level is above 50%. There are probably two independent reasons why evidence becomes weaker in those positions. One reason is phonetic. Elements other than the initial consonant are much more susceptible to lenition processes, weakenings that may cause the sounds to be lost entirely or merge into the same values as other sounds (Hock 1991:83), making cross-linguistic phoneme matching more difficult. The

TABLE 16 Ringe's English–German Tests in Different Word Positions

Table[1]	Environment	Cells[2]	Found[3]	Needed[4]
8	$C_{1.1}$[a]	289	16	8
10	$C_{1.2}$[b]	48	3	4
12	$C_{2.1}$[c]	324	11	9
14	$C_{2.2}$[d]	66	0	4
15	final rime	36	1	3

Note: Tests were performed with the entire Swadesh 100-word list.
[1] Table number from Ringe (1992).
[2] Number of cells (consonant pairs) in the table.
[3] Number of cells with count significantly ($p \leq .01$) above expectation by binomial theorem.
[4] Number of cells that would need to have significantly high counts in order to correctly demonstrate that English and German are related, by Ringe's binomial correction.
[a] Word-initial consonant.
[b] Second consonant of word-initial cluster.
[c] First consonant after first vowel.
[d] Second consonant of cluster immediately following the first vowel.

other reason is that many words are not long enough to have all the positions in question. Concepts for which the word in either language was too short were omitted from the comparisons; the N columns in the table tell how many words remained. As those columns indicate, quite a few words lacked sufficient material past the the vowel. One could of course assign a special null value for those short words, but that amounts to using word length itself as a comparandum, which may not be a good idea, for reasons to be discussed later.

Nevertheless, there is the possibility that for some languages, tests on the initial consonant alone might not be ideal or even satisfactory. Baxter and Manaster Ramer (1996:378) called attention to the case of those Australian languages that have consistently dropped word-initial consonants (a process called aphaeresis). Obviously, attempts to match such a language with a related language that has not undergone aphaeresis will fail. Even in less exotic cases, one could well imagine that the evidence of word-initial consonants may be too weak to constitute conclusive evidence of connection alone, but perhaps evidence from other parts of the word could in some way buttress the evidence, and cumulatively push the test into the positive zone.

Although Ringe ndid not give explicit advice about how to integrate the results of the different tests, they are apparently intended to be cu-

TABLE 17 Tests in Different Word Positions

Languages		$C_{1.1}$		$C_{1.2}$		V_1		$C_{2.1}$		$C_{2.2}$		Cogn[1]
		p	N	p	N	p	N	p	N	p	N	
English	German	.000	100	.000	85	.009	73	.000	37	.834	4	.593
French	Latin	.000	100	.108	82	.004	78	.000	30	1.00	2	.565
English	Latin	.000	100	.136	83	.167	73	.190	34	1.00	1	.292
German	Latin	.000	100	.094	85	.800	72	.620	38	1.00	2	.290
English	French	.000	100	.936	81	.226	70	.014	20	1.00	1	.285
French	German	.001	100	.714	82	.860	69	.002	18	1.00	1	.253
Albanian	Latin	.002	100	.692	81	.271	78	.092	45	1.00	2	.230
Albanian	French	.005	100	.700	79	.095	74	.465	24	1.00	1	.203
Albanian	German	.300	100	.974	81	.500	71	.160	33	1.00	2	.152
Albanian	English	.062	100	.878	82	.190	71	.764	34	1.00	1	.130
Albanian	Turkish	.015	100	1.00	85	.529	80	.589	48	.602	6	.005
English	Hawaiian	.441	100	1.00	86	.238	80	.593	41	1.00	5	.005
Albanian	Hawaiian	.012	100	1.00	85	.064	86	.840	50	1.00	8	.000
Albanian	Navajo	.896	100	1.00	87	.854	86	.293	46	1.00	0	.000
English	Navajo	.630	100	1.00	88	.895	80	.839	38	1.00	0	.000
English	Turkish	.117	100	.445	84	.923	73	.273	38	1.00	0	.000
French	Hawaiian	.247	100	1.00	82	.292	81	.902	30	1.00	4	.000
French	Navajo	.111	100	1.00	84	.420	81	.204	30	1.00	0	.000
French	Turkish	.209	100	.772	81	.782	75	.980	28	1.00	2	.000
German	Hawaiian	.534	100	1.00	86	.732	78	.618	45	1.00	7	.000
German	Navajo	.526	100	1.00	88	.582	78	.937	38	1.00	0	.000
German	Turkish	.710	100	.452	84	.871	72	.579	42	1.00	2	.000
Hawaiian	Latin	.188	100	1.00	86	.801	87	.387	58	1.00	9	.000
Hawaiian	Navajo	.052	100	1.00	98	.110	100	.790	63	1.00	0	.000
Hawaiian	Turkish	.335	100	1.00	91	.465	91	.605	71	1.00	19	.000
Latin	Navajo	.002	100	1.00	88	.646	87	.122	56	1.00	0	.000
Latin	Turkish	.076	100	.249	85	.737	81	.165	54	1.00	5	.000
Navajo	Turkish	.417	100	1.00	93	.563	91	.372	62	1.00	0	.000

Note: Table gives p statistic for χ^2 tests in different environments (defined as in Table 17), as well as N, the number of word pairs considered.
[1] Closeness judgments from Table 10.

mulative in force. Ringe (1992:10, fn. 17) stated: "Of course one does not rely on a single set of word-initial consonant matchings to prove or disprove language relationships!". And later in the same work he counted the number of concepts that have more than one phoneme match that is significantly more frequent than chance. For example, he adduced as additional positive evidence of an English–German relationship the fact that their word lists have three word pairs that share four significantly frequent pairs, such as *sand:Sand* /sand/:/zand-/ for 'sand', which has the significant pairs /s/:/z/, /a/:/a/, /n/:/n/, and /d/:/d/. Of course these were found in four different sets of tests. On the other hand, Ringe (1993) proved the relationships between French and Albanian and between French and English by considering only the word-initial consonants, so it is hard to know exactly which method he preferred. But I believe a fair appraisal is that he considered any of these tests to be potentially sufficient grounds for establishing language connections, and that the more tests that agree, the better.

However, such assumptions are unfounded unless one knows something about the dependency relations between the tests. If different tests measure the same thing in a redundant fashion, then an extra test tells nothing more than what one started out with. An example of such a problem can be seen in the English–German tests. If an English word has /t/ as the second element of a word-initial consonant cluster $(C_{1.2})$, then the previous phoneme $(C_{1.1})$ has to be /s/. In Standard German, if a word has /t/ in that position, then the previous letter has to be /ʃ/. Therefore noting that a pair like *stone:Stein* /ston/:/ʃtain/ has the /t/:/t/ match at $C_{1.2}$ in effect gives for free the information that the word has a /s/:/ʃ/ match at $C_{1.1}$ as well. So from an informational point of view, the contribution that /s/:/ʃ/ makes toward the solution is zero, once one knows that one has a /t/:/t/ match as the second consonants of initial clusters. One must not take either of the two tests to be independent corroboration of the other. The problem of dependence may not be as severe in other environments, but the fact that restrictions on phonotactics might not be absolute elsewhere just hides the problem and makes it more insidious. Even in English, the identity of the coda is probabilistically constrained by the identity of the preceding vowel and onset, even though few such effects are absolute (Kessler and Treiman 1997). Consequently, pairs could make relatively different contributions, depending on what other sounds appear in the word.

One could devise methodologies to make sure that the various tests are independent. For example, standard loglinear techniques (Wickens 1989) could be used to test for direct associations between the first consonants of one language and the first consonants of another, while factoring out any indirect contributions of the second consonants in either language. Then one would know that one could calculate the overall $1 - p$ level of the family of tests—that is, the probability that success in none of the tests is due to chance—by multiplying the $1 - p$ levels of the various tests. So, for example, if one runs two independent tests at the $p \leq .01$ level, as Ringe suggests, then if either test succeeds, then one can conclude that the languages are connected with a likelihood of $(1 - .01)(1 - .01)$, i.e., .9802. Conversely, p, the likelihood that the correspondences are due to chance, is .0198. On the other hand, if both tests succeed, then the overall significance level plummets to .0001.

It does not appear, however, that it is really worthwhile to attempt to achieve independence between the various tests. For one thing, one is normally more concerned with the upper limit of p than with its lower limit. The question one really wants to answer is, if only one test out of many shows historical connection between languages, then what is the overall probability that that is due to chance? The Bonferroni inequality states that, in the worst possible case, that number is equal to the sum of those probabilities for each test. So in the current example, that number is $p \leq .02$, just slightly greater than the number (.0198) that was calculated under the special condition that the tests are independent. For six tests of $p \leq .01$, the Bonferroni inequality gives a limit of $p \leq .06$, again scarcely distinguishable from the special case of $p \leq .059$. What is the cost of proving independence so as to get such a small increase in precision? This turns out to be rather massive, when one considers how much data one is trying to squeeze out of 100 words. Already the cell frequencies in a two dimensional table comparing initial consonants are very small. To control for the second consonants, one would have to add two more dimensions to the table, making the numbers that much smaller. That may be thinkable if one limits one's concern to those two syllabic positions; but if one wishes to make each of six different tests independent, then the table will become so huge that virtually all of the attested frequencies will be 0 or 1. In fact, the six tests Ringe discusses are sufficient to distinguish virtually all the words in a list of 100 basic words in any language. Not many words will share the same first consonant, second consonant, vowel, first postvocalic consonant, second postvocalic consonant, and second rime. How much less so then would one expect any recurrent match between two languages. It would be virtually impossible to demonstrate significance in such a case.

The difference between the target of $p \leq .01$ and the more defensible and still respectable $p \leq .06$ may seem like a minor quibble until one considers that those six tests are not the only possible or desirable ones. When one takes into account the possibility that fairly simple changes can totally destroy evidence in one position (such as aphaeresis destroying agreements between initial segments), it becomes evident that practitioners will be tempted to add new tests, such as between word-initial consonants in one language and the first postvocalic consonant in another. How many tests then can a linguist potentially apply? Ordinary researchers are going to mention only the tests that work, making it difficult for other scholars to interpret the validity of their findings; Ringe is exceptional in doing otherwise. The more candidate tests that are tried, the lower the significance of the overall results.

The easiest valid corrective if one wants to report significance at the $p \leq .01$ level would be to institute a fixed suite of N tests, which are applied in all situations; compute a per-test p level of p/N; then report success only if one of the tests succeeds at that level. For example, if there are ten tests, then each of them needs to be run at the $p \leq .001$ level. Alternatively, one could take the best of the p values, multiply it by N, and report that number. Of course, if some tests are clearly inapplicable because of the nature of either of the languages taken independently, then it seems that it would be safe enough to drop that test from the suite. For example, testing the second consonant in word-initial clusters is clearly pointless in languages like Hawaiian, which have no consonant clusters. But it should go without saying that a test could be thrown out only because of objective qualities of a candidate language, not because prior hypotheses about language relatedness make a particular comparison look unlikely, and certainly not because running the test gives bad results.

Another theoretical possibility, if one can restrict the number of conditions sufficiently, is to run one big test where the relevant factors are already multiplied out for each language. Such a test would report one p number, which would not need to be interpreted in relation to other such results. For example, if it turns out to be best to just look at the initial consonant and the first vowel of the words in each language, one could do an analysis where the rows list all those word-initial two-phone sequences (CV-) of one language, and the rows list the CV- sequences of the other language. That would directly test that there is some association between the languages in either or both of those two phones. Such a test would have the advantage that it could detect associations even if they occurred between the consonant in one language and the vowel in the other. Of course, lots of data is necessary before this will work; such a large table will mostly be filled with 0's and 1's if one is only looking at 100 or 200 words. I tried this with English–German comparisons, using 162 words, somewhat more than Ringe's 100 comparanda. The 162 words produced a contingency table of dimensions 106×115.

Of the 12,190 cells, all had counts of 0 and 1 except for five cells that had a count of 2 (/wo/:/va/, /fa/:/fe/, /si/:/zi/, /æ/:/a/, /wai/:/vai/). Despite these unfavorable conditions, the χ^2 test did reveal a connection between English and German, at $p = .025$. But that p statistic does not meet Ringe's standard of .01, and shows several orders of magnitude less certainty than the test on the initial consonant alone. When one considers that the relationship between English and German is much closer than most relationships that a linguist would bother to test, it is clear that a CV- comparison is not going to prove useful in general unless many more words are available. More importantly, such a dramatic drop in power means that one would be better off doing several tests at a more restrictive p threshold than doing one test that is as weak as this.

The optimal test suite is perhaps a topic for future research. There are at least three possible approaches to figuring out what the best general test suite is. One could run tests across languages already known to be related, and see what works best. From Table 17, one can see already the beginnings of such work. For this set, at least, the word-initial consonant is by far the best comparison, and the second element of a postvocalic consonant cluster is the worst. Another approach would be to study the literature on reported correspondences, to see which positions figure most prominently cross-linguistically. Lastly, one could consider primary facts about acoustics and articulation. It is reasonable, for example, to expect there to be less rapid change in consonants that release into vowels, because they are acoustically easier to distinguish, than in consonants that do not; therefore in general a prevocalic consonant environment should be better than a postvocalic one. A factor that may be somewhat more difficult to control is stress. It is clear that the stressed vowel and the consonant preceding it will be better conserved than other vowels; on the other hand, stress patterns themselves are subject to a great deal of change. That may be a good reason for multiple tests. Word-initial C and V, as well as stressed C and V, if any, cover a lot of different cases.

There are other ways to test more than one phoneme in the word at a time. That idea will be developed in a later chapter, when I turn my attention to improving the power of already valid statistical tests. My concern here, rather, is to point out the potential pitfalls in the existing practice of running multiple tests, and to suggest minimal sufficient corrections. In a nutshell, the experimenter must select which parts of the words to compare, without taking into account any prior knowledge of the hypothesized linguistic relationships. Multiple tests must be compensated for by proportionally increasing reported overall p or, equivalently, by proportionally lowering the p sought in each test. From a practical standpoint, one is probably best off minimizing the number of tests while selecting them for maximum power. Current knowledge suggests that the best tests are on consonants near the beginning of the word, especially the word-initial one.

5

Size of the Word Lists

Ross (1950) advocated using one thousand concepts (word pairs), but did not recommend a specific list, and his own tabulations were performed with smaller lists, ranging from 533 to 391 words. Villemin (1983) used 215 words. Ringe (1992) used 100 words. Are linguists getting lazier? Or might smaller word lists be just as good as longer ones, or even better?

Ringe (1992:55–64) discussed whether there would be any benefit to extending his wordlists beyond the Swadesh list of one hundred words. He decided there would not be any benefit, on both theoretical and experimental grounds. His theoretical arguments were first, that the Swadesh 100 words were more basic and hence superior to longer word lists; and second, that no matter what size vocabulary one uses, the relative frequency of related to unrelated words will remain the same, and that the number of matches expected by chance is proportional to the number of words. The experiment consisted of showing that when English and Latin initial consonants are compared, the number of consonant pairs that show up significantly more frequently than chance is 7, whether 100 words are used or 200 (Table 11).

I am not sure I fully understand Ringe's second theoretical point. In a binomial test of the type he uses, if one increases the number of trials and gets the same proportional increase in the number of events, the significance level drops quickly. For example, the odds of getting at least 80% heads by chance is .055 with 10 coin tosses, but .006 with 20 tosses. Or, looking at it the other way, if one doubles the number of coin tosses, one does not have to double the number of heads needed to reach significance; the proportional requirement drops. Surely that should also work for vocabulary. If one doubles the number of words, then to a rough approximation one should get double the counts in each cell. Therefore if there is a connection between the languages, some cells that were formerly on the threshold of significance should be pushed over the limit. The result should, on average, be a higher number of significant cells in the test that uses twice as many words.

But it is perhaps pointless and maybe even confusing to dwell on the impact of word list size on Ringe's significance tests, because after all I have already advocated using a different type of analysis. Certainly the χ^2 metric I have been discussing benefits from extra data. As illustrated in the Gallup poll example, increasing the amount of data will, all things being equal, increase the significance level of a test. In fact the χ^2 will on average grow as fast as the number of words, and the significance level always goes down when the χ^2 goes up. On purely mathematical grounds, therefore, it seems clear that the more words we look at, the better. And that reasoning holds for virtually all significance tests.

Another reason to prefer more data, this one perhaps more social than mathematical, stems from the fact that one is going to miss out on some data if one only considers 100 words. For example, the Swadesh 100 word list is pretty good on revealing the correspondences for initial consonants in English and German, but the data simply did not have any instances of the /j/:/j/ correspondence, as in the word pair *year:Jahr*. Things work out pretty well in the case of those languages, but if the test had ruled that there was insufficient evidence for deciding that the languages were related, and it had never been shown that data, then historical linguists would have reason to doubt the test. Certainly such reasoning can be taken too far, and would lead to error if it impelled one to bring in correspondences that were not selected blindly. But as long as the criteria are not biased with respect to the research hypotheses, more inclusive bodies of data are both mathematically and socially preferable to small data sets.

Of course, this is not to say that increasing the number of words is guaranteed to improve the results in every case. On the contrary, random forces are at work. Especially if rather small numbers of words are being used, one could find that in individual cases the additional words make no difference, or even push the results in an unexpected direction. There can also be systematic forces at work. I have verified, for example, Ringe's claim that increasing the size of the English–Latin word list from the Swadesh 100 list to the Swadesh 200 list makes no practical difference. For both word lists, the χ^2 metric using a permutation test with one million randomizations found that the significance level was less than 1 in a million. One possible conclusion is that when languages are closely related, p values may be so small that it is impractical (and scarcely useful) to compute the difference between them.

But that may not be the only reason for the failure of an increase in word list size to give consistently better results. Table 18 shows what happens when the χ^2 test is run over all language pairs, using either the Swadesh 100 and/or the Swadesh 200 word lists. One might have expected the larger data set to give consistently smaller numbers for the related languages, and larger numbers for the unrelated languages. But no such pattern emerges. Just about half of the results are better with the Swadesh 200 set, about half are worse.

But if this demonstration shows anything, it is not the difference (or lack thereof) between word-list sizes per se, but between Swadesh 100 and Swadesh 200. Neither of those are randomly selected word lists. Here is one fairly clear example of how those differences could matter. Both are designed to be list of words of high cultural universality, but the Swadesh 100 (a subset of Swadesh 200) is meant to be even more universal than the fuller list (Swadesh 1955). If the words on the Swadesh 200 list really are less universal than the words on the shorter list, they are more likely to be borrowed. Therefore the Swadesh 200 list could make related languages seem less related, and unrelated languages seem more related. The reason why this is so will be discussed in more detail later. Here the point is simply to illustrate the general statistical principle that if the items on the short and long list are not chosen randomly, a bias could well creep in that will throw off the results. Table 18 does support the idea that Ringe was correct in concluding that, within the context of his general methodology (comparing word-initial consonants), the Swadesh 200 list seems to be no better than the Swadesh 100 list. But the more general conclusion, that word-list size is not an important factor, is premature.

Table 19 focuses more specifically on the size issue. In this experiment, the basic population for both samples was the Swadesh 200 list. From that set, for each pair of languages, ten random samples of size 50 were drawn and tested, and the resulting p values averaged. The same was done for samples of size 100, and the results are contrasted in the table. Here the differences are very clear. It can be seen that languages known to be connected receive much lower p values when 100 words are selected rather than 50, without adversely affecting the findings for the unconnected languages. Looked at in terms of the actual decisions that the tests would support at the .05 significance level, only the two closest connections, those with cognate levels over 56%, were uncovered when 50 words were used, whereas five pairs, with cognate levels going down to 23%, were discovered when 100 words were used.

This experiment demonstrates clearly that sample size does matter. All things being equal, it pays to have more words in the sample. On the other hand, there are certain natural limits. When the p estimate is already very low, it may be impossible to get lower within the confines of

TABLE 18 Swadesh 100 vs. Swadesh 200 Word Lists

Languages		Swadesh 100	Swadesh 200	Cognates[1]
English	German	.000	.000	.593
French	Latin	.000	.000	.565
English	Latin	.000	.000	.292
German	Latin	.000	.000	.290
English	French	.000	.030	.285
French	German	.001	.011	.253
Albanian	Latin	.002	.000	.230
Albanian	French	.005	.006	.203
Albanian	German	.300	.093	.152
Albanian	English	.062	.020	.130
Albanian	Turkish	.015	.797	.005
English	Hawaiian	.441	.083	.005
Albanian	Hawaiian	.012	.021	.000
Albanian	Navajo	.896	.855	.000
English	Navajo	.630	.105	.000
English	Turkish	.117	.078	.000
French	Hawaiian	.247	.278	.000
French	Navajo	.111	.702	.000
French	Turkish	.209	.329	.000
German	Hawaiian	.534	.351	.000
German	Navajo	.526	.233	.000
German	Turkish	.710	.681	.000
Hawaiian	Latin	.188	.384	.000
Hawaiian	Navajo	.052	.256	.000
Hawaiian	Turkish	.335	.834	.000
Latin	Navajo	.002	.002	.000
Latin	Turkish	.076	.304	.000
Navajo	Turkish	.417	.201	.000

Note: All comparisons use χ^2 metric on initial consonants.
[1] Closeness judgments from Table 10.

TABLE 19 Effect of Randomly Doubling Sample Size

Languages		50 Words	100 Words	Cognates[1]
English	German	.000	.000	.593
French	Latin	.000	.000	.565
English	Latin	.076	.018	.292
German	Latin	.170	.028	.290
English	French	.204	.100	.285
French	German	.240	.064	.253
Albanian	Latin	.232	.044	.230
Albanian	French	.398	.125	.203
Albanian	German	.442	.266	.152
Albanian	English	.346	.190	.130
Albanian	Turkish	.639	.631	.005
English	Hawaiian	.389	.393	.005
Albanian	Hawaiian	.329	.207	.000
Albanian	Navajo	.610	.526	.000
English	Navajo	.456	.344	.000
English	Turkish	.519	.337	.000
French	Hawaiian	.402	.382	.000
French	Navajo	.526	.506	.000
French	Turkish	.440	.424	.000
German	Hawaiian	.469	.536	.000
German	Navajo	.404	.337	.000
German	Turkish	.666	.434	.000
Hawaiian	Latin	.621	.503	.000
Hawaiian	Navajo	.493	.482	.000
Hawaiian	Turkish	.314	.701	.000
Latin	Navajo	.314	.052	.000
Latin	Turkish	.452	.301	.000
Navajo	Turkish	.474	.266	.000

Note: All comparisons use initial consonants; 50 or 100 concept pairs from the Swadesh 200 word list were then randomly tested using the χ^2 metric.
[1]Closeness judgments from Table 10.

TABLE 20 Closeness Judgments, Swadesh 200 vs. 100

Languages		Swadesh 200	Swadesh 100
English	German	.593	.740
French	Latin	.565	.660
English	Latin	.292	.360
German	Latin	.290	.380
English	French	.285	.330
French	German	.253	.320
Albanian	Latin	.230	.235
Albanian	French	.203	.210
Albanian	German	.152	.190
Albanian	English	.130	.180
Albanian	Turkish	.005	.000
English	Hawaiian	.005	.000

Notes: Human judgments as to how closely related languages are, expressed as the ratio of the Swadesh word pairs that are cognate; numbers closer to 1 represent closer connections.

the methodology. For example, when doing 10,000 iterations in a Monte Carlo test of significance, one cannot improve a result of .0001, and one would want to round off a decimal place anyway lest one give false impressions about the precision.

The more important consideration, however, would appear to be the properties of the words in consideration. Some words may simply be better candidates for this kind of study, and on the face, it appears that the Swadesh 100 words may on the average be more suitable than the Swadesh 200 words, as Ringe conjectured.

To verify that hypothesis, I recomputed closeness judgments as in Table 10, but this time separately computed results over words found only in the Swadesh 100 list. Table 20 summarizes the results for the language pairs that have any connections at all in the Swadesh 200 list. For all the related languages, the closeness judgment (the proportion of words that are known cognates) is higher for the Swadesh 100 list. That is to say, for these languages, morphemes are retained among the words in the Swadesh 100 list more frequently than in the Swadesh 200 list as a whole.

To take a concrete example of what the problem is, assume that one is working with the Swadesh 100 list and wants more data. Therefore one goes to the larger Swadesh 200 list and picks, say, the concept 'dirty'. That is more data, which in itself is good. But it appears to be the case that cultures are always thinking up new ways of saying 'dirty'. It is just not a very stable word (Swadesh 1955). Even if two languages are historically connected, then unless the divergence is very recent indeed, it is very likely that at least one of the languages will have innovated a new word for 'dirty'. What makes this bad is that this case is indistinguishable from the situation where the languages never shared a word for 'dirty' in the first place, that is, are not historically connected. If indeed the words for 'dirty' are not cognate, then the additional data point does not help in the comparison. It is not even neutral noise in the comparison. It actually leads one to the conclusion that the two languages are not historically connected.

Does this mean that one should work only with the Swadesh 100 list? The answer is too close to call definitively. On the one hand, there is the clear evidence that the Swadesh 100 words are more stable, as well as an empirical demonstration that adding the extra Swadesh 200 words does not seem to make much of a difference. Those factors would legitimize a choice to restrict the size of one's word lists, especially if there are time constraints to consider. On the other hand, there is the undeniable fact that size itself does matter, and it does not seem that the extra words do any harm. One should also take into account that this demonstration was performed here only among a few Indo-European languages, and it does not constitute proof that other culture groups will replace their vocabulary in the same way. If the relative stability of the Swadesh 100 list turns out not to hold up for a particular group of languages, then it would be disadvantageous to discard the other words.

An ideal solution would take into account solid numerical cross-cultural evidence about which concepts tend to have the most stable words. To a certain degree, such evidence may be easier to come by than one would imagine. The Swadesh lists have been widely used by glottochronologists to estimate the time depth at which related languages separated. That enterprise calls for the linguist to count how many words on the lists are cognate between the languages under consideration, and very often the detailed data is published. Thus one could count whether concept A tends universally to be retained as a cognate x times as often as concept B, and the word lists could be prioritized based on that information. I have not heard of any such large-scale survey of surveys, but fortunately several researchers have approached the question directly, either by looking at actual retention rates diachronically, or by seeing which words tend to be found most widely across sets of related languages. Swadesh himself (1955), in a study heavily biased toward European languages, gave the retention rate of all 200 words. Dyen et al. (1967) did the same for 89 Austronesian languages. Kruskal et al. (1973)

looked at Indo-European, Philippine, and Cushitic languages, and compared the word stabilities across those families. The correlation was not high, but was significant. Oswalt (1975), working across many language groups, found certain universal trends, such as the fact that verbs of life support, like 'drink', are retained more than other verbs. Dolgopolsky (1986) did a similar study for 140 languages of Europe and Asia. Taken together, these studies do suggest that it is meaningful to rank concepts with respect to their universal retention rate, which directly correlates to their power and usefulness in statistical studies of language connections. And a good start has already been made on gathering the numbers. On the other hand, the work also shows that there is a great deal of variance cross-linguistically. There is a rather high chance that a concept that is stable in the history of one language group will be very unstable in another language group, and vice versa. When it comes time to testing the probability of a connection between two specific languages, knowing that concept X has a retention rate a few percentage points higher than concept Y across several language families might not help very much if one knows that individual languages vary widely from that average.

It is interesting to note that such findings call into question one of the key assumptions of standard glottochronological computation, that all replacements of vocabulary count equally. But differential replacement rates do not affect the validity of my tests for determining language connection. The tests are valid no matter what the rates of replacement are for the different words. A p below .05 still means that the data are not likely to be due to chance. However, it goes without saying that the *power* of the test does diminish as the rates of replacement of the vocabulary items goes up. If one is restricted to using N words, one is of course better off using words that have the greatest likelihood of surviving as cognates.

6

Precision and Lumping

Virtually all critics of demonstrations such as Greenberg and Ruhlen's (1992) 'suck' (see Chapter 1) have condemned the use of imprecise equations in both the semantic and phoneme domains (e.g., Campbell 1999, Fox 1995, Rosenfelder 1998b). On a superficial level, such criticism is puzzling, for the 'suck' comparanda appear, if anything, to be more similar to each other than are the data often encountered in standard manuals. Semantic shifts from 'suck milk' to 'breast' or 'drink' or even 'throat' in Greenberg and Ruhlen's data set hardly seem more farfetched than shifts from 'beech tree' to 'book' or from 'blessed' to 'silly' in the history of English. Phonetic changes such as Guamo *mirko* from **mälgi* seem much smaller than the universally accepted derivation of English *five* from Proto-Indo-European *penk^we*.

On a more careful inspection, however, the analogy with Proto-Indo-European scholarship is simply not relevant. As Nichols (1996) pointed out, the study of a language family proceeds in stages. First one demonstrates that the languages are connected. After that is established, one can proceed to using that knowledge to uncover surprising facts about the history of those languages. The overwhelming majority of etymologies that one reads in the standard handbooks of comparative linguistics rely crucially on the prior knowledge that the languages are related, and on sound correspondences that are otherwise well established. Prehistoric etymologies with sizable semantic shifts, far from being part of the proof that languages are related, are in fact the fruits of a prior proof.

But regardless of whether there are analogies with standard linguistic practice, one can still wonder whether Greenberg and Ruhlen's demonstrations are probative. The above-cited negative arguments are conclusive and do not need to be repeated here. The crux of the matter is that one needs to take into account the universe of all possible words from which a fuzzy semantic match is drawn. If one lets oneself look anywhere in the vocabulary for a word that is semantically close to a comparandum until one finds one that is phonetically close, there are often dozens of candidates one can consider. Finding such a match becomes orders of magnitude easier than finding a (reasonably) exact semantic match and phonetics that involve recurring sound correspondences. Critics have convincingly demonstrated this both with numbers (Rosenfelder 1998b) and by demonstrating that anybody can easily develop extensive word lists with comparable levels of semantic and phonetic matching between any sets of languages, no matter how unlikely their relationship, and even when the facts contradict secure etymologies. For example, Rosenfelder (1998a) gave long lists of words that are similar in Chinese, Quechua, and English.

And yet, there is no a priori reason why one could not demonstrate language connections using morpheme lists and imprecise matchings on the semantic or phonetic levels, or both. The great difficulty with the current attempts is that one is not given the wherewithal to evaluate just how easy it would have been to come up with the proferred data by chance. Such difficulties would be overcome if one employed the general methodology I am advocating in this book. If the researcher defined in advance the list of concepts that were going to be compared; defined in advance the criteria for accepting a semantic or phonetic match; and then evaluated the positive data along with the negative data in an experiment that provided for significance testing; then the results could be much more convincing. There is nothing in standard statistical methodology that calls for an exact match. It is only necessary that what qualifies for a match be defined in advance and applied equitably and blindly across all candidate words.

But the fact that semantic or phonetic imprecision is theoretically usable does not mean that it is desirable to use it. If imprecision is going to be no better than exact matches, the effort required to precisely define imprecise (or "fuzzy") matches would be wasted. This chapter explores whether fuzzy matching may be useful in the context of the general family of methodologies being developed in this book.

6.1 Semantic Fuzziness

Ringe (1992:64–67) claimed that when one pairs words from two different languages on the basis of their meaning, the significance tests will necessarily be more precise the more exact the matching criterion is. Any introduction of fuzziness will make it harder to prove relatedness. Ringe's model for such fuzziness is to add word pairs that cover all possible combinations of mappings for approximate synonyms. Thus he proposes 'skin' and 'bark' as a pair of concepts that might be considered to be mutually liable to etymological replacement; for example, that a word originally meaning 'skin' may change its meaning to 'bark'. To cover the possibility that such a change may have taken place in only one of a pair of historically related languages, Ringe would suggest adding the matchings 'bark':'skin' and 'skin':'bark', in addition to the already-obtained 'bark':'bark' and 'skin':'skin'. For example, for English and German, this would give the pairs *bark:Haut* and *skin:Rinde* in addition to the more precisely matched pairs *bark:Rinde* and *skin:Haut*. All four of these pairs would be included in the contingency table that one constructs for pairs of phonemes between the two languages.

Such an approach has the admirable property that it avoids the problem of a posteriori matching. Naïve attempts to demonstrate connections between languages very often founder when the proponent points out resemblances between pairs of words that have some semantic connection, but fails to take into account the potentially vast number of word pairs that are just as close to each other semantically, but which were ignored. Ringe's proposal takes that problem into consideration. And it is hard to see that it has the particular drawback that Ringe (1992:66) himself claimed it has. He stated that this procedure simply increases N, the number of comparisons, without increasing the number of cognates, and therefore can only result in a reduced significance measure. However, that claim is clearly not true. It is certainly possible that in a particular language pair only the match 'skin':'bark' will prove to be related when 'skin':'skin' and 'bark':'bark' do not; it is easy enough to imagine that when two languages diverge, one of them might start broadening the application of the old word for skin to integuments in general, and later borrow the French word for human skin in particular. And it is certainly theoretically possible to run into a situation where such inexact matches add enough phoneme matches that they will demonstrate historical connection, despite the dross introduced by the other comparisons.

The following imaginary scenario demonstrates that counting inexact matches could make a difference. The 100-word list expands to 200 comparisons because all of the words are paired not only with their exact semantic match, but also with one near-matching word, which is also involved in the reverse matching, as discussed for 'skin' and 'bark' above. Assume that semantic shifts have taken place so that in each of the four resulting matches, one of the near-matches is historically connected and indeed, the words have the same initial consonant; but that none of the other 3 pairs are connected, including the exact semantic matches. Such a neat case may be unlikely, but certainly large amounts of semantic shifting is quite common. Could the mathematics show that the languages are connected? Certainly: 50 phonetic matches out of 200 is not a bad showing at all, especially if one considers that any noise in the other 150 can only contribute to the count. Imagine for example that in each language, half the words begin with /p/, the other half with /t/. Under the conditions just described, 25 of the words will have a /p/:/p/ match, 25 will have a /t/:/t/ match. For the other 150 pairs, there is no historical connection, so the matchings will be distributed evenly among the four possibilities of matching phonemes: approximately 37.5 each of /p/:/p/, /t/:/t/, /p/:/t/ and /t/:/p/. In the end, one would have a 2×2 contingency table with about 62 in the matching cells (/p/:/p/ and /t/:/t/) and 38 in the nonmatching cells. Such a table has an over-all χ^2 of 11.5, a number which would be significant at $p < .001$ (using the χ^2 distribution, df=1). In contrast, in the same situation, the original test of 100 exactly matching words would have given a very low χ^2, approaching 0. Of course that is an extreme and artificial case, but it does serve to falsify Ringe's claim that the mathematics are such that comparing near-synonyms can never be as good as comparing only exact matches (1992:67).

Nevertheless, I agree with Ringe's intuition that the general procedure he outlined would not be effective. One problem is that there is far too much historical contingency involved in semantic shifts for one to be able to set up pairs of non-synonyms that would be nearly as likely to co-occur as exact synonyms. That is, for any single word such as that for 'skin' in any particular language A, that word in language B which it is most likely to correspond to would be the word for 'skin'. It is almost certainly the case that comparing 'skin' to any other specific word will fail, on the average. And it should be kept in mind that such failure is not harmless, but effectively dilutes the counts for recurrent phoneme pairs, easily leading to potential failure to identify true historical connections between languages.

Perhaps more to the point, the procedure of repeating words violates the basic assumptions of the tests. The point is not to develop a heuristic that, more or less, will give a higher score if languages are connected. The point is to give a statistically rigorous analysis of the likelihood of connection. Ross's original insight was that the arbitrariness of meaning-sound mappings translates into the idea of statistical independence. That allows the idea of historical connection to be expressed as statistical association between two variables, namely, initial consonants in one language vs. initial consonants in the other. The whole scheme is destroyed if words are repeated in the two lists. Certainly a χ^2 test no longer gives a useful metric.

It is conceivable that a more complicated statistical approach could compensate for such deviations. But as soon as one departs from the well-accepted principle of Saussurian arbitrariness, it is not at all clear what numbers would be accurate and acceptable. In principle, a statistical model could be elaborated if one knew just how likely it would be that a word meaning 'skin' would shift to mean 'bark' after two languages diverge, and so forth for the rest of the vocabulary. If it were just a matter of figuring out the histories of words and tallying up the results, the problem would be daunting enough. Most specific shifts such as 'skin' to 'bark' (as opposed to general types of shifts, such as metaphorical extension) are low enough in frequency that one would probably have to research hundreds of languages before one would have a clue as to what the differential frequencies were. It is even harder to figure out if the shifts in question were truly independent events, or whether they might have been encouraged by some sort of contact, or indeed, whether they reflect a situation in a parent language (perhaps one is dealing with an original pair of homonyms, or a single hypernym). But finally, even if one figures out that in any arbitrary language the word for 'skin' is likely to shift to mean 'bark' 0.004 times per century, exactly how does one extrapolate this information to deal with comparisons of unknown time depth? And would such numbers be big enough to contribute appreciably to the overall equation?

A darker outcome of ruminating about semantic change is that it leads one to wonder whether Ross's original model was valid or not. It takes into account the well-known possibility of sound change, but on its face it appears to imply that meanings do not change, which is patently false. Fortunately, however, while semantic change makes the problem of identifying language connections more problematic, it does not invalidate the model. All that this method requires one to believe is that if semantically matched words show more phonetic matchings than (most) other word pairings, then the languages are likely to be connected. The direction of that implication means that the method is not invalidated by the fact that words are likely to drift away in meaning from their matched words. That only reduces the power of the test, meaning that some language connections will go unidentified.

6.2 Phonetic Lumping

Ringe (1992:67–70) made much the same sort of argument as he did for semantic lumping when he rejected the idea that lumping together phonemes can ever do any good. He considered for example the possibility of having table rows and columns not for individual phonemes, but for broader classes such as liquids or velars. His theoretical argument against this was the claim that higher observed frequencies will always be offset by higher expected frequencies. His experimental demonstration was that when one tries this, English and Navajo still are not proved to be related.

It is not clear that a method should be rejected just because it fails to prove that unrelated languages are related. As for the theoretical argument, he is of course stating a general tendency, which will not necessarily hold in all circumstances. Certainly lumping will normally have the effect of weakening the statistical argument, and that is one reason why I advocated moving beyond that way of compensating for under-populated table cell when applying χ^2 tests. But equally certainly, there must be situations in which such lumping would give positive results. Imagine for example a situation where a language splits into two languages, one of which undergoes a palatalization of consonants before /i/, the other of which undergoes a labialization of consonants before /u/. Then matching up pairs of words like /tʲina/:/tina/, /tuna/:/tʷuna/, and /tapa/:/tapa/ will result in three separate $C_{1.1}$ pairings. If that is repeated across the board, one can see that one will get many fewer correspondences than one would have if one had lumped together all phonemes sharing the same primary articulation, i.e., treating /t/, /tʲ/ and /tʷ/ as the same consonant. In general, if the lumping algorithm manages to lump together phonemes that are the result of phoneme splits that took place after the languages separated, and minimize other lumpings, then it is possible to get true positives that would otherwise not be found.

But this is not to say that one should do such a thing. In fact it would be a very bad thing to pick a lumping algorithm based on observed qualities of the two languages under consideration—that would of course be yet another example of the fallacy of trying to prove connectedness by demonstrating the rarity of carefully selected similarities. This does however raise the general question of whether it would be a good thing to call for some specific lumping that should be employed across the board. That is a much harder question, which may have no a priori answer. It seems clear enough that arbitrarily selected lumpings (such as lumping together /p/ and /t/) will simply reduce the power of the tests. But might it be useful to lump together sounds that cross-linguistically have a relatively high probability of being products of conditioned change from the same parent sound? For example, if one has good reason to believe that secondary articulations such as palatalization in consonants very often represent phonemic split conditioned by an adjacent vowel, would one be better off systematically ignoring or "undoing" palatalization by lumping together palatalized and unpalatalized variants of consonants in all the tests? The results would clearly depend on circumstances. If there are palatalizations that have other sources (perhaps /tʲ/ comes from /kʲ/), or if the palatalizations predate the separation of the languages under consideration, or even if they postdate the separation but independently took place under the same conditions, then the lumping will only add noise, leading to a weakening of the estimate of language relatedness. In other cases, where one of the languages underwent a palatalization process that the lumping rule would undo, then the lumping would quite possibly correctly add to the estimate of language relatedness. These considerations mean that the choice of whether or not to lump would depend in part on the application to which one wanted to apply the test. When comparing dissimilar languages and looking for any hint at all of historical connectedness, then it may very well be useful to lump together phonemes that are likely to have the same origin. The general answer is more empirical than mathematical. If on the whole the application of a lumping algorithm in randomly selected cases of known relationship or non-relationship tends to improve the desirability of the results, then one might wish to apply the algorithm for other pairs of languages. Candidate lumpings could be compiled from counting cases in the literature where secondary splits lead to the introduction of new phonemes, or could possibly be derived from the data themselves by randomly combining phonemes.

6.3   Semantic Lumping

It is interesting to note that Ringe came up with two separate approaches to two different types of precision problems. For the phonetic case, he considered lumping phonemes together; for the semantic case, he considered the possibility of adding additional word pairs which are permutations of existing pairs. Equally well one could consider lumping in the semantic case, thereby avoiding the statistical chaos of adding all the extra nonindependent word pairs. If 'skin' and 'bark' are frequent etymological matches cross-linguistically, then one way of approaching the problem is to have one concept, call it 'integument', which would subsume the notions of 'skin' and 'bark'. Here the same problems of precision would apply, but in addition there is the problem of selecting the appropriate word. In a language where the same word means both 'skin' and 'bark' (such as Hawaiian), the collapsing of the categories is a welcome fit between the test and reality. In languages where there are different words (such as English), then one is faced with the problem of whether to use the word for 'skin' or the word for 'bark' when constructing the word pairs. One must of course guard against selecting the word that makes the test come out best. If one selects between the two randomly, then one is almost certainly on the whole increasing the chances of adding noise to the data and getting a less powerful test. In the cases where the words for 'skin' and/or 'bark' do align, as between English and Danish, one is introducing the chance that one will instead match up, for example, English *skin* with Danish *bark* and English *bark* with Danish *skind*.

This sort of lumping of concepts would no doubt be a net win only for concepts that have a relatively high probability of being lumped together. But even the Swadesh lists have terms that are prime candidates for lumping. For example, both 'this' and 'that' are on the list, causing problems in languages like spoken French that do not normally distinguish demonstratives based on proximity, as well as in languages like German that have three-way distinctions (*dieser* 'this', *der* 'that', *jener* 'that yonder'). One suspects that it would have been better if the list simply called for 'demonstrative', or 'this/that', and advised the user on how to select from multiple candidates, for example, by selecting the term that can be unmarked for distance, or to use the word appropriate for the more remote distance.

In a larger sense, though, the Swadesh list already does a lot of lumping. That is, after all, unavoidable in any list of natural language words. If the list calls for the word for 'fat', it is implying that all languages have a single word that denominates exactly the same set of objects that are called 'fat' in English. Of course everybody knows this is not so. Virtually no language comparison has ever been made where the researcher did not have to decide between words for human fat versus that of food animals versus that of olives, or animal fat in situ versus that used as food versus that used as an ingredient in food, or some other such qualification. Of course when the researcher is making such decisions on the fly, it is very difficult to provide guidance that will guard against biased choosing (in an English–Yiddish study, would one be tempted to choose Yiddish *fets* because it sounds like English *fat*, or conversely, to choose *shmalts* because it does not?). From the standpoint of most people who have worked with the Swadesh list, it would appear that the problem is not lack of lumping, but the presence of too much lumping already. On the other hand, if one is too specific ("the fat taken from the shanks of a three-year-old boar and used for frying rice"), one increases the risk of choosing concepts that are not familiar in a particular culture, and for which they are therefore likely to have a term that is borrowed or recently derived, or indeed, not have any term at all.

What is needed in an ideal test is a level of semantic specificity that balances the need for nonarbitrariness with the need to find concepts that are universal and terms that are not subject to or likely to be the product of replacement. This may be a complex undertaking. It is in part an empirical question: What are the collocations of objects that are most likely cross-culturally to be denominated by words that are commonly used in everyday discourse, are not of transparent composition, and are least likely to be borrowed and replaced? Which of these are (nearly) universal? And how should one decide between competing terms when there is no universal agreement? The answer to the latter will in part be logical and a priori (e.g., pick the most specific word, or the word used most often in daily discourse, or the simplest, or the shortest), and in part a posteriori (if a culture has no general word for 'fat', prefer the one that applies to cooking fat taken from common food animals, because that is known to be more stable than terms for human fat).

From a statistical point of view, semantic lumping is valid, provided of course that the range of concepts to be lumped, and an objective way of deciding between words that express concepts within that range, are determined in advance and without bias. However, I have chosen not to lengthen this chapter with experiments of my own for two reasons. First, coming up with principles for semantic lumping is a complicated, time-consuming enterprise, particularly if one wishes to do a creditable job of identifying concepts that really are especially likely to be subsumed by a single word cross-linguistically. It would prove nothing if I decided to collapse 'skin' and 'bark' after experience with the very language set I am testing on; such decisions really should be made on the basis of research with hundreds of languages not in my test suite. A second reason for demurring is an educated hunch that any gain from even well-informed lumping would not be very impressive. The Swadesh 100 list is already the product of at least five years of refinement under conditions of extensive use with diverse non-Indo-European languages. To a large extent it already represents a well considered judgment as to the proper level of specificity for investigating words cross-linguistically. As far as I am concerned, a difficult experiment is not necessarily off-putting, nor is an easy experiment with a low likelihood of payoff, but a difficult experiment that is not likely to work is another matter entirely. Nevertheless, those who adjudge the odds differently are encouraged to give it a try; this general methodology is conducive to such experiments.

7

Nonarbitrary Vocabulary

It is easy to lose sight of exactly what is being measured by the tests under consideration. Because the historical linguistics literature is full of discussions about whether languages are genetically related or not, it is tempting to use a shorthand and say that when the tests have low p values, they are showing that the languages are related. While that is one useful conclusion that the tests can support if used very carefully, it is important to keep in mind that what they are actually testing is whether there is any pattern at all connecting the vocabularies of the two languages.

Positives (i.e., low p values) can be found if some universal property of human language tends to make words in different languages be similar along some dimension measured by the test. It is, for example, commonly noted that onomatopoetic words may well be expected to be rather similar from language to language, because each language is trying to a certain extent to imitate the natural sound (Campbell 1999:320, Hock 1991:558). It is clear therefore that words such as *kerplop* or *bow-wow* should be left off the word list. But other concepts may also have natural linguistic expressions. For example, it is possible that certain words expressive of emotions may tend toward having similar forms in various languages: for example, words like *ah*, *oh*, *ouch*. And it is often noted that certain nursery words like *mama* and *papa* have a very good chance of being similar in unconnected languages (Murdock 1959, Jakobson 1971). Naturally, animal cries, exclamations, and nursery words are kept out of word lists such as those of Swadesh precisely out of the fear that similarity between them will be due to factors other than linguistic connection. Less obviously, perhaps, these things can sometimes act to mask what would otherwise be a recurrent correspondence, and so actually raise p values. Campbell (1996) gave the following example. Proto-Mayan /ts'/ normally remains unchanged in the daughter languages, leading to recurrent /ts'/:/ts'/ matches in daughter languages such as Tojolabal and Jacaltec. But while Proto-Mayan /ts'uh-/ 'drip' remains unchanged in Jacaltec, in Tojolabal it is deflected to /t'uh-/, almost certainly because that sounds more like a drip. That breaks the regularity of the /ts'/:/ts'/ correspondence and therefore makes the language relationship harder to identify.

Excluding obvious onomatopoeia is one thing, and by and large, pretty easy. However, it must be kept in mind that the origins of vocabulary are obscure. It may well be possible that some mainstream words have their origin in such universal expressions. Students of sound symbolism claim that stop phonemes are used disproportionately often for denoting abrupt acts, continuants for continuous acts, nasals for reverberating sounds, acute sounds for small size and sharpness, grave sounds for large size and smoothness (Hinton et al. 1994:9–10). It hardly seems out of the question that the Indo-European word for 'mother', typified by Latin *mater* (where *-ter* is a suffix of familial relationship, as in *pater* 'father'), may have its origin in nursery expressions like *ma* or *mama*. Nor does it seem unlikely that the Navajo word *-má* 'mother' has the same origin, especially since /m/ is a rare initial (Young and Morgan 1992, 395 gave as its etymology "PA *-han, mother", which would contradict this point, but which seems etymologically unintelligible; more likely they implicitly meant to contrast the two forms). Comparing Latin and Navajo using the Swadesh 200 word list, which includes the concept 'mother', would yield a $C_{1.1}$ match for /m/:/m/, and indeed a V_1 match for /a/:/a/. That one case may be more or less harmless, but the cumulative weight of several such cases could end up lowering the p value reported for a language pair. If a low p is to be taken as evidence of a special connection between languages, then such cases are trouble. The question is to what extent such cases exist in human language in the core adult vocabulary. Is 'mother' a special case, or the tip of the iceberg?

The everyday experience of linguists and polyglots tells them that vocabulary is almost entirely arbitrary. But there is also the insidious possibility that universal tendencies in language are more subtle than people are capable of easily perceiving. What if words for sharp things were just 20% more likely to contain an acute vowel like /i/ than the typical word is, prescinding from questions of language connectedness? If there were more than one word for sharp things in the word list, then their inclusion would have the general effect of lowering the p values one gets from tests, because the number of matches like /i/:/i/ would be higher than one expects from chance combinations. That is, any such universals can have the effect of making researchers claim that languages are historically connected, when in fact the evidence does not support that conclusion. To be sure, I doubt that such effects are strong, but the fact that /m/ is such a widespread initial for 'mother' (Murdock 1959) forces one to admit that such universals are possible and therefore worth looking into. Ideally, word lists for these kinds of study should be subjected to statistical tests across many languages, to determine whether there is any skew in their phonological properties over and above what can be attributed to particular language connections. Words with appreciable skew would be removed from the lists.

Unfortunately, a rigorous examination of the Arbitrariness Hypothesis may be impossible in principle. If related languages are being considered, unusually high frequencies of correspondences for words like 'mother' could mean simply that that word is particularly resistant to replacement. Such resilience, of course, is usually considered a desirable property, especially when one is particularly interested in investigating the possibility of relatedness between languages. What one needs to do instead is look for concepts that have unusually similar vocabulary across unrelated languages. But, as discussed earlier, no pair of languages is known to be unrelated. Even if one did a large-scale survey of vocabulary across widely separated languages, and found a set of words that were more similar to each other than are other words, it would be impossible to state whether one has succeeded in uncovering linguistic universals that contradict the Arbitrariness Hypothesis, or whether one has found a core Proto-World vocabulary that is unusually resistant to being replaced by neologisms or loans, or whether one has found early and widely traveled loan words.

A more promising approach would entail the careful study of the history of individual languages. If one finds that certain words consistently seem to resist sound changes, or appear after a certain period of time to change back to their original form, then one might logically have grounds for suspecting that universals are at work. For example, the English word *papa* could not be cognate to Latin *papa* or Greek *páppas* (Homer *Odyssey* 6.57) by normal descent from a common Proto-Indo-European ancestor. If it were, one would have expected the various sound changes to have turned it into something like **fave*. The fact that the form is now *papa* could be taken as demonstrating that there is a universal trend for that word to be formed from the syllable /pa/. Of course, there are usually other possible explanations that are hard to rule out. In many cases such a form could just be a borrowing back from another language. In this particular case, the word *papa* is probably influenced by the French word. The possibility of such borrowings makes it almost impossible to determine whether a word is a universal or a borrowing. Fortunately, one is often just as anxious to get rid of loans as to get rid of universals, so a large-scale effort to identify such items could pay off even if those two cases are hard to distinguish; one way or the other, there is something funny about *papa*. The unfortunate side of the coin is the fact that such an effort really would have to be large scale. It is difficult enough to figure out the history of a language family well enough to be able to formulate rules that would help one to determine whether a particular word history is irregular; even more, to do that over hundreds of words; and much more, to do that for dozens or hundreds of languages so that one can generalize that particular concepts are suspect and should be omitted from word lists.

It is also conceivable that experimentation may help resolve the issue to a certain extent, but the problems are complex. Most experimentation has addressed the issue of whether speakers have consistent judgments about direct connections between sound and meaning, regardless of the origin of those judgments. For example, Berlin (1994) reported that American students were able to distinguish bird names from fish names in an unknown language (Huambisa) with 58% accuracy. That fact alone is interesting enough, whether the explanation turns out to be onomatopoeia or sound symbolism based on such factors as animal size or modalities of movement. But if it turns out that the effect is due to the fact that the two languages contain within their structure certain codings that the speakers internalize, then one has gotten no further than in examining the languages themselves. There is no way of knowing whether the symbolic code is universal or the result of some linguistic contact, perhaps in the remote past.

There does however seem to remain open the possibility of demonstrating that speakers are sensitive to sound-symbolic codes even though the languages they are familiar with do not deploy that code themselves. But the methodology required for doing so may be complex, for it is not clear how one can definitively prove that a language has no perceptual deployment of a code. For example, it is often pointed out that the English words *big* and *small* contravene the claimed universal that words for 'big' should have grave, sonorous vowels, and words for 'small' should have acute vowels. Therefore, one might argue, if a English speaker reports in experiments that words with grave vowels sound like they should denote big objects, it must be because of a psychological universal, and not the effect of the English vocabulary. But the English speaker may be picking up patterns instead from arrays of words like *teeny-weeny, itsy-bitsy, little*, and *large, humongous, huge*, which do conform to the claim. And one would also have to consider words for large objects and small objects, and somehow weigh the salience of these various objects. That is a daunting task.

Until attacks on the Arbitrariness Hypothesis are finally resolved, one must be content with generalizations and observations that have been made in the literature. In general it is agreed that the following types of concepts may well be subject to certain universal pressures, and so should probably be omitted from studies that rely on language arbitrariness:

1. Representations of paralinguistic sounds and expressive interjections, such as *oomph* and *ouch*.

2. Onomatopoeia: imitations of natural sounds, such as *bang* and *meow*.

3. Items that are not normally used in a syntactically structured sentence, but instead are normally complete utterances in themselves. This includes words only used as interjections.

4. Words that include a sound or sequence that is unique or otherwise very rare in the language, such as *ptui* (which is supposed to represent a sound where the tongue is placed against the lip), *uh-uh* (which includes nasalized vowels and a glottal stop), and *boink* (/ŋ/ plus noncoronal consonant sequences do not occur word-finally after diphthongs or long vowels in the core vocabulary). An example of a more ordinary word is the aforementioned Navajo *'amá*, where /m/ is rare. In many languages, reduplication or partial reduplication is also diagnostic of a possible sound-symbolic form (Bloomfield 1984, Hinton et al. 1994:9), for example, English *ding-dong*, Albanian *nënë* 'mother'. That diagnostic would of course not apply in the many languages that used reduplication as a regular derivational or syntactic process.

5. Ideophones. Many ideophones are clearly onomatopoetic and can be excluded on that ground. But even when they are not onomatopoetic, there seems to be an intimate connection between sound and meaning. This is evidenced in part by the fact that ideophones often fall in families where small changes in the pronunciation correspond to partial changes in the meaning. For example, Bahnar /dəbuuŋ/ 'ridge of an immense roof', /dəbooŋ/ 'ridge of a large roof', /dəbɔɔŋ/ 'ridge of a small roof' (Diffloth 1994). (Note by the way that this is another example of a language where the alleged universal correlating size with sonority is reversed.) Even if the sound–meaning correspondence is not universal, the patterning would produce other statistical problems, as will be discussed later. One also gets the impression that most ideophones are not extremely old anyway, and so would be of little value in diagnosing whether languages are connected at any profound time depth.

6. Terms used mostly by children, or by and with infants, for example, *mama, papa, caca, booboo*. If these are not actually universal forms, it appears that they are highly likely to be borrowed, because of certain undeniable universalistic pressures, namely the need or desire to have a working vocabulary that is simple enough for the youngest children to vocalize. It would probably be reasonable to routinely discard a few terms such as 'mother', 'breast' and 'father', even if they are not nursery forms, because there is always the possibility that they derived from nursery forms.

7. Well-documented instances of sound-symbolic words. Correlations between the size of an object and the frequency (pitch) of sounds in the denoting word seem to be the best case made for a universal sound-symbolic code: High frequencies (including acute and/or voiceless segments like /i/ or /t/) suggest smallness, low frequencies (including grave and/or voiced segments like /a/ or /g/) suggest largeness. There is good argumentation for why this particular coding should be universal (Ohala 1994), although of course the conventional nature of language can override it, as in the case of Bahnar above. It is hard to say just how deeply this particular effect is embedded in languages, but it would be wise to consider carefully whether one should include particular words where the code seems to be in effect, and to systematically exclude from word lists concepts like 'large' and 'small'.

8. Spontaneous words like *dork*, nonce words *(slithy)*, and words of irregular formation, such as blends *(swipe)*, are also best avoided for reasons of potential sound symbolism, but it is of course already rather obvious on other grounds that they would not be very useful for investigations of deep connections between languages.

9. Words that seem to defy sound laws. If a word defies a common change such as /p/ > /f/ in Germanic, that usually suggests that it is a loanword, or that there is some unknown condition to the sound change, but it may mean that the word "changed back" because the new form is no longer motivated. As an example, baby birds still *peep* /piːp/, even though the English Great Vowel Shift changed the Middle English word /piːp-/ to /paip/ *pipe* (Anttila 1972:86).

It may be imagined that such problems will not arise in the Swadesh lists. These lists have after all been constructed for the express purpose of investigating how closely languages are related by noting how rapidly the vocabulary is replaced. It would not serve the interest of glottochronologists if the list contained motivated items. And to the eye the list contains "normal" words, nothing like *dork* or *achoo*. However, a closer inspection suggests many potential pitfalls. I checked the etymological sources for all 1600 words in the word lists (the Swadesh 200 list for 8 languages), and found not a few suspicious cases. Granted, it is difficult to know for sure whether an item is motivated. But for the statistical purpose of proving language connections, the burden of proof is the opposite of what one might be used to. If one considers a word to be potentially motivated, and therefore discards it from the tests, at worst one is slightly reducing the power of the test. If one however neglects to remove a motivated word, then the test becomes invalid, in the sense that one may spuriously find that languages are connected on the basis of inappropriate data. Therefore it is important to discard words that *might* have some universal motivation. The following cases impress me as being possibilities worth considering.

- 'Bad' (Swadesh 200). The Albanian word is *keq*, which has an obscure etymology, but is often assigned to Greek *kakos* as a loan (Huld 1984). Whether borrowed or native, this strongly suggests the common nursery word, /kaka/.

- 'Belly' (Swadesh 200). The German word *Bauch* sounds arbitrary enough, but this was *būh* in Old High German, and suggests a PIE root *bhū-*. Not only is *ū* an unusual root vowel in PIE, but the word would be perfect onomatopoeia for a blowing sound, and the belly is that which puffs up (Kluge 1995). If this etymology is accepted (note that the English word *belly* has the same Old English source as *bellows*), it points out two dangers. One is that the motivating metaphor may not be immediately obvious to modern speakers. The other is that the onomatopoeia may be obscured by sound changes. Here, the diphthongization of long high vowels in German makes the modern word less obviously onomatopoetic for blowing than does the Old High German word. The Navajo word for 'belly' is *'abid*, stem *-bid*. Young and Morgan (1992:65) also note that *bid* is onomatopoeia for "a hollow thumping sound, as that produced by patting a dog's stomach". They also provide a perfectly regular etymology for the 'belly' word from a Proto-Athapaskan *-wɪ't'* 'be gluttonous', so it is difficult for a nonexpert to weigh which came first, or to decide whether *-wɪ't'* itself sounds onomatopoetic. But it does not seem inconceivable that there is some onomatopoetic motivation in the word.

- 'Bird' (Swadesh 100). Of the possible Navajo translations, I follow Ringe (1992) in using *tsídii*, which refers to smaller birds. Young and Morgan (1992:1000) derive this as nominalization of *tsíd*, a chirping sound. Perhaps this highlights the potential problem that onomatopoeia in foreign languages may not sound particularly representational to people unfamiliar with the language. In hindsight, however, it does seem fairly appropriate. All the sounds are acute, and the tone vowel is high, all of which makes a natural representation for a high-pitched sound.

- 'Blow' (Swadesh 200). The Hawaiian word is *puhi*. The first syllable at least sounds like excellent onomatopoeia for blowing, and even if the reconstructed Proto-Oceanic /s/ in *pusi* (Ross 1995) does not sound as natural as the Hawaiian /h/, it is at least compatible with the idea that continuants are natural sounds for continuous actions. Compare English *puff*. Swadesh himself (1955) withdrew this word from his list because of the observation that words for 'blow' tend to have labials or sibilants, and rounded vowels.

- 'Bone' (Swadesh 100). The German word *Knochen* is difficult to etymologize, unless one follows Kluge (1995) in deriving it from the sound of cracking one's joints (in German, *knacken*; cf. also the English onomatopoetic word *knock*).

- 'Breast' (Swadesh 100). The nursery words that are taken as universals for 'mother' appear just as readily to designate the mother's breast. Arguably labial sounds are a very natural designation for the site of nursing. The Latin word *mamma* meant either 'mother' (informally) or 'breast', and is nowadays widely accepted as coming from the nursery word (Pokorny 1959:694). The Turkish word *meme* has the same labials and an uncharacteristic reduplication. Note that Greenberg and Ruhlen's (1992) 'suck' etymology contains several examples of words for 'breast' that begin with /m/. Labials or lip-rounded vowels are also found in the Hawaiian word *ū* (Proto-Oceanic *susu*, Blust 1993) and the Navajo word *-be'*.

- 'Dirty' (Swadesh 200). The Navajo word in the word list is *baa'ih*. Its degenerate conjugation suggests that it is a fairly recent derivative of the current interjection of disgust, *'ih* (Young and Morgan 1992:241). Compare English *icky*.

- 'Fall' (Swadesh 200). The Navajo word selected is *naashtłíísh*, from the root *tłíízh*. That also turns out to be the onomatopoeia for the sound of falling timber (Young and Morgan 1992:572). This is particularly hard to hear, and one might suspect that the interjection comes from the word, like English *timber!*; but Young and Morgan claim that the word is imitative. Compare English *crash*.

- 'Father' (Swadesh 200). There are several words with labial sounds, which suggest a nursery origin. The Navajo *-taa'* has no obvious etymology, and would seem to be a simple nursery word like English *da, dad.* The Turkish word is *baba,* which has unusual reduplication; Albanian *babë* is usually thought to be borrowed from it. The Latin, French, English, and German words would reflect a Proto-Indo-European *pH₂ter,* that is, root *p;* nowadays, this is generally accepted as having an origin in the nursery (Pokorny 1959:829). Swadesh (1955) also noted problems with this word cross-linguistically. This case is instructive in that the *pHter* word has been considered the formal word for probably thousands of years; the presence of nursery by-forms such as *papa* can serve as a decoy distracting researchers from the fact that the formal word has a nursery origin as well. The problem is exacerbated by sound changes. The French word for example now has the vowel /ɛ/ instead of the more basic /a/, and the English and German words have substituted /f/ for /p/. Considered completely in isolation, it may not have been obvious that German *Vater* is ultimately a nursery word.

- 'Few' (Swadesh 200). The German word *wenige* would seem to be ultimately a cry of lamentation, PIE *wai* (Kluge 1995, Pokorny 1959:1111).

- 'Fight' (Swadesh 200). The Latin word *pugnat* would come from a root *pug,* which sounds like a natural word for *poking* or *punching.*

- 'Fire' (Swadesh 100). The English and German *(Feuer)* words would go back ultimately to a Proto-Indo-European root like *pew,* which has been referred to the sound of air movements caused by fire (Kluge 1995).

- 'Fly' (Swadesh 100). Albanian has several words, such as *flutur-, flu-, flatr-;* cf. also *fletë* 'wing'. These are shockingly close to English words like *flitter, fly, flutter,* and yet should have no etymological connection (English initial /f/ should come from Proto-Indo-European /p/, which would remain /p/ in Albanian; Beekes 1995). Çabej (1982:196) considered the Albanian words to be spontaneous, motivated creations; Bloomfield (1984) considered *flitter* and *flutter* to be symbolically motivated words, and noted that that sort of word has some small chance of appearing in different languages. This raises a question: Even if the Germanic words (English *fly,* German *fliegen*) have etymologies that are not onomatopoetic (in this case, a Proto-Indo-European root *pleu-* 'flow'), can one be sure that sound-symbolic considerations did not influence speakers to select those words from all competing Indo-European words that could have served as etyma for a word expressing the concept 'fly'?

- 'Hit' (Swadesh 200). Bloch and von Wartburg (1964) considered the French word *frappe* to be onomatopoeia. Certainly words ending in a stop are very suggestive of hitting; consider also German *schlag-*, from a Proto-Indo-European *slak*.

- 'Laugh' (Swadesh 200). The English and German words are derivable from a Proto-Indo-European *klak*, whose unusual root vowel supports Kluge's estimation (1995) that this is onomatopoetic. The Hawaiian word is *'aka*, which would reflect a Proto-Polynesian *kata*. Swadesh (1955) also noted problems with this word being motivated in many languages when he withdrew it from his list.

- 'Mother' (Swadesh 200). The nursery origin of Albanian *nënë* is apparent in its atypical reduplication. Similarly for the atypical geminate in Turkish *anne*. Most of the other languages have the expected /m/ sounds: Navajo *-má* (despite the fact that it is otherwise an unusual phoneme), Proto-Indo-European *ma* surviving in Latin, French, English and German. Swadesh (1955) also noted problems.

- 'Mouth' (Swadesh 100). The French *bouche* would appear to have the same origin as German *Bauch* 'belly', discussed above: that which puffs up (Ernout and Meillet 1979, "bucca"). The Hawaiian word is *waha*, which has been traced back to a Proto-Malayo-Polynesian *baqbaq* (Blust 1993). In general labial sounds for words meaning 'mouth' or 'lips' have such obvious symbolic potential that they should be scrutinized very carefully.

- 'Nose' (Swadesh 200). The 'nose' words in the eight languages under discussion are not usually considered sound-symbolic and I did not treat them as motivated in this study, but perhaps I should have. The Indo-European languages all reflect a PIE naH_2s, which begins with a nasal consonant. A nasal sound would be a very obvious component of a word for 'nose'. The Navajo word *'áchį́į́h* contains a nasalized vowel, which may be as good as starting with a nasal consonant. Less compelling are the final nasals in Turkish *burun* and Proto-Oceanic *isuŋ*, the source of the Hawaiian *ihu*.

- 'Path' (Swadesh 100). The English and German *(Pfad)* words do not properly connect with the Proto-Indo-European root *pent*, which have led many people to imagine unlikely borrowings from Iranian. A simpler solution, attributed by Kluge (1995) to Sommer, is to posit recurrent onomatopoetic formations. Cf. English *pitter-patter*, an expressive formation often applied to walking. Perhaps more alarmingly, once one accepts the idea that a Proto-Germanic *pat-* for 'path' may be onomatopoetic, this raises the suspicion that the Proto-Indo-European root *ped-* for 'foot' may have the very same motivation.

- 'Rotten' (Swadesh 200). Latin *putridus*, French *pourri*, and German *faul* would all share a Proto-Indo-European base of *pu*, which is a rather obvious interjection of disgust, cf. English *pooh, pfui*. Note how sound changes have disguised the German form (also English *foul*).

- 'Sleep' (Swadesh 100). The Navajo root *ghaazh* is supposed to originally mean 'make a bubbling noise' (Young and Morgan 1992:234). Thus the root is onomatopoetic.

- 'Small' (Swadesh 100). Although this word is meant to be a canonical case of sound symbolism, the word seems arbitrary in most of the eight languages. But French *petit* comes from a Late Latin form like *pitito* (Bloch and von Wartburg 1964), which clearly seems to be an expressive encoding according to the rule "acute = small". The Hawaiian word is *iki*; its Proto-Polynesian ancestor, *'iti*, is even more expressive.

- 'Spit' (Swadesh 200). Etymologists by and large see onomatopoetic sources for most of these words.

- 'Squeeze' (Swadesh 200). The English word has emerged from a tremendous mishmash of words such as *squash, quease, squiss, squish, squize, quash*. Such a proliferation of variants is a strong indication that speakers at least find something expressive in the sounds. The Hawaiian word is *'uwī*, which might have gone unnoticed if it did not also mean 'squeal'.

- 'Suck' (Swadesh 200). The Albanian word *thith* is the same stem as *thithë* 'nipple'. The reduplication and the fact that the latter alternates with *cicë* and *sise* clearly points this out to be a nursery word. The English, French, German and Latin words all go back to forms like *sug* or *suk*, which is an obvious representation of the sound of sucking or slurping something in. The Pre-Proto-Athapaskan root *t'ut'* also strikes the ear as onomatopoetic.

- 'That' (Swadesh 100). The basic morpheme for a remote demonstrative is *'á* in Navajo and an unexplained *a-* in Albanian. Arguably there may be a certain natural pressure to select a very short vocalization for the demonstrative, and the most sonorous vowel may be naturally representative of a great distance (Ruhlen 1994:12).

- 'Wind' (Swadesh 200). The Proto-Indo-European root underlying the Latin, French, German and English words is approximately

$H_2 w\bar{e}$, which is a credible representation of the sound of the wind blowing.

All in all, including some transitive cases (e.g., the French and perhaps the Navajo words for 'wipe' are derived from the motivated word for 'suck') I count approximately 30 concepts that had some overt credible problem with nonarbitrariness. About a third of them appeared in the Swadesh 100 words. That is, the Swadesh 200 additions have about twice as many problems with motivated words as do the original 100. In addition to the aforementioned problem, that Swadesh 200 words are more likely to disappear over time, this is another reason why work with the full Swadesh 200 list may not give results as good as those afforded by the shorter list, despite the fact that they constitute more data.

Whether 30 problematic entries out of 200 should be considered bad is a moot question. One mitigating factor is that not all of the identified words are likely to cause much of a problem. There might not be many languages in the world where the word for 'few' derives from the natural sound of crying. And one would have to be tone deaf to miss words as clearly motivated as French *petit*. But many of the other cases seem much more credible as potentially dangerous universals. More alarmingly, the list points out how difficult it can be to recognize incontrovertible cases. Unfamiliar sound systems may make onomatopoeia virtually unrecognizable, and sound changes may turn originally natural sounds into something quite arbitrary.

In the end, it is difficult to decide on the impact of possibly nonarbitrary words. One sober judge could decide that nonarbitrariness is well constrained in the vocabulary among certain rather obvious classes of words, and that it would be rather generous to even concede the advisability of discarding from the Swadesh lists a few words, perhaps 'mother', 'father', 'large', and 'small'. For the rest, one can trust the linguist to identify and discard the rare problems. And a few mistakes could be relatively harmless, because a few motivated words are unlikely to lead to recurrent sound correspondences. But another equally sober judge could decide that the problem is essentially boundless. After all, the idea is to make tests that are as sensitive as possible to any deviation from absolute arbitrariness in sound-meaning correspondence. So theoretically, quite tiny doses of nonarbitrariness can throw powerful tests off completely, with false positives. If there is any tendency at all toward onomatopoeia in the core linguistic vocabulary, that is, if people have even a small tendency toward being influenced in naming things and actions by the natural sounds associated with them, then to the extent that that tendency is not recognized and compensated for in all particulars, any finding of linguistic connection can be spurious. Likewise, even if the only important universal case of sound symbolism turned out to be the Frequency Coding for size, could one possibly hope to root out every case where a low tone or grave phoneme happens to appear in a word that denotes big things?

For the sake of argument I will take an intermediate position in this study. While I admit that any findings of distant linguistic connections must necessarily remain suspect until such a time as nonarbitrariness in words can either be disproved or factored out of the equation, I will for the sake of argument trust that diligence can suffice to factor out all cases of nonarbitrariness from the word lists. In all further statistical tests, I will discard words that are likely to have a motivated source. Table 21 shows how the tests come out when I discard concepts for which either of the two languages under consideration has a nonarbitrary form. As can be seen, the number of words discarded is moderate (ranging from 10 to 21), and the difference in results is minuscule. In most cases, the related language for which the p values had room for improvement do go down when motivated words are discarded, but the magnitude of the change is for the most part small, and there is one exception: The Albanian–English connection has a higher p. There is no noteworthy trend among the unrelated pairs of languages.

7.1 Grammatical Elements

Many researchers are enthusiastic about comparing grammatical elements such as inflectional affixes in statistical studies such as this (Cowan 1962, Collinder 1947, Ringe 1992). Belief that morphology can be especially probative of language relationships is inherited from the traditional methodology of historical linguistics. For example, Meillet (1926:91) taught that comparing words was by itself suggestive evidence of linguistic relationship, but not absolute proof in the absence of morphological evidence. This theory has broad support today (Nichols 1996, Poser and Campbell 1992) and is widely put into practice. Sapir (1911) was hesitant to pronounce that languages were related on the basis of the lexicon alone. And among the evidence that Hittite is an Indo-European language, for example, the general consensus is that the most convincing is not word lists per se, but rather, morphological alternations like nominative *wādar* versus genitive *wedenaš* 'water'; participles in *-nt-*; and pronominal alternations like *kuiš* 'who' versus *kuit* 'what' (Poser and Campbell 1992, favorably citing Hrozný 1917 and several other Hittitologists). Even Sir William Jones is often claimed as a true father by adherents of the comparative method such as Nichols, not because there

TABLE 21 Effect of Discarding Motivated Vocabulary

Languages		Discarding		No Discards		Cognates[1]
		p	N	p	N	
English	German	.000	185	.000	199	.593
French	Latin	.000	184	.000	199	.565
English	Latin	.000	186	.000	199	.292
German	Latin	.000	185	.000	200	.290
English	French	.027	182	.030	199	.285
French	German	.001	180	.011	199	.253
Albanian	Latin	.000	185	.000	200	.230
Albanian	French	.004	179	.006	199	.203
Albanian	German	.053	180	.093	200	.152
Albanian	English	.051	182	.020	199	.130
Albanian	Turkish	.803	188	.797	200	.005
English	Hawaiian	.073	185	.083	199	.005
Albanian	Hawaiian	.045	183	.021	200	.000
Albanian	Navajo	.693	183	.855	200	.000
English	Navajo	.218	181	.105	199	.000
English	Turkish	.077	188	.078	199	.000
French	Hawaiian	.396	182	.278	199	.000
French	Navajo	.606	178	.702	199	.000
French	Turkish	.281	185	.329	199	.000
German	Hawaiian	.308	182	.351	200	.000
German	Navajo	.178	180	.233	200	.000
German	Turkish	.703	186	.681	200	.000
Hawaiian	Latin	.465	187	.384	200	.000
Hawaiian	Navajo	.261	183	.256	200	.000
Hawaiian	Turkish	.791	190	.834	200	.000
Latin	Navajo	.001	185	.002	200	.000
Latin	Turkish	.354	192	.304	200	.000
Navajo	Turkish	.202	188	.201	200	.000

Note: All comparisons use initial consonants; the Swadesh 200 word list; and the χ^2 metric. N columns show the number of concepts remaining after discarding concepts with unusable words (motivated or with zero phonology) in either language.
[1]Closeness judgments from Table 10.

was any evidence that he worked with recurrent sound correspondences, but because he included grammatical elements among his comparanda.

It might first be useful to see why researchers believe morphological evidence to be especially probative. The most often stated reason is that such elements are least likely to be borrowed (Meillet 1926:91, Ringe 1992:77, Cowan 1962). That belief is largely correct, though the generalization one sometimes hears, that grammatical morphemes are never borrowed, or so rarely borrowed that one can safely ignore that possibility, is clearly wrong. Thomason and Kaufman (1991) have developed a well-nuanced scale for relating type and degree of borrowing with the intensity of contact between cultures. At the most casual level of contact, only content words are borrowed; function words such as conjunctions may be borrowed at the next more intense stage of contact; at an intermediate stage, adpositions and derivational affixes may be borrowed, and inflectional affixes may remain attached to borrowed words; and under strong cultural pressure, even inflectional affixes may be borrowed and attached to native words. They also find several instances of speakers carrying over inflectional endings from their old language when they learn a new one, as when an entire culture shifts imperfectly to the language of their conquerors. So looking at grammatical particles does protect somewhat against loans, but one must not think that is absolute protection. But why is protection against loans important at all? The issue of loans in general will be addressed in detail in the next chapter, but for now suffice it to say that the motivation of historical linguists has always been that they wish to explore the question of genetic relatedness in the strict sense. Therefore the fact that the Hittite present participle is -nt- is taken as more important than a lexical comparison, because it is more likely to be an inherited form. However, if one is interested in the broader question of whether two languages are historically connected in general, the fact of borrowing is much less important, although it will be seen later that in some respects the tests will be more powerful if certain types of loans can be avoided.

Another reason for preferring morphology is that the structural patterns that call for the application of a morphological element remove an additional amount of chance from the equation. To take the simplest case again, the fact that there is a participle in -nt- means a little bit more than the fact that, say, the word for 'night' is nekuz. In addition to the consideration that, say, Latin and Hittite have picked similar (or, rather, phonetically corresponding) morphemes for these two concepts (Latin -nt- and noct-), there is also the consideration that the languages have in common the very idea of having the grammatical category of participle, and that they express that category by means of adding a suffix to a verb stem. In contrast, the fact that they both have a word for 'night' is less of a surprise. Similarly, the alternation of *kuiš* 'who', *kuit* 'what' is somewhat more important than just the fact that the interrogative pronoun is *ku-* (Latin *qu-*) and that the animate is in *-š* (cf. Latin *quis*) and the inanimate in *-t* (corresponding to Latin *quid*). There is also the factor that the languages agree in having an animacy distinction for this word in the first place, and that the difference is expressed by means of suffixes, and furthermore, that the dental suffix expresses the neuter only for pronouns, not for nouns. The case is even stronger for some of the other comparanda. The case of 'water' is interesting not only because the two stems agree with the two stems seen elsewhere throughout Indo-European (e.g., *wādar* with English *water*, *wedenaš* with Swedish *vatten*), but also because of the very fact that the stem has such an alternation at all.

The most probative of all types of morphological evidence is held to be shared irregularities (Meillet 1925:27). Consider for example the third person indicative present of the verb 'be' in the singular and plural: Latin *est, sunt*; German *ist, sind*. In addition to the fact that the words for 'is' agree so well in these two languages, there is also the consideration that the plural has a peculiar irregularity in both languages: The root vowel disappears. Not only is it odd that the two languages have this special behavior in the plural, but it is also a striking coincidence that the irregularity, which occurs nowhere else in the languages, is found in the same verb.

My opinion is that the above-cited reasoning to the effect that shared morphology in general and shared morphological irregularities in particular reduce the likelihood of chance is good, although the magnitude of the importance may be overstated. Sometimes our natural satisfaction at seeing patterns may cause us to be overly enthusiastic when we discover data that fall together into a coherent explanatory pattern. The matter of the *r/n* stems is a case in point. The discovery that Hittite has an entire declension of words that vary through their inflection by having *r* in some cases and *n* in others is a wonderful explanation of the fact that Indo-European languages differed among themselves in whether the word for 'water' (and one or two other words such as 'fire') had an /r/ or an /n/; one may reasonably hypothesize that the Hittite situation reflects the facts of Proto-Indo-European, and that the other daughter languages generalized one or the other of those forms. But that excitement should not distract the researcher from considering the broader picture. The *r/n* declension explains very few words. Elsewhere the Indo-European languages are full of words that agree in the first few phonemes but not in the rest. The very idea is elevated to the status of theory. The standard Benvenistean root theory (Benveniste 1935, chap. 9) holds that the basic Proto-Indo-European root was a simple monosyllable, to which could be added, in ways that seem almost random to us, various extensions which, by and large, have uncertain function and meaning. Given such a situation, the odds that a heteroclisis uncovered in a new language would agree with such an alternation in one or two words in some other Indo-European languages are probably not as low as we would like to believe. It certainly does not seem on its face to be the sort of thing that proves the Indo-European nature of Hittite all by itself.

In a more general vein, scholars often attach a somewhat mysterious value to the very notion of paradigmaticity. My impression from reading arguments such as those of Nichols (1996) is that demonstrating that languages like Latin and Greek share the same endings for nouns when cross-classified by gender and case, e.g., the six endings for the nominative and accusative feminine, masculine, and neuter, means much more than it would mean if some six lexical words were found to match. Of course the fact that the languages both have cases and three genders means something in itself, more so than the fact that a language might have a word for 'water' and 'fire', but I believe the claim goes beyond this. For example, Nichols approvingly cited Meillet as saying that correspondence among the first five numerals would also be probative evidence because it is paradigmatic, and of course having words for the concepts 'one' through 'five' is not itself so unusual cross-linguistically that one would make anything of it. The paradigmaticity itself must be adding something. Although this sounds on its face like a claim about some numinous linguistic quality, I suspect what we really have here is a principle of statistical inference making. Namely, the fact that the first five numbers "go together" means that we have less freedom to pick and choose our data than if we made a claim about the five words 'water', 'fire', 'tree', 'dig' and 'eye'; the probability that we could find some correspondence by chance would be much higher in the latter case. In the absence of a mathematical statistical methodology, the rule of thumb about seeking paradigmaticity has much to recommend it. On the other hand, the recommendation of standard statistical methodology, that one specify the comparanda in advance, is much more rigorous than allowing people to bring up any paradigm they come across, and so it completely supersedes this requirement for paradigmaticity. And insofar as one of the great benefits of morphological information was the idea that it forms significant paradigms more readily than lexical words, a statistical methodology benefits correspondingly less from considering morphology.

Furthermore, as I discussed in the introduction, there can be evidence that is clear and even convincing, but which might nevertheless not fit into the sort of statistical study being undertaken here. By far the most serious problem with this evidence is that it does not belong to the class of completely arbitrary phenomena. It will be recalled that in order to demonstrate whether a congruity between two systems is chance or not, one needs to know the prior probabilities of the phenomena. Unfortunately, the shape of general lexical vocabulary might be the only situation in which such probabilities can be theoretically determined; standard theory teaches that lexical choice is completely arbitrary. Everything else is impossible to judge. We do not know for sure whether OVS languages are rare because that word order is unlikely to occur during glottogenesis, or whether they are rare because most nations that had OVS languages were wiped out in prehistoric warfare, to take two extreme possibilities. In the same way, we do not know what number to place on the probability that a language will have a participial verb form, or that it will use suffixation, or that suppletion or heteroclisis will occur in a particular high-frequency word.

The case of shared irregularities is even farther beyond our grasp. In the case of *est:sunt::ist:sind*, one would need to know the probability that a language would evolve in such a way that there could be an alternation involving loss of root vowels (in principle that is not terribly surprising; in this case it no doubt reflects a simple shift in word accent to the suffix in the plural); that that alternation would be found between the singular and plural of verbs; and that it would be restricted to the particular verb meaning 'be'. This last point is perhaps easiest to grasp, precisely because we have a general idea that irregularities and suppletions tend to be found precisely in the most frequently used words. Consider what would happen if one were to apply the lexical-correspondence test to the question of shared irregularities, marking a correspondence based on whether the languages have, say, a suppletion in the inflectional paradigm for a particular word. Even for unconnected languages, such marks would not be distributed at random, but would be concentrated in high-frequency words. But it is precisely such a departure from random distribution that is taken as indicative of connections between language. As in many other cases discussed in this book, such a connection would be real, but that connection is not language contact or shared descent, but rather a shared universal. Because the exact probabilities of that universal are unknown, they cannot be factored out.

Thus it is unfortunately necessary to exclude from a rigorous statistical test of this nature all of the really interesting information about morphology. But one might still wonder whether one could at least include the identity of certain morphological elements, for example, the bare fact that the exponent of the grammatical function of participle is -*nt*- in Hittite. Even then, the problem of coming up with an unbiased test suite would be daunting, though perhaps not insuperable. I have repeatedly pointed out the fallacy in basing a test on the most striking evidence. One needs to make the test as general as possible. The observation for heteroclisis in 'water' would have to be broadened to a test for heteroclisis across all words; the existence of a participle would have to be broadened to the existence of various grammatical categories in general. Of course, as already pointed out, such things can be very difficult, because there are so many complications, such as the fact that languages vary so widely in so many different ways. One language may not have the category at all; another might conflate it in different ways (e.g., a form might be used both as a participle and an infinitive, or one language may distinguish several tenses and voices of participles while another might have only one); another might collapse categories so that one cannot disentangle morphemes; one language may use prefixation, another infixation, another may use a separate word or some special word ordering. Such problems can be overcome if one writes the selection criteria carefully enough and without bias, although there is a danger that if one is too specific one could end up in situations where one discards virtually all evidence. For example, if one insists on considering only suffixes, one would draw a blank in languages like Navajo that mostly prefix, or languages like Chinese that have few affixes at all. There is also the consideration that grammatical morphemes change and are reorganized in many ways due to analogical and paradigmatic factors that are much stronger than those affecting typical lexical roots (Koch 1996). Thus even for languages that are related, one can expect that matches between a specific suffix and a specific grammatical function will typically be less stable than for lexemes. It was considerations such as these that led Swadesh to exclude almost all grammatical words from his refined list of 100 concepts (Swadesh 1955).

There is one additional consideration that is often overlooked, and which explains why morphology is being discussed in this chapter on nonarbitrary vocabulary: Affixes do not vary as freely as lexical words (Justeson and Stephens 1980). Such elements tend to be very short, for example, while lexical words, to the contrary, often have requirements for minimal length. Affixes tend to emphasize less-marked phonemes such as dentals (Meillet 1926). Consequently people tend to exaggerate the importance of matching of affixes; the fact that the exponent of a particular grammatical category happens to be /n/ in two languages is not a terribly striking coincidence. But more specifically in the context of these tests for historical connection, there would be the possibility that the phonological differences between lexical and grammatical particles could lead to the spurious finding of recurrences. For example, if lexical words were more likely to begin with a /b/ and grammatical elements more likely to begin with a /t/, then one is somewhat more likely to find /t/:/t/ matches, for no other reason than that they are somewhat more highly expected cross-linguistically in the grammatical section of the lexicon.

I do not mean to thoroughly quash the notion of comparing grammatical morphemes. Certainly the idea that some kinds of grammatical correspondences are especially probative is correct, and there is much room for invoking them in studies that do not require strict knowledge of the probabilities. Even within the context of these statistical tests, one could study affixes by isolating them in a separate test from which lexical items are excluded; this option is revisited in the later section on stratificational approaches. Firm believers in morphological evidence are even invited to run these tests using only grammatical particles and affixes, and totally ignoring lexemes, provided, of course, that the comparanda are picked in advance, and clear instructions are given for choosing in an unbiased manner which exponent should be selected in complicated cases.

To be absolutely clear where my reticence lies, let me reiterate that the trouble with using grammatical elements lies essentially in the extreme difficulty of coming up in advance with objective lists of grammatical categories that can be objectively matched with morphemes in languages of radically different typologies across the world. That is, the problem is that they are not as easily amenable to statistical treatment of any type as are full meaning-bearing words. The problem does not lie in the fact that the methodology expounded in this book deals with sound correspondences. Current thinking about the comparative method sometimes leaves one with the impression that reasoning about grammar and working out recurrent sound correspondences are disjoint exercises. Nothing could be farther from the truth. Even the great proponents of morphological data such as Meillet insisted that what one compares across languages is the exponent of the grammatical category, that is, a morpheme. Comparing grammar in the abstract, without reference to the morphemes, is an exercise in typology. And, of course, demonstrating that two morphemes are historically connected amounts to showing that they participate in regular sound correspondences.

8

Historical Connection vs. Relatedness: The Albanian Fallacy

In the previous chapter, I discussed a situation where problems with the data can confuse the interpretation of any finding of significant connection between languages. If any of the words are motivated by nonconventional criteria such as onomatopoeia, the connection that is uncovered may reflect natural facts about the universe rather than particular genetic relationships between languages.

Probably a much more important reason why the tests do not measure language relatedness alone is the fact of vocabulary borrowing. All languages to a greater or lesser extent adopt words from other languages. Though for some languages, such as Navajo, this seems to be quite a limited occurrence, at least in historical times, for others it seems to be rather common. It is sometimes claimed however that the Swadesh lists minimize the risk of including borrowed items; that is, the words are purported to be rarely borrowed. It seems intuitively likely that the words in the Swadesh lists are, on the whole, less likely to be borrowed than other words one might pick at random from the dictionary. There does not seem to be any reason why a language would borrow words like *egg* and *dog* from some other language, but one can think of compelling reasons why they might borrow words along with novel cultural concepts such as *telephone* and *miniskirt*. But the Swadesh words are demonstrably borrowed much too often to allow one to believe that the fact of borrowing could not significantly lower p scores when comparing languages. That is to say, a p score that is lower than one's threshold could well be due entirely to borrowing, and not at all to language relatedness in the traditional sense; and such a state of affairs would probably be quite common. Assuming universals are kept under control as discussed above, positive test results demonstrate that languages have some historical connection, but that could quite well come from vocabulary borrowing and not from relatedness.

I have checked the etymologies of the words on the Swadesh 100 and 200 lists in standard sources for each of the languages in the test set. In case of disagreement among authorities, I have tried to judge who has the best evidence. In case of doubt, I leaned in the direction of classifying a word as a loan, because it is safer to err in that direction. However, I am sure that the overwhelming majority of these judgments are reliable. The information is presented in full in the appendix; each word thought to be borrowed is followed by the designation *Loan*. However, the appendix is arranged by concept, which makes it difficult to get an idea of how pervasive loans are in particular languages. In the following discussion and tables, loans will be presented on a language-by-language basis.

Table 22 lists the Albanian words that are likely to be loans. I cite all words in their normal orthography, and mark whether the word is found in the Swadesh 100 list (S100) or only on the Swadesh 200 list (S200). It is evident that Albanian has borrowed much vocabulary. It will be recalled that even though Albania was part of the Roman empire, it did not give up its native language, as was common in the Celtic countries, and so Albanian is not a Romance language, but it is genetically related to it as a common descendant from Proto-Indo-European. It did however borrow heavily from Latin. Of the loans I identified, 16 are in the Swadesh 100 list, and 25 are only in the Swadesh 200 list.

Table 23 shows that English has also borrowed heavily in its basic vocabulary. The Old Norse loans are due to the extensive Scandinavian settlement in England in the 9th and 10th centuries. The Old French borrowings reflect French cultural ascendancy in the Middle Ages. I count 11 loans in the S100 list and 20 in the S200 list.

The French word list contains quite a number of loans as well. The fact that the bulk of its vocabulary descends naturally from Latin sometimes obscures the fact that it also contains many loans from Latin. For example, the word for river was originally *flum*, but that was replaced by a learned borrowing from Latin, resulting in *fleuve*. Table 24 shows 9 borrowings in the S100 list and 18 in the S200 list.

German tends not to borrow as heavily as some other languages, but table 25 lists 3 Swadesh 100 loans and 4 among the Swadesh 200 words.

Table 26 shows a few borrowings into Hawaiian. Borrowings into Latin and Navajo are also fairly negligible for this list. Navajo is in fact celebrated for its adamant resistance to incorporating foreign loan words. Turkish on the other hand has borrowed moderately heavily, especially from Persian and Arabic, as Table 27 shows. There are 6 words in the Swadesh 100 list, and 16 in the Swadesh 200 list.

TABLE 22 Borrowings into Albanian in Swadesh Lists

Gloss	Swadesh	Word	Source
and	S200	*e*	Latin *et*
animal	S200	*kafshë*	Latin *causa* 'thing'
back	S200	*shpinë*	Latin *spīna* 'backbone'
bad	S200	*keq*	Greek *kakos*
bird	S100	*zog*	?
child	S200	*fëmiljë*	Latin *familia* 'family'
come	S100	*vjen*	Latin *venit*
count	S200	*numëron*	Latin *numerus* 'number'
dog	S100	*qen*	Latin *canis*
far	S200	*larg*	Latin *large* 'broadly'
father	S200	*babë*	Turkish *baba*
few	S200	*pak*	Latin *pauci*
fight	S200	*lufton*	Latin *lucta*
fish	S100	*peshk*	Latin *piscis*
flower	S200	*lule*	Latin *lilium*
foot	S100	*këmbë*	Latin *camba* 'leg'
fruit	S200	*pemë*	Latin *pomum*
good	S100	*mirë*	Latin *mirus* 'wonderful'
grease	S100	*dhjamë*	Greek *dēmos*
green	S100	*gjelbër*	Latin *galbinus* 'pale green'
hair	S100	*flok*	Latin *floccus* 'flock of wool'
lake	S200	*liqen*	Latin *lacus* or Greek *lekanē*
many	S100	*shumë*	Latin *summa* 'mass'
mouth	S100	*gojë*	Latin *gula* 'throat'
narrow	S200	*ngushtë*	Latin *angustus*
neck	S100	*qafë*	Turkish *kafa*
old	S200	*vjetër*	Latin *veterem*
red	S100	*kuq*	Latin *cocceus* 'scarlet'
rub	S200	*fërkon*	Latin *fricat*
sand	S100	*rërë*	Latin *arena*
short	S200	*shkurtër*	Latin *curtus*
sing	S200	*këndon*	Latin *cantat*
sky	S200	*qiell*	Latin *caelum*
smell	S200	*marr erë*	Latin *aer*
smoke	S100	*tym*	OCS *timijasati* 'burn incense'
straight	S200	*drejtë*	Latin *directus*
think	S200	*mendon*	Latin *mentem* 'mind'
true	S200	*vërtetë*	Latin *verus*
wind	S200	*erë*	Latin *aer*
woods	S200	*pyll*	Latin **padulem* 'swamp'
yellow	S100	*verdhë*	Latin *viridis* 'green'

Table 23 Borrowings into English in Swadesh Lists

Gloss	Swadesh	Source
animal	S200	Latin.
bark	S100	Old Norse.
because	S200	Native *be-* + French *cause*
count	S200	Old French *conter*
cut	S200	Probably dial. Swedish *kuta*
die	S100	Prob. Old Norse *deyja*
dig	S200	Prob. ult. from Dutch
dirty	S200	Prob. Old Norse *drit* 'excrement'
egg	S100	Old Norse
flower	S200	Old French *flour*, from Latin *florem*
fog	S200	Prob. Scandinavian
fruit	S200	Old French, from Latin *fructus*
give	S100	Infl. by Old Norse
grease	S100	Old French *graisse*
hair	S100	Old French *haire* 'haircloth'
hit	S200	Old Norse *hitta* 'hit upon'
human	S100	Old French *humain*
husband	S200	Old Norse *húsbóndi*
lake	S200	Old French *lac*
mountain	S100	Old French *montaigne*
push	S200	Old French *pousser*
river	S200	Old French *rivere*
root	S100	Old Norse *rót*
rotten	S200	Old Norse *rotinn*
round	S100	Old French *rund*
skin	S100	Old Norse *skinn*
sky	S200	Old Norse *ský*
split	S200	Middle Dutch *splitten*
they	S200	Old Norse *θei-*
vomit	S200	Latin
wing	S200	Old Norse *vængir* 'wings'

TABLE 24 Borrowings into French in Swadesh Lists

Gloss	Swadesh	Word	Source
animal	S200	*animal*	Latin
breathe	S200	*respire*	Latin *respirat*
burn	S100	*brûle*	Infl. by Frankish *brōjan*
claw	S100	*griffe*	Frankish
cut	S200	*coupe*	Greek *kolaphos* 'slap'
dig	S100	*creuse*	Perhaps Celtic
dirty	S200	*sale*	Frankish *salo*
fall	S200	*tombe*	Frankish *tûmon*
fog	S200	*brouillard*	Germanic
guts	S200	*intestins*	Latin *intestina*
hit	S200	*frappe*	Germanic
human	S100	*humain*	Latin *humanus*
lake	S200	*lac*	Latin *lacus*
left	S200	*gauche*	Frankish
liver	S100	*foie*	Infl. by Greek *sykōton* 'with figs'
river	S200	*fleuve*	Latin *fluvius*
scratch	S200	*gratte*	Germanic
spit	S200	*crache*	Germanic
stand	S100	*debout*	Frankish **bōtan* 'shove'
stick	S100	*bâton*	Greek *bastazo* 'carry a burden'
stone	S100	*pierre*	Greek *petra*
think	S200	*pense*	Latin *pensat* 'weigh, judge'
throw	S200	*lance*	Celtic 'lance'
tie	S200	*attache*	Frankish *stakka* 'stake'
vomit	S200	*vomit*	Latin *vomit*
white	S100	*blanc*	Germanic
woods	S200	*bois*	Germanic

TABLE 25 Borrowings into German in Swadesh Lists

Gloss	Swadesh	Word	Source
fight	S200	*kämpft*	Lat. *campus* 'field'
fruit	S200	*Frucht*	Lat. *fructus*
grease	S100	*Fett*	Low German
head	S100	*Kopf*	Latin *cuppa* 'cup'
round	S100	*rund*	Old French *rund*
short	S200	*kurz*	Latin *curtus*
spit	S200	*spuckt*	Low German

TABLE 26 Borrowings into Hawaiian in Swadesh Lists

Gloss	Swadesh	Word	Source
grease	S100	ʻaila	English oil
snake	S200	naheka	Hebrew nahaš and/or Eng. snake
wipe	S200	kāwele	English towel

TABLE 27 Borrowings into Turkish in Swadesh Lists

Gloss	Swadesh	Word	Source
and	S200	ve	Arabic wa
animal	S200	hayvan	Persian, from Arabic haywān
bad	S200	kötü	Armenian godi
breathe	S200	nefes alıyor	Arabic nafas
dirty	S200	pis	Persian pīs
dull	S200	kör	Persian kūr 'blind'
fire	S100	ateş	Persian ātaš
fruit	S200	meyva	Persian maiva
human	S100	adam	Arabic ādam
husband	S200	koca	Persian xoǰa 'eunuch'?
if	S200	eğer	Persian agar
liver	S100	ciğer	Persian jigar
many	S100	çok	Possibly Armenian jok 'crowd'
old	S200	ihtiyar	Arabic ixtiyār 'choice'
river	S200	nehir	Arabic nahr
seed	S100	tohum	Persian tuxm
some	S200	bazı	Arabic baʻd
stab	S200	hançerliyor	Arabic xanjar 'dagger'
wind	S200	rüzgâr	Persian rūzgār
woman	S100	kadın	Sogdian xwātūn 'queen'
worm	S200	solucan	Armenian sołun 'creeper'
year	S200	sene	Arabic

I indulge in this listing in order to emphasize that borrowing of words in the Swadesh lists is by no means rare. This demonstration is necessary because there is an idea in the air that it has been proved that words on the Swadesh 100 list are so rarely borrowed that it is safe to ignore the problems of loans when using that list—maybe there will be one or two borrowings, but surely not enough to skew the overall results. Certainly that cannot have been proved, as these lists make clear—languages can have many borrowings among the Swadesh 100 list. In fact one wonders how the idea could have got started, when even English has a large number of loans in those lists. True, it is sometimes held that English is a special case, that it has an extremely high number of loans because of its special history, England having been subjected to the Danelaw and the Norman invasion. But invasion, and intimate language contact in general, really is not all that unusual in this world, unless one chooses to believe that the Frankish influence on French (Table 24) and the Latin influence on Albanian (Table 22) were also unusual coincidences.

Including these loan words in these lists is by no means an unqualified error. One can think of a couple of good reasons for keeping them in. If the researcher is mostly concerned with the role of chance versus the role of historical connection, loans are of interest as one sort of historical connection. In many cases where one is investigating languages not known to be related, the first task is to see if there is anything in the perceived similarities at all; one can reserve for later the question of just what the exact history of the similarities is. The second reason for including loans is that most of the time one really is in the dark about whether words are loans or not. At some point in prehistory, those questions become unanswerable. So arguably one should just as well keep the loans in, and hope the methodology proves robust enough to cope with them.

There are also disadvantages to including loans. The most obvious problem is that it is easy to think that if correspondences are not due to chance, then that means the languages are related. For example, Villemin (1983) concluded via statistical tests that Japanese and Korean were related, but without carefully excluding words that were loans from Chinese into both languages. Ringe too fell into that trap at one point. Greenberg (1993) challenged Ringe, saying that his statistical test (Ringe 1992) was not powerful enough to detect some weak but well-established relationships, such as that between English and French. Ringe rejoindered (1993) that his technique was indeed up to the task, and published the results. As has been shown, the significance testing involved was questionable, and ultimately we are asked to accept as an empirical rule of thumb rather than by statistical reasoning that the three significant cells in the contingency table are indicative of a connection between the languages. But just as importantly, it appears that Ringe did not discard loans. In fact his test considers as crucial evidence the fact that there are three recurrent matches between English /r/ and French /r/, but one of these is *round:rond*, where the English word is borrowed from Old French. Ringe then proceeded to up the ante and show that the test can find the relationship between Albanian and French, but again he failed to discard loans. We have just seen that both Albanian and French have borrowed heavily from Latin, and most of the rest of the French vocabulary is descended from Latin. Part of the demonstration, for example, depended on there being three /p/:/p/ matches; one of these was 'fish', Albanian *peshk*:French *poisson*, where *peshk* is usually considered a loan from Latin. The implications are clear. Even though the tests were capable of showing a historical connection between French and Albanian, they did not prove that that connection is due to their being related (that the correspondences are due to common inheritance from Proto-Indo-European). Ringe's demonstration therefore rather missed the point of Greenberg's challenge. Historical contact between the vocabularies of the two language groups is obvious upon inspection; it is genetic relationship that took so long to establish, in part precisely because of several confusing layers of loans. The most expert of linguists can easily fall prey to the Albanian Fallacy of thinking that low p levels necessarily imply historical relatedness.

The other problem with loans is a little less obvious. Loans can raise p values when one is comparing historically connected languages, if at least one of them has borrowed from some third party. For example, if one were testing to see whether French and Spanish were related, and the word list included many Germanic borrowings into French and many Arabic borrowings into Spanish, then the loans are only going to obscure the Romance inheritance between those two languages. For 'fog', Spanish retains the Latin word *nebula* as *niebla*; if French had not borrowed a Germanic root for this concept (*brouillard*), then the Latin derivative *nielle* (now meaning 'smut') might still have meant 'fog', giving us a historically correct /n/:/n/ match, which would have joined with other pairs like Spanish *negro*, French *noir* 'black' to increase the number of recurrent correspondences. Instead, one gets an /n/:/b/ match, which will recur only by accident. Thus the evidence for a connection between the two languages is diminished. In practice, of course, the Romance inheritance of both French and Spanish is overwhelming, but it is easy to extrapolate how that effect could cause problems in language pairs that have diverged more.

On the other hand, it may be argued that the problem of loans is minimal. Even if the first prong of attack fails (the claim that words on these vocabulary lists are rarely borrowed between languages), there are other less subtle reasons why loans between two languages need not greatly increase the measure of the test that those two languages are related. One reason is that even when words are borrowed between ancestors of two languages, they may not end up expressing the same idea. This is obvious enough even with recent loans, such as French *smoking* 'dinner jacket'. This is even more true in the case of Albanian, where loans from Romance began almost two millennia ago: *gjelbër* 'green' comes from the same source as French *jaune* 'yellow'; *verdhë* 'yellow' matches *vert* 'green'; and so forth. In fact only three of the many Romance loan words in the Swadesh 100 Albanian list (Table 22) match up semantically as well as etymologically with the Modern French correlates on the word list (*vjen:vient* 'comes', *qen:chien* 'dog', *peshk:poisson* 'fish'). One could argue that such a small number of cognates will make little difference in the statistics.

The other, even more subtle, reason why loan words may not necessarily radically increase the measure of linguistic relationship is due to the fact that the test counts recurrent matches, not absolute similarity. Because both languages will change after the time of their separation, a given ancestral sound may well have different reflexes depending on whether the correspondence stems from the time of language separation or from any of several later stages. To take a textbook example, the canonical match between English and French when the English is /h/ in a certain environment would be /ʃ/, as in *head:chef* /ʃɛf/. But when the same root was borrowed into English from an earlier stage of French, English ended up with *chief*, which in fact more closely corresponds in meaning to French *chef*: /tʃ/:/ʃ/. When both English and French borrowed from Latin, they ended up with equations like *capital* = *capital* /k/ = /k/. The identical Latin morpheme is involved in all these cases. Thus a series of three borrowings involving the same root would just give three distinct correspondences. In themselves they would not lower the p values, because, in principle, only recurrences count. (Actually the statistical methods under discussion so far do not technically measure recurrence, and the distinction will be discussed in more detail later. But for purposes of the discussion, it is clear that three separate correspondences would almost certainly end up being weaker evidence than would three instances of the same phoneme match.)

But at best both of these arguments only apply if there is a very small number of loan words in the word list, and perhaps not even then. Already in the Swadesh 100 list one sees Albanian–French *vjen:vient* 'come', where the /v/:/v/ match reinforces the /v/:/v/ matches one ex-

pects as normal reflexes of Proto-Indo-European */w/ (Beekes 1995).
Indeed, when one moves up to the Swadesh 200 list, one begins to see
recurrent matches with French among the Latin loan words themselves.
To the above /v/:/v/ match is added *vjetër:vieux* 'old', and there is
also *pak:peu* 'few' and *peshk:poisson* 'fish', where according to Beekes
(1995) /p/:/p/ is also an inherited correspondence. And fundamentally
it does not really matter whether the number of loans is small or large.
The purpose of these statistical tests is to quantify a probability, and
the χ^2 technique works only to the extent that there is no other factor
influencing the distribution of phonemes among word senses beyond the
hypothesized factor. So if loan words are a factor at all, then it is incorrect to say that the test gives a mathematical quantification of the
probability that the languages are related.

The χ^2 technique has no way of mathematically allowing for any
such extraneous forces. On the other hand, there is in principle nothing wrong with the simple expedient of removing from the lists all word
senses involving known or suspected loans before running the tests. If
one encounters the word *ideopolitik* in an Albanian word list, it is a safe
bet that a loan is at hand, and the word can be discarded. At the same
time, one must keep in mind that not all layers of loan words are quite
so obvious. Several of the items I presented above as Albanian loans
from Romance have in fact been the object of lengthy contention as to
whether they are loans or direct heirs of Proto-Indo-European, and such
issues are by no means peculiar to Albanian. Many proposed language
relationships were subsequently rejected by showing that resemblances
and correspondences were actually due to loan words. Poser and Campbell (1992) pointed out that Sir William Jones was misled by Aramaic
loans in Pahlavi, classifying it as Semitic instead of Indo-European; by
Arabic loans in Malay, classifying it as Semitic instead of Austronesian; by Sanskrit loans in Sumatran, classifying it as Indo-European
instead of Austronesian; and by Sanskrit loans in Tibetan, classifying it
as Indo-European instead of Sino-Tibetan. They also noted that Armenian was generally believed to be an Iranian language, again because of
loan words, until 1875. Perhaps the best one can hope for is to systematically err on the side of the null hypothesis. If one wants to demonstrate
that there is a relationship between languages, one should aggressively
reject all suspected loans. And certainly any results would have to be
prefaced with the important proviso that the results are valid only if all
loan words have been correctly removed from the lists.

Table 28 shows what happens when the related languages in the
test set are measured with the Swadesh 200 list, both including and
omitting known loan words. I do not include here pairs of unrelated lan-

guages, because none of the unrelated languages in my sample borrowed heavily one from the other, at least not until modern times. Therefore the better result in all of these comparisons would be a lower p value. Sadly, the only nontrivial changes caused by omitting loan words seem to be for the worse. The results are less desirable, but, for the researcher investigating genetic relationships, more honest. For the related languages, including loans sometimes vastly inflates the estimate that the languages are related. The case of the English–French comparison is perhaps easiest to understand, because English has borrowed so heavily from French; here, keeping loan words out of the mix causes us to lose pairings like *flower:fleur*, *fruit:fruit*, and so forth. So many loans clearly increase the number of recurring correspondences. It may be less obvious why the comparisons of Albanian with the Germanic languages come out stronger when loans are included, seeing that there were few direct loans between Albanian and those languages. But consider for example the fact that Albanian has borrowed the words *peshk* 'fish' and *pemë* 'fruit' from Latin. These two concepts make a recurring /f/:/p/ correspondence by themselves, which is reinforced by the fact that /f/:p/ is already a true correspondence between English and native Albanian words, as in *five:pesë*. One might also note that eliminating loans reduces the amount of data, so that fact alone may tend to lower p values.

In this sample I do not see evidence of the opposite effect, where loans from a third language impede the recognition of related languages. It might be the case that French loans into English lower its tested relationship with German, and that Germanic loans into French lower its relationship with Latin, but the probability of chance in those two cases is already so low that no effect is visible. At any rate, the borrowings all come from languages that are related anyway. Effects of third-party loans would be expected to be much greater if any of the languages under consideration had extensive loans from non-Indo-European languages.

Amidst all the concern about whether the test is valid for determining genetic relationship between languages, it should not be forgotten that in many cases the issue is irrelevant. Often the debate is not whether perceived similarity between two languages is a matter of genetic relationship in the classical sense, but whether similarities are due to linguistic connection at all. When linguists debate whether similarities can be due to chance, loan words arguably belong among the factors that contribute to non-chance similarities. If one bears in mind that the Ross procedure basically tests for the probability of linguistic connection, not genetic relationship, the thorny problem of loans is, on the whole, much less important.

Nevertheless, because of concerns about confusing genetic relation-

TABLE 28 Effect of Omitting and Including Loan Words

Languages		Omit		Include	
		p	N	p	N
English	German	.000	153	.000	185
French	Latin	.000	162	.000	184
German	Latin	.000	181	.000	185
English	Latin	.000	157	.000	186
English	French	.088	141	.027	182
French	German	.001	153	.001	180
Albanian	French	.002	125	.004	179
Albanian	Latin	.000	148	.000	185
Albanian	German	.094	142	.053	180
Albanian	English	.194	123	.051	182

Notes: Comparisons use χ^2 metric with word-initial consonants, using the Swadesh 200 list, omitting motivated words.

ship with historical connection, in addition to the many ways in which loan words can obfuscate true relations, I will omit known or suspected loan words from further tests in this book. Note that the cognate judgments presented in Table 10 is now less appropriate, because that table included loans. Known loan words are omitted from the new tabulations presented in Table 29.

TABLE 29 Closeness Judgments for Language Connections, Omitting Loans

Languages		Cognates	N
Albanian	English	.143	136
Albanian	French	.191	136
Albanian	German	.157	156
Albanian	Hawaiian	.000	157
Albanian	Latin	.186	159
Albanian	Navajo	.000	159
Albanian	Turkish	.000	144
English	French	.280	152
English	German	.664	165
English	Hawaiian	.000	167
English	Latin	.293	169
English	Navajo	.000	169
English	Turkish	.000	153
French	German	.272	167
French	Hawaiian	.000	170
French	Latin	.613	173
French	Navajo	.000	173
French	Turkish	.000	157
German	Hawaiian	.000	191
German	Latin	.295	193
German	Navajo	.000	193
German	Turkish	.000	172
Hawaiian	Latin	.000	197
Hawaiian	Navajo	.000	197
Hawaiian	Turkish	.000	175
Latin	Navajo	.000	200
Latin	Turkish	.000	178
Navajo	Turkish	.000	178

Note: Human judgments as to how closely connected languages are, expressed as the ratio of the Swadesh word pairs that are cognate; numbers closer to 1 represent closer connections. N tells how many words were compared.

9

Language-Internal Cognates

The core assumption behind the language connection tests that makes the whole thing viable is our faith that the connection between word sense and word pronunciation is completely arbitrary. I have already discussed above one challenge to that assumption: namely, that certain concepts such as 'mother' may actually have natural pronunciations. But there is another challenge that is of a much greater magnitude. Within a language, words that have associated senses tend to have similar forms.

This may seem controversial, but it is in fact very obvious even when one looks at the Swadesh lists at the grossest level. Multiple appearances of the same word are a surprisingly frequent occurrence. In Albanian, *ai* translates both 'he' and 'that'; *burrë* is both 'man' and 'husband'; *grua* is both 'woman' and 'wife'; *në* is both 'at' and 'in'. In English, *you* is both singular and plural. In French, *droit* means both 'right' and 'straight', and *femme* is both 'woman' and 'wife'. In German, *See* translates both 'sea' and 'lake'. In Hawaiian, *'ike* translates both 'see' and 'know'; *hele* 'go' and 'come'; *'ili* 'bark' and 'skin'; *hua* 'fruit' and 'egg'; *kāne* 'man' and 'husband'; *lā* 'sun' and 'day'; *lā'au* 'tree' and 'stick'; *lepo* 'earth' and 'dirty'; *moe* 'sleep' and 'lie'; *nui* 'big' and 'many'; *wahine* 'woman' and 'wife'. In Latin, the same word *fodit* translates 'dig' and 'stab'. In Navajo, *łeezh* means 'dust' and 'earth'; *'at'a'* means 'feather' and 'wing'; *hastiin* means 'man' and 'husband'; *nihí* means both 'we' and 'you' plural; *tsin* means both 'tree' and 'stick'; *yázhí* is glossed as both 'small' and 'short'; *yígháah* is both 'go' and 'come'. In Turkish *doğru* is both 'true' and 'straight'; *o* is both 'he' and 'that'.

It will be noted that some of these repetitions are repeated across languages. Consider the case of Albanian and Hawaiian (Table 30). When it comes time to tallying recurrent correspondences, it will be seen that the phoneme matching /b/:/k/ appears twice, and the matching /g/:/w/ appears twice, and there are still 196 other pairs of words to count. The

TABLE 30 Recurrent Word Repetitions in Albanian and Hawaiian

Concept	Albanian	Hawaiian	$C_{1.1}$
husband	*burrë*	*kāne*	/b/:/k/
man	*burrë*	*kāne*	/b/:/k/
wife	*grua*	*wahine*	/g/:/w/
woman	*grua*	*wahine*	/g/:/w/

recurrent phoneme counts are adding up because of the fact that Albanian and Hawaiian both use the same word for adults of a given sex and adult spouses of a given sex. But interesting as that fact may be, it does not have the same status as would the determination that Albanian and Hawaiian has recurrent /b/:/k/ matches on concepts that have no semantic connection with each other. It is quite common for languages to conflate the concepts 'man/husband' and 'woman/wife' (Campbell 1999:272). Therefore the additional words cannot count as independent evidence as to whether there are enough consonant pairings to decide between chance and historical connection. For there is a tertium quid: motivated connections between the words in a given language.

For statistical purposes, motivated connections masquerade as historical connection because they are not products of mere chance. But in fact, motivated connections do not depend on actual historical connection. The human faculty that sees such a close connection between 'husband' and 'man' that it will use the same word for the same concept is universal and is very likely to produce the same effect time and again all around the world. Similarly, the fact that Hawaiian and Navajo can use the same word for 'tree' and 'stick' is not a random occurrence, for it is not equally likely that all pairs of concepts will be subsumed under the same word. The choice to subsume two concepts is not arbitrary in the Saussurian sense of the word, but is motivated by how the speakers view their world and how they manipulate semantic concepts. And these qualities are to a significant extent universal. People of different cultures think alike to a very close approximation.

If one had an excellent model of human cognition, or at least strong quantitative control over how processes like polysemy work across cultures, one might be able to use to advantage the fact that two languages have one word that means both 'fruit' and 'egg'. As it stands, however, the information merely interferes with the task of computing statistical correlations between languages. By and large, the presence of repeating words will tend to sharply inflate the estimation that two languages are connected, if indeed they both happen to share the same polysemy.

Campbell (1999:182) made a similar point about how polysemy would artificially lower time of separation for two languages in a glottochronological study.

All but one instance of the repeating word need to be removed from the word list. Exactly how that should be done is an interesting question. One must guard against removing words in such a way that they will bias the test. If for example one were to remove the German word *See* from the 'lake' concept so that it would remain glossed by the 'sea' concept, doing so because one knows that that would provide a better match with English *sea*, that would be a serious case of experimenter bias. It seems safest therefore to remove such words one language at a time, without considering the language to which it is to be compared. Ideally one would retain the word in the meaning that is the core, or original meaning, because on average that will be the meaning most likely to be found on cognates. However, determining core or original meaning can be maddeningly difficult in the absence of historical records; which came first, the fruit or the egg? In some cases, the literature provides guidance as to the natural direction of semantic shifts (for example Sweetser 1990, Traugott 1989, Wilkins 1996). In other cases, the safest and least troublesome approach is probably to delete one of the words at random, or to always delete the one whose gloss comes first in alphabetical order.

Unfortunately the problem does not stop with cases of pure polysemy. There are many other instances of nonarbitrary connection between elements of a language's vocabulary. Consider, for example, the data in Table 31. Assuming that one takes the obvious approach and removes the inflectional prefixes from the Navajo words, the presence of the two concepts 'fly' and 'wing' on the Swadesh lists results in a recurring correspondence of /f/:/t'/. This is very similar to the table previously considered. The only difference is that in this case it is not the entire word that repeats, just the root. Such patterning happens all of the time; in fact, it is extremely rare that a new word is invented that does not have some sort of phonetic similarity with one or more other words to which it is semantically connected in some way. Just as with polysemy, there is no reason to believe that languages' ideas of semantic connectivity are completely independent from each other. To the contrary, there is every reason to believe that they are highly correlated. If something in the Albanian mind sees a connection between wings and flying, to the end that the one word is based on the other, there is no reason to be surprised that the Navajo mind would see the same connection and likewise come up with a pair of words that are similar to each other, as indeed it did. So while the mapping behind any given concept and pronunciation may be arbitrary in itself from a universal

TABLE 31 Recurrent Root Repetitions in Albanian and Navajo

Concept	Albanian	Navajo	$C_{1.1}$
fly	fluturon	yi-t'ah	/f/:/t'/
wing	fletë	'a-t'a'	/f/:/t'/

point of view (prescinding from 'mother' and such), the mapping of the next concept to a pronunciation is partly dependent on the mapping of that previous concept. That failure of independence can completely undermine the validity of tests founded on Saussure's arbitrariness hypothesis. The presence of the recurring /f/:/t'/ correspondence simply does not count as very much evidence for a connection between Albanian and Navajo.

That sort of correspondence happens all the time, and the use of the Swadesh lists does little to help. Swadesh (1955) trimmed his list from 200 words to 100 in part because he had noticed that some words were particularly susceptible to such problems, but he admitted that several potential problems remain, and certainly one cannot discard all words that have any possibility of being derivationally related in any arbitrary language. It is therefore necessary for the linguist to root out such cases by hand when compiling the word list. Just how difficult will this turn out to be in practice?

The clear and incontrovertible cases are productive processes within the language; e.g., from German *Gatte* 'husband', one derives *Gattin* 'wife' with the feminine suffix. Others may not be productive, but any native speaker would recognize the connection; e.g., *Fluss* 'river' from *fließen* 'flow'. Some may require some etymological expertise to uncover; e.g., it is by no means obvious that French *dans* 'in' is composed of the initial morphemes of *debout* 'standing' and *enfle* 'swell'. Many correspondences are even quite unclear to historical linguists: Some see a connection between English *and* and *in*, others do not. Unfortunately, there is no necessary connection between etymological obscurity and the likelihood that some other language will reprise the semantic and morphological connection. For example, any connection between Latin *caput* 'head' and *capillus* 'hair' is etymologically very tenuous; few linguists would want to stake their career on saying they are related at all. And yet the question needs to be decided, for if they are connected in any way, then matching them up with Navajo *-tsiighá* 'hair' and *-tsii'* 'head' would result in spurious recurrences of the /k/:/ts/ match.

Because our interest is in the history of languages, a historical approach to derivation is the most appropriate. Words need to be expunged if they devolve from the same historical etymon. I've explicitly followed this approach in this book, and to document those decisions, in the appendix I have provided etymologies for most of the words used in these analyses. Strictly speaking, the synchronic state of the grammar is irrelevant. It is on a historical basis, for example, that I grouped English *what* and *who*, even though their (American) pronunciations are quite different (/wɑt/ and /hu/); historically, both continue the Old English interrogative stem /hw-/. But it must be conceded that a historical requirement often puts researchers at a serious disadvantage. Even for the Indo-European languages, which have been studied thoroughly for two centuries and which have unusually strong documentation stretching back thousands of years, there are very many uncertain etymologies. For most of the languages of the world, that sort of detailed information is simply not available. It is indeed daunting to realize that failing to take out a word that is historically connected to some other word on the list technically invalidates the whole study, when synchronically there is no limit on how far apart words with the same etymon may have drifted both phonetically and semantically. Could a researcher investigating a language without a long-documented history and many well-studied related languages possibly recognize that words as dissimilar as *augment, augur, auxiliary, author, wax, waist,* and *nickname* all devolve from the same etymon (Pokorny 1959:84)?

Even if one knows the history, derivational processes are often irregular and can present problems of analysis. The Latin word for 'that (man)' was originally *olle*, but in the Classical period it became *ille*, apparently by analogy with the near-synonym *is* 'he'. While *ille* and *is* fundamentally have different etymons, nevertheless the recurrence of initial /i/ in words of similar semantics would cause a problem if the language Latin is being compared with has related words for 'that' and 'he' (e.g., Turkish *o*). Words in sound-symbolic families like *spot, blot, blotch, botch,* and *splotch* or *squash, quease, squiss, squish, squize,* and *quash* are clearly not independent of each other even though there are no clean etymologies decomposing these words into root plus affix. Especially intriguing are folk-etymological formations like *crayfish* and *sockeye*, where we have clear etymologies that have nothing to do with fish or socks or eyes. But even if *crayfish* is originally a monomorphemic word that has nothing to do with *fish*, it would seem rash to say that the two words now have no connection. If English speakers saw something fish-like in this animal, other speakers could as well. Even more subtly, the recent shift in meaning of the word *buxom* to mean 'large-breasted' means that people now connect it with the word *bosom*. To people interested in deep linguistic relationships, a connection between *buxom* and *bosom* is absurd, and there is certainly no general derivational pattern of creating negatives by infixing a /k/. But for our purposes, the Arbitrariness Hypothesis is violated just as certainly. The reason we need to avoid repeating morphemes in the word list is not because they share a remote etymological connection, but because any similarity in sound that is due to a connection in meaning, or vice versa, is fundamentally not arbitrary and is likely to be repeated in some other language at greater than chance levels.

The implications of such exotic derivations for morpheme identification are appalling. In the absence of perfect historical knowledge, the fact that such word formations can appear without regular patterning and indeed without any motivation beyond an offhand analogy based on sound or meaning implies that the conservative approach to word-list selection should be to throw out any word that remotely sounds like another and has any remote semantic connection to it. But of course there is no limit to such a procedure. Every word has some, even if infinitesimal, degree of phonetic and semantic similarity to every other word. Worse, such associations can be almost indiscernible if they are inspired by unfamiliar cultural contexts or intermediated by several steps of semantic change. Worst, there seems to be little if any guidance of a statistical nature to be had. If one is comparing words based on initial consonants, then one should of course pay special attention to words that start with the same consonant in the same language. However, a maximally conservative approach, namely, omitting words with the same consonant if they have any conceivable semantic connection, would guard perfectly against erroneously connecting languages, but it would also guard perfectly against ever finding any connections at all. The best advice one can offer is to exploit every source of information in weeding out one's word list: synchronic analysis, historical etymology, comparison with related languages, and knowledge of what concepts tend to be derivationally connected in the languages of the world.

Fortunately, the other side of the coin is that the more obscure the semantic connections are between words within a language, the less likely they are to actually turn up obscurely in another language. If, as has been suggested, *child* and *cloud* actually have some deep etymological connection in English, it does not seem very likely that one some other language will also have an obscure etymological connection between these words. Whereas the theoretist may despair that it is technically impossible to make sure that all language-internal cognates have been removed without completely crippling the test, the pragmatist can find plenty of room to argue that after one has done one's best to weed the list, the likelihood of actual disaster appears to be small.

The appendix lists the language-internal cognates that were found for each word. The net is cast rather wide here, and I do include some speculative etymologies that may not be accepted by all researchers. In particular, the Indo-European etymologies generally include the often rather speculative connections proposed in Pokorny (1959). The appendix is arranged by concept, so it is a bit difficult to see at times how pervasive the problem might be on a language-by-language basis. The following lists may be useful as a finding aid. For each language, they list words whose roots have been traced back to the same etymon by at least some linguists.

Cognates within Albanian:

1. *në* 'at, in'
2. *bark* 'belly', *mban* 'hold', *burrë* 'husband', 'man'
3. *fryn* 'blow', *marr frymë* 'breathe'
4. *pa mprehtë* 'dull', *mprehtë* 'sharp'
5. *bie* 'fall', *borë* 'snow'
6. *fluturon* 'fly', *fletë* 'wing'
7. *këmbë* 'foot', *rri më këmbë* 'stand'
8. *plotë* 'full', *pyll* 'woods'
9. *shkon* 'go', *shteg* 'path'
10. *zorrë* 'guts', *gojë* 'mouth'
11. *ai* 'he, that', *aty* 'there', *ata* 'they'
12. *këtu* 'here', *ky* 'this'
13. *vjetër* 'old', *vit* 'year'
14. *kripë* 'salt', *shkurtër* 'short'
15. *marr erë* 'smell', *erë* 'wind'
16. *ç'* 'what', *kush* 'who'
17. *grua* 'wife, woman'

Cognates within English:

1. *and, in*
2. *belly, blood, blow, flower*
3. *breast, breathe, burn*
4. *child, claw, cloud*
5. *dull, dust*
6. *ear, eye*
7. *eat, tooth*
8. *flow, fly, full*
9. *grass, green*

10. *he, here*
11. *new, now*
12. *say, see*
13. *sharp, short*
14. *that, there, they, this*
15. *thin, think*
16. *tree, true*
17. *wash, water, wet*
18. *what, who*
19. *wide, with*
20. *wife, woman*
21. *you* (sing., pl.)

Cognates within French:

1. *écorce* 'bark', *court* 'short'
2. *oiseau* 'bird', *œuf* 'egg'
3. *vente* 'blow', *vent* 'wind'
4. *brûle* 'burn', *brouillard* 'fog'
5. *coupe* 'cut', *beaucoup de* 'many'
6. *poussière* 'dust', *pousse* 'push'
7. *loin* 'far', *(est) allongé* 'lie', *long* 'long'
8. *fleur* 'flower', *feuille* 'leaf', *fleuve* 'river', *enfle* 'swell'
9. *gèle* 'freeze', *glace* 'ice'
10. *plein* 'full', *pluie* 'rain'
11. *intestins* 'guts', *ici* 'here', *dans* 'in'
12. *il* 'he', *là* 'there', *ils* 'they'
13. *entend* 'hear', *tient* 'hold', *maintenant* 'now'
14. *rit* 'laugh', *ver* 'worm'
15. *près de* 'near', *presse* 'squeeze'
16. *droit* 'right, straight'
17. *tranchant* 'sharp', *trois* 'three'
18. *quelques* 'some', *quoi* 'what', *qui* 'who'
19. *suce* 'suck', *essuie* 'wipe'
20. *cela* 'that', *ceci* 'this'
21. *épais* 'thick', *pense* 'think'
22. *femme* 'wife, woman'

Cognates within German:

1. *und* 'and', *in* 'in'

2. *an* 'at', *nahe* 'near'
3. *schlecht* 'bad', *Leber* 'liver'
4. *Vogel* 'bird', *fließt* 'flow', *fliegt* 'fly', *voll* 'full', *viele* 'many', *Fluss* 'river', *Flügel* 'wing'
5. *Blut* 'blood', *bläst* 'blow', *Blume* 'flower', *Blatt* 'leaf'
6. *Brust* 'breast', *brennt* 'burn'
7. *Ohr* 'ear', *Auge* 'eye'
8. *ißt* 'eat', *Zahn* 'tooth'
9. *weit* 'far', *Weib* 'woman'
10. *gut* 'good', *Gatte* 'husband', *Gattin* 'wife'
11. *Gras* 'grass', *grün* 'green'
12. *Mensch* 'human', *Mann* 'man'
13. *wenn* 'if', *was* 'what', *wer* 'who'
14. *See* 'lake', *See* 'sea'
15. *neu* 'new', *nun* 'now'
16. *eins* 'one', *einige* 'some'
17. *stößt* 'push', *Stock* 'stick'
18. *rund* 'round', *gerade* 'straight'
19. *reibt* 'rub', *wirft* 'throw', *Wurm* 'worm'
20. *sagt* 'say', *sieht* 'see'
21. *scharf* 'sharp', *kurz* 'short'
22. *riecht* 'smell', *Rauch* 'smoke'
23. *glatt* 'smooth', *gelb* 'yellow'
24. *das* 'that', *da* 'there', *dieses* 'this'
25. *dünn* 'thin', *denkt* 'think'
26. *wäscht* 'wash', *Wasser* 'water'

Cognates within Hawaiian:

1. *'ili* 'bark, skin'
2. *nui* 'big, many'
3. *hele* 'come, go'
4. *lā* 'day, sun'
5. *lepo* 'dirty, earth'
6. *hua* 'egg, fruit'
7. *maka* 'eye', *makahiki* 'year'
8. *makua kāne* 'father', *makuahine* 'mother'
9. *kaka'ikahi* 'few', *kahi* 'one', *kekahi* 'some'
10. *lima* 'five, hand'
11. *ho'opa'a i ka hau* 'freeze', *pa'a* 'hold', *pa'akai* 'salt'

12. *lauoho* 'hair', *lau* 'leaf', *lā'au* 'stick, tree'
13. *kāne* 'husband, man', *keiki* 'child', *kanaka* 'person'
14. *hau* 'ice', *hau kea* 'snow'
15. *loko* 'in, lake'
16. *'ike* 'know, see'
17. *moe* 'lie, sleep'
18. *'ōlelo* 'say', *lelo* 'tongue'
19. *kēlā* 'that', *kēia* 'this'
20. *wahine* 'wife, woman'
21. *'oe* 'thou', *'oukou* 'you'

Cognates within Latin:

1. *quod* 'because', *quid* 'what', *quis* 'who'
2. *avis* 'bird', *ovom* 'egg'
3. *mordet* 'bite', *moritur* 'die'
4. *flat* 'blow', *fluit* 'flow', *flos* 'flower', *folium* 'leaf', *flumen* 'river'
5. *puer* 'child', *pauci* 'few', *parvos* 'small'
6. *nubes* 'cloud', *nebula* 'fog'
7. *secat* 'cut', *scit* 'know'
8. *fodit* 'dig, stab'
9. *auris* 'ear', *audit* 'hear'
10. *edit* 'eat', *dens* 'tooth'
11. *procul* 'far', *prope* 'near'
12. *gelat* 'freeze', *glacies* 'ice'
13. *plenus* 'full', *pluvia* 'rain'
14. *intestina* 'guts', *in* 'in'
15. *capilla* 'hair', *caput* 'head'
16. *is* 'he', *unus* 'one', *ibi* 'there'
17. *hic* 'here', *hoc* 'this'
18. *ferit* 'hit', *fricat* 'rub'
19. *tenet* 'hold', *tenuis* 'thin'
20. *ridet* 'laugh', *vermis* 'worm'
21. *iacet* 'lie', *iacit* 'throw'
22. *novos* 'new', *nunc* 'now'
23. *alius* 'other', *aliqui* 'some'
24. *trudit* 'push', *terget* 'wipe'
25. *umidus* 'wet', *uxor* 'wife'

Cognates within Navajo:

1. *t'áá 'áłtso* 'all', *'áníłtso* 'big'
2. *łeeshch'ih* 'ashes', *łeezh* 'dust, earth'
3. *'akásht'óózh* 'bark', *'akágí* 'skin'
4. *háálá* 'because', *ha'át'íísh* 'what', *háí* 'who'
5. *ních'i* 'blow', *níłch'i* 'wind'
6. *ts'in* 'bone', *ńdiists'in* 'hit'
7. *'akéshgaan* 'claw', *'akee'* 'foot', *'akétl'óól* 'root'
8. *yígháah* 'come, go'
9. *'aháshgéésh* 'cut', *'ałtániishgéésh* 'split'
10. *jj́* 'day', *'ooljéé'* 'moon'
11. *hahashgééd* 'dig', *bighá'níshgééd* 'stab'
12. *doo deení da* 'dull', *deení* 'sharp'
13. *'at'a'* 'feather, wing', *yit'ah* 'fly'
14. *t'áá díkwíí* 'few', *kwe'é* 'here'
15. *yishtin* 'freeze', *tin* 'ice'
16. *'atsiighá* 'hair', *'atsii'* 'head'
17. *bí* 'he', *daabí* 'they'
18. *hastiin* 'husband, man'
19. *bééhózin* 'know', *nízin* 'think'
20. *tooh siyínígíí* 'lake', *tooh* 'river', *tónteel* 'sea', *tó* 'water'
21. *'at'ąą'* 'leaf', *'áłt'ą́'í* 'thin'
22. *łáa'ii* 'one', *łah* 'other', *ła'* 'some'
23. *yázhí* 'short, small'
24. *dilkǫǫh* 'smooth', *'ashkǫ́ǫ́h* 'swim'
25. *tsin* 'stick, tree', *tsintah* 'woods'
26. *'éi* 'that', *'áadi* 'there'
27. *ni* 'thou', *nihí* 'we, you'

Cognates within Turkish:

1. *çünkü* 'because', *içinde* 'in'
2. *tırnak* 'claw', *tırmalıyor* 'scratch'
3. *kesiyor* 'cut', *keskin* 'sharp'
4. *gün* 'day', *güneş* 'sun'
5. *ölüyor* 'die', *öldürüyor* 'kill'
6. *toz* 'dust', *toprak* 'earth'
7. *düşüyor* 'fall', *düşünüyor* 'think'
8. *uzak* 'far', *uzun* 'long'
9. *yürüyor* 'go', *yürek* 'heart'
10. *o* 'he, that', *orada* 'there', *onlar* 'they'

11. *baş* 'head', *başka* 'other'

12. *burada* 'here', *ben* 'I', *bu* 'this', *biz* 'we'

13. *yakın* 'near', *yeni* 'new'

14. *bir* 'one', *ile* 'with'

15. *doğru* 'straight, true'

16. *sen* 'thou', *siz* 'you'

9.1 Item Deletion

As was true for the case of the repetition of whole words, the most straightforward approach to the problem of repeating morphemes is to discard one of the items in question. In this case, however, there occasionally arises the situation where one word is clearly derived from the other, such as German *Gattin* 'wife' from *Gatte* 'husband'. When this happens, it is probably preferable to delete the derived form, because the base will most likely be closer in meaning to any cognate found in a related language.

Table 32 shows that deleting words with repeating morphemes can make an important difference. When tests for linguistic connection were run between all pairs of languages, it was found that allowing repeats increased the test's assessment of the likelihood that the languages were related most of the time. More importantly, for several of the tests, the difference pushed the test over the .05 decision point. Albanian appears related to Hawaiian, and Latin to Navajo, only if the lists include repeating morphemes. Much of the explanation for these findings can be seen in the lists of internal cognates themselves (recall Table 30 for example). And those matches are in addition to random matches found elsewhere in the vocabulary; one or two recurrences due to repeating morphemes can push total tallies over the edge to significance. A more insidious danger is that repeating morphemes can increase the estimation that truly related languages are related. Connections between English and French appear quite a bit stronger when language-internal cognates are included. Because they are indeed connected and low p values are desirable, it can be easy to overlook a methodological flaw.

But surely not all of the decrease in significance is due to the problem of repeating morphemes. As the N columns in the table make clear, by the time one has discarded words with repeating morphemes, on top of loans and motivated words, one has discarded a good deal of the comparanda, indeed up to half of the concepts in this tableau. The decrease in data points may also contribute to the decline in significance in the tests. Is there a way to minimize this decrease? At first it might appear that it would be sufficient to eliminate one of a pair of concepts only

TABLE 32 Effect of Excluding Repeating Morphemes

Languages		Exclude		Include		Cognates[1]
		p	N	p	N	
English	German	.000	116	.000	153	.664
French	Latin	.000	132	.000	162	.613
German	Latin	.000	135	.000	181	.295
English	Latin	.020	123	.000	157	.293
English	French	.131	108	.088	141	.280
French	German	.001	116	.001	153	.272
Albanian	French	.004	106	.002	125	.191
Albanian	Latin	.000	116	.000	148	.186
Albanian	German	.061	118	.094	142	.157
Albanian	English	.316	99	.194	123	.143
Albanian	Hawaiian	.117	119	.044	143	.000
Albanian	Navajo	.573	115	.637	145	.000
Albanian	Turkish	.285	118	.193	137	.000
English	Hawaiian	.265	115	.132	153	.000
English	Navajo	.353	115	.702	152	.000
English	Turkish	.045	116	.004	147	.000
French	Hawaiian	.628	126	.369	158	.000
French	Navajo	.377	123	.571	157	.000
French	Turkish	.266	122	.112	150	.000
German	Hawaiian	.243	128	.387	175	.000
German	Navajo	.758	124	.243	175	.000
German	Turkish	.788	125	.663	165	.000
Hawaiian	Latin	.506	141	.431	184	.000
Hawaiian	Navajo	.256	138	.255	180	.000
Hawaiian	Turkish	.673	137	.663	169	.000
Latin	Navajo	.125	135	.001	185	.000
Latin	Turkish	.066	134	.093	174	.000
Navajo	Turkish	.360	135	.145	171	.000

Notes: All comparisons use the Swadesh 200 word list, and apply the χ^2 metric to initial consonants. Known loan words and motivated words are omitted.
[1] Closeness judgments from Table 29.

when both languages share the decision to associate the same pair of concepts with the same (language-specific) sound, such as when Navajo and Hawaiian are caught using the same word for *tree* and *stick*. But actually the problem is broader than that. Any sharing of sounds between word senses within a language will causes problems if the basis of that sharing is accessible cross-linguistically, even if the language it is being matched with does not happen to avail itself of that same sharing. This is of course a restating of the fact that the χ^2 statistic is not a useful measure of independence if the data points it is based on are not independent along the relevant dimensions. Because one is measuring whether the pronunciation between two languages is independent, given a fixed word sense, then any interdependence of pronunciation and word sense within a given language can spuriously increase that measure.

A synthetic example may help develop this intuition. Imagine that language A only has the sounds R, X, and Y, and that language B has the sounds G, X, and Y, and that the three sounds of each language are equally common. If one had a 90-concept list, one would expect each of the nine cells representing each possible trans-language phone match (3×3) to have about 10 entries. But now imagine this wrinkle. The list contains 45 concepts for red things, and 45 concepts for green things. In language A, the names for red things always begin with the sound R, but the names for other things are free to begin with any other sound. In language B, the names for green things always begin with the sound G, but the names for other things are free to begin with any other sound. Here the languages do not share the same propensity: language A only groups red things, B only groups green things. And yet it is clear that when one counts how many times each of the nine possible pairings actually occur, there will be a wildly different distribution from the expected 10 per cell: 4 of the cells will share the count for all 90 items, evenly among themselves, and the other 5 cells will have no items in them. If one were to compute the χ^2 for that table based on the assumption that the expected frequencies are 10 per cell, it would prove to be tremendously significant ($\chi^2 = 90, p < .001$). And yet that high χ^2 is due entirely to the individual patterning within each language. If there is any sense in which it is true that the high value is due to similarity between the two languages, it lies in the fact that they both like to include color designations in the names of things. But since the semantic category of color is universal, and the propensity to similarly denominate semantically similar objects is universal, this particular similarity is an invalid indicator of historical connection between languages.

Of course this thought experiment is extreme, but it is meant to be extreme for the sake of clarification, and it is taking to extremes true

TABLE 33   Color-Based Skew in Two Hypothetical Languages

B	A	
	G	X	Y
---	---	---	---
R	0	22.5	22.5
X	22.5	0	0
Y	22.5	0	0

tendencies of languages. English itself offers a case not too dissimilar in principle: All demonstratives, and indeed many grammatical words in general, begin with /ð/. No lexical words begin with /ð/. No grammatical words begin with /θ/. So the more grammatical words on the word list, the more problems will arise. And the Swadesh 200 list does include grammatical words, no doubt from an effort to find concepts that are universally recognized.

In one respect, though, the thought experiment was one-sided: 'Red' and 'green' come from the same semantic domain and are mutually exclusive. Would the same problem occur if neither of the two concepts implied the other or the absence of the other? Not exactly, but a more subtle problem can arise. Imagine one language names red things with words having the prefix R, and another prefixes the names of large things with L. In itself that would not cause a problem, unless the word lists disproportionately emphasized large red objects. Then there would appear to be a lot of $R : L$ matches, whether or not they have any common historical origin. There may not be any particular reason for the word list to have such an imbalance, but, crucially, there is no particular reason why it should not. And it is easy enough to imagine imbalances that a Swadesh-type list might have. In its effort to achieve cultural universality, such a list might easily overemphasize, say, parts of the body, or the natural environment.

So it does not appear sufficient to simply discard words with recurring morphemes only when the repetition holds across both langagues. The above argumentation shows that that can occasionally lead to the finding of spurious connections, although, to be sure, the magnitude of the danger is probably fairly low. The opposite possibility, however, is much more obvious and dangerous. If in one language concepts 1 and 2 are represented by the same morpheme, but not in the other, then one is guaranteed to not find a valid sound recurrence across those two concepts. That will probably artificially lower the estimation that the two languages are connected.

The most helpful solution is probably obvious: The researcher should discard from each list concepts that share the same morpheme. Or, to put it a bit finer, discard all but one of any set of concepts that share the same morpheme in the part of the word that is under investigation. In comparisons involving English, it is sufficient to eliminate either *this* or *that*, provided the decision is made in some way that is not biased by the pronunciation of the word in the other language. One could for example, eliminate a word that is actually a derivative of the other. Or, in a pinch, always eliminate the word whose Swadesh gloss is alphabetically prior.

9.2 Trimming Words

It is also a good idea to routinely reduce words to single morphemes. Strictly speaking, that might not be absolutely necessary. If *blueberry* appears in an English word list, it should be clear enough whether *blue* or *berry* also appears (in which case the 'blueberry' concept should be discarded). But the problem of polymorphemic words may go deeper. In some languages, there may well be some property of complex words themselves that manifests itself in a phonetic property that is being tested. For example, if one's metric were to include some measure of word length (which is not a good idea, as will be demonstrated later), one would run up against the fact that complex words would usually be longer than single morphemes. Or if, more reasonably, one were comparing word-internal phonemes, one might find that compound words have special linking units, such as the Greek /o/ (e.g., *cheir-o-manteia* 'palm-reading', lit. 'hand-prophecy') or Latin /i/ (e.g., *pedisequus* 'servant', lit. 'foot-follower'). In comparing Greek and Latin, one would get several /o/:/i/ matches among compounds. That fact alone would not be a problem if the choice of whether a concept will be expressed by a simple or complex word in some language is arbitrary. But that assumption is not justified. Languages tend to correlate on this property. Certainly in everyday experience, languages more often agree on whether a concept is named by a simple or compound word, than one would expect by chance. So if two languages have, say, specific linking vowels for compounds, then even if those vowels are different and the languages are not related, the presence of several compounds in the word lists would result in an increased number of recurrences for that pair of vowels, which would spuriously increase the likelihood of reporting that the languages are related.

Allomorphic variation can also be a difficult wrinkle. Computations can easily be thrown off by changes that were conditioned by adjacent sounds or by differences in the metrical context in which a morpheme finds itself. It is easy to see how that can make it harder to identify connections between words, as in the aforementioned case of English *who* and *what*, where originally both began with /hw/, but then /w/ was dropped in clusters before round vowels (*who*) and later /h/ was dropped before remaining /w/ (*what*). Somewhat more subtly and therefore more dangerously, adjacent sounds can have the affect of making different morphemes appear more similar. In many languages, for example, a front vowel often has the effect of making an adjacent velar consonant become coronal (palatalization). For example, the French word *cent* /sã/ is derived from the Latin word *centum* /kɛntʊm/, both meaning 'hundred'. But the opposite is not a general pattern; coronals do not often become velars next to back vowels. Now imagine a situation where in such a language morphemes in a universal category X tended to be followed by a morpheme beginning with a front vowel. In general, then, words of that category would end in coronals more often than one might otherwise expect. This would skew the distribution of expected frequencies in the language, and be particularly problematic if the other language had any sort of skew influenced by the same property.

The distinction between parts of speech may well turn out to be such a universal category. Many languages differentiate nouns and verbs morphologically. If those differences had any effect at all on the rest of the word, then that would potentially skew frequencies away from the expectation. French may be a concrete example of this. Consider for example the case of root-final Latin /k/. Through palatalization, /k/ has regularly become Modern French /z/ before original /e/ or /i/ (e.g., *facimus* 'we do' > *faisons*). Before other vowels it generally lenited to /ɣ/, which at most left a trace as *i* (e.g., *baca* 'berry' > *baie*, Price 1971). Now those vowels were not distributed randomly across inflectional categories. Most verb stems, especially in the core vocabulary, ended in /e/ or /i/, some ended in /a/, and none ended in /o/ or /u/. Most noun stems ended in /a/ or /o/, although some had /e/ and /i/. So it would seem possible that root-final *i* would end up particularly common in nouns, and /z/ especially common in verbs. I have not actually counted the French roots to verify that possibility, and in fact so much restructuring has gone on in French conjugations that I would be rather surprised if the numbers came out solid, but the general idea that such imbalances can easily arise should not be in doubt. One could imagine the problem if another language (Anti-French) had nouns mostly ending in palatalizing /i/ and verbs ending in nonpalatalizing /u/. Then when matched against Old French, it would appear that French coronals pair up with Anti-French velars significantly more often than expected by chance. But that would tell us nothing about the possibilities of linguistic connection between French and Anti-French. Palatalization before front vowels and using suffixes to distinguish nouns from verbs are both universally available features of language.

In the case of French, one could easily correct for that possibility. In fact, in real life, one probably would not want to compare French directly with anything, but instead use Latin, or even Proto-Indo-European reconstructions, thus working from a stage preceding such palatalization. But it is obvious that such steps can only undo the very surface of the problem. At some time depth (which will be quite recent for most languages) one will be at a stage where one cannot reliably determine what the antecedent form is, but nevertheless the language may still suffer from such a distributional skew. Even in a situation where it was clear that verbs tended to end in, say, /s/ more than nouns, and that that was obviously due to the palatalizing effect of a following verbal marker, there might be no principled way to take such a skew into account in a χ^2 test unless there were some clear evidence as to how to properly reconstruct the antecedent. After all, the fact that the count for /s/ is elevated suggests that there was some merger. Perhaps original /k/ became /s/ in that environment, and original /s/ remained unchanged. The only proper way to correct for the skew would be to undo the change and list the original /k/ verbs under /k/ instead of /s/, but there may be no evidence to tell one which verbs were which.

The best advice I can suggest is that the investigator run statistical tests over the set of morphemes, to see if there is any discernible skew along categories known to be particularly susceptible to such alternations. If so, then the investigator will know to be particularly careful to look for any internal or comparative evidence for reconstructing forms antecedent to the processes that caused the skew in question.

In my own investigations using the Swadesh lists, I have experimented with correlating word shapes with the part of speech called for by the Swadesh lists: *verb, adjective, linker, limiter, noun.* (*Linker* is my omnibus term for grammatical categories such as conjunctions and prepositions; *limiters* are words like demonstratives and numerals, which restrict the reference of a word without modifying it.) Then I tried a simple χ^2 test between the phonemes of the words and each of those categories. For example, one test would see whether there was a connection between the identity of the final phoneme of the word, and whether or not it was a noun. One simply builds up a contingency table where the rows list each of the possible final consonants, and the columns are *Noun* and *Not Noun*. Then one goes through the Swadesh 200 list and tallies how many times words with each final phoneme are nouns or not. If there is an imbalance that comes out reasonably significant by the χ^2 test, then one may wish to investigate whether there is a causal connection. As a further refinement, I designed the computer program to keep track of which phonemes are especially strongly associated with a particular part of speech.

Because the outcome of these studies may prove useful not only for the narrow task of deciding on the validity of running language-connection tests with the Swadesh word lists, but also for illustrating the sort of problems that may arise in trying to reject repeating elements from the word lists, I present the results in some detail. I begin by comparing the results across all words at the full word level. That is, I am here considering the entire word spelled out in italics after the language name in the entries in the appendix, not just the part that I have determined to be the root.

Table 34 shows the significance of the correlations between the phonology of the words and the parts of speech. One may object that the methodology is not sound, because I am inappropriately running many χ^2 tests over the same data. I am not however using quantitative methods in this case to prove anything; these tests are simply exploratory aids. Running separate tests is simply one of the most straightforward ways to highlight potential problems.

The *Initial* rows in the table correspond essentially to the sort of tests that have been under discussion up till now, except that I treat the vowels separately instead of lumping them all together as *0*, the absence of an onset. It will be noticed that in general these numbers are not particularly significant, with the major exception of Navajo. For most other languages, the *Final* rows are most significant, that is, the results of comparing part of speech to the final phoneme of the word. This difference is not surprising when one considers that Navajo has prefixing inflection, whereas the other languages have mostly suffixes. The odd man out, Hawaiian, does not typically inflect words at all. These patterns strongly suggest that inflection may be the causal factor for differences in those two rows.

Table 35 shows some typical results of performing these tests. It tells what connections between phoneme and part of speech appear particular noteworthy at the beginning of words in Navajo. After each phoneme, the sign tells whether the phoneme is positively associated with the part of speech. For example, the display shows that /ʔ-/ appears particularly often at the beginning of nouns, but that /n-/ is particularly rare. The following number gives some idea of the significance of the connection, that is, low numbers show stronger connections. Some of these figures, to be sure, are based only on a very few words. It is, for example, probably not worth worrying about figures involving linking words, since there are only a few such words in the lists. But the fact that initial /ʔ-/ is so strongly associated with nouns attracts immediate attention. And

TABLE 34 Correlations Between Phonology and Part of Speech

Language	Position	Verb	Adjective	Limiter	Link	Noun
				Role		
Albanian						
	Initial	.291	.917	.298	.151	.515
	Final	.000	.425	.502	.076	.061
	Anywhere	.008	.282	.195	.752	.094
English						
	Initial	.183	.379	.011	.008	.789
	Final	.000	.002	.006	.414	.004
	Anywhere	.000	.256	.055	.874	.228
French						
	Initial	.658	.443	.808	.374	.232
	Final	.000	.151	.115	.010	.000
	Anywhere	.000	.154	.005	.261	.003
German						
	Initial	.667	.440	.039	.001	.345
	Final	.000	.050	.020	.085	.000
	Anywhere	.000	.211	.135	.951	.000
Hawaiian						
	Initial	.003	.167	.359	.010	.464
	Final	.244	.205	.126	.474	.957
	Anywhere	.359	.030	.114	.948	.474
Latin						
	Initial	.200	.416	.129	.002	.159
	Final	.000	.000	.000	.001	.000
	Anywhere	.000	.000	.025	.985	.000
Navajo						
	Initial	.001	.651	.010	.192	.000
	Final	.015	.002	.163	.020	.100
	Anywhere	.000	.019	.555	.981	.000
Turkish						
	Initial	.280	.804	.118	.353	.026
	Final	.000	.151	.070	.000	.000
	Anywhere	.000	.400	.694	.250	.000

TABLE 35 Significant Correlations Between Part of Speech and Initial
Sounds in Navajo Words

Part of speech	Sound	p
Verb	/j/	+0.000
Adjective	/d/	+0.021
Limiter	/t'/	+0.008
Link	/ʔ/	−0.025
	/b/	+0.033
	/d/	+0.034
	/h/	+0.035
Noun	/ʔ/	+0.000
	/n/	−0.003
	/d/	−0.017
	/j/	−0.031
	/t/	+0.042

the reason, fortunately, is obvious. In Navajo, many nouns (those that
are not perceived as first class, independent objects in the universe) are
obligatorily inflected to show possessor. In my word lists, they have been
entered in the most neutral such inflection, the *'a-* /ʔa-/ meaning *someone's*: *'anághah* '(someone's) back', *'akásht'óózh* '(something's) bark',
'abid '(someone's) belly', etc.

In comparison, Table 36 is the display for the end of Latin words. It
can be seen that /-t/ is a common ending of verbs, which is not surprising
when one considers that verbs were entered in the third person singular.
For nouns, /-a/ and /-m/ were listed as common endings; these reflect
the common nominative singular endings *-a* of first declension nouns
and *-um* of neuter second declension nouns. For adjectives, /-s/ was
identified as the most typical ending; this reflects the fact that the most
common types of adjectives, the *-o/-ā* stems, were always entered in the
nominative masculine singular, which ends in *-us*.

Similar results obtain for other languages. For English, for example,
the program finds a strong association between verbs and the endings
/-s/ and /-z/. For Turkish, verbs end disproportionately in /-r/. It will
take only a little thought to see that if languages are compared in a way
that takes any appreciable account of the end of the word, undesirable
effects can take place. English and Latin will be linked because there are
many recurrent /s/:/t/ and /z/:/t/ pairs. That may sound like a good
thing, until one considers that either of those languages will also appear
to be very closely connected to Turkish, with many /s/:/r/ and /z/:/r/
pairs linking English and Turkish.

TABLE 36 Significant Correlations Between Part of Speech and Final Sounds in Latin Words

Part of speech	Sound	p
Verb	/s/	−0.000
	/t/	+0.000
	/a/	−0.022
Adjective	/s/	+0.000
	/t/	−0.001
Limiter	/k/	+0.001
	/iː/	+0.009
	/t/	−0.011
	/e/	+0.027
Link	/d/	+0.001
	/iː/	+0.043
Noun	/t/	−0.000
	/a/	+0.000
	/m/	+0.008

The lesson is not to avoid comparing the ends of words, because, as was seen for Navajo, it is not unknown for languages to inflect at the beginning of words instead. In fact, inflection in the middle of a word happens as well. In Navajo verbs, the leading material is often arguably derivational, and truly inflectional morphemes such as person agreement come between those derivational particles and the word-final root. (The *Anywhere* rows in the tables are meant to pick up such cases, although they are hard to interpret because "anywhere" is not as precisely defined as the absolute beginning or end of a word. They should also not be treated as completely reliable statistically, because phonotactic dependencies within a word violate the usual assumptions of independence. If one were really doing significance testing, one would not consider multiple phonemes from the same word in the same χ^2 test; see the discussion on page 60.) Rather than avoid the end, beginning, or middle of words, the linguist must be careful to first strip inflectional morphemes off of words before comparing them. This advice may seem obvious; Ringe (1992, 1993) did accordingly without feeling the need to justify it. But it is important as an example illustrating the potential pitfalls involved in leaving recurring morphemes in the word lists. It is also helpful to start this discussion at an easy level, considering inflections on words, because the next steps are less intuitive but equally important.

Table 37 shows what happens when the tests correlating phonemes with part of speech are run after stripping away inflectional endings.

This is apparently the level which Ringe (1992) worked with. One's first reaction to the table is a sense of disappointment that so many strong correlations remain. The very low p scores in so many of the entries means, in effect, that one can still to a very large extent predict part of speech from the shape of the word stem.

It may be useful to first examine a well-known language. The data indicate that after stripping off inflectional endings, Latin nouns, adjectives, and verbs are still fairly distinctive at the tail end. The detailed phoneme pairings reveal that verb stems end disproportionately often in /-aː/, /-e/, /-eː/, and /-iː/. Adjectives tend to end in /-o/, and nouns in /-a/. The historical linguist or classicist will immediately recognize these as the most distinctive stem vowels (except that I have applied a synchronic analysis giving the first declension a stem vowel length identical to that of the second, rather than giving it its historical length), as were mentioned in the semi-hypothetical discussion above about French. Similar results can be found with other languages. French verb stems end often in /-e/, reflecting the size and distinctiveness of the *-er* conjugation. Albanian verb stems tend to end in /a/ or /o/; the latter is equivalent to the productive /-aː/ of Latin.

One might immediately argue that I engaged in some trickery by not removing the stem vowels along with the inflectional endings. After all, native speakers tend to think of that vowel as part of the ending; most French speakers will see *vomit* as stem *vom* plus ending *it*, not as *vomi* plus *t*. And I am certainly not inclined to argue the point, because my conclusion is congruent. Stem vowels must be eradicated, because they too readily identify an item's part of speech. But unfortunately the problem goes much farther. For one thing, some remaining correlations have nothing to do with stem vowels, at least not in any direct sense. In German, those vowels have essentially disappeared, and most verb stems end in a consonant or a long vowel. But then many nouns are identifiable because their stem ends in /-ə/, an option not available to verbs. Those final schwas have to be eradicated as well, for the same reason that theme vowels must go: not because they are significant morphemes, but just because they are strongly associated with part of speech, a universally available grammatical category.

In view of such findings, I made one last radical attempt to stamp out correlations between phonemes and part of speech by trimming back words, using several different principles. First, all meaningful elements were cut back, except for the root. For example, the causative *h-* prefixes of many Hawaiian verbs, and the final *-i* found on many transitive verbs were stripped off. Secondly, certain gross phonological principles were applied, such as stripping off all final schwas. Thirdly, all identifiable

TABLE 37 Correlations Between Stem Phonology and Part of Speech

Language	Position	Verb	Adjective	Lim	Link	Noun
Albanian						
	Initial	.440	.921	.310	.171	.444
	Final	.000	.602	.700	.125	.001
	Anywhere	.227	.367	.069	.786	.181
English						
	Initial	.183	.379	.011	.008	.789
	Final	.003	.769	.025	.666	.113
	Anywhere	.014	.792	.089	.868	.819
French						
	Initial	.748	.526	.324	.802	.190
	Final	.000	.002	.064	.162	.000
	Anywhere	.001	.289	.014	.997	.014
German						
	Initial	.499	.498	.039	.000	.348
	Final	.012	.103	.071	.085	.013
	Anywhere	.001	.258	.359	.924	.000
Hawaiian						
	Initial	.003	.259	.274	.026	.408
	Final	.318	.167	.105	.403	.969
	Anywhere	.559	.020	.066	.954	.416
Latin						
	Initial	.163	.467	.136	.002	.176
	Final	.000	.000	.172	.002	.000
	Anywhere	.000	.004	.076	.991	.001
Navajo						
	Initial	.002	.894	.279	.081	.000
	Final	.022	.003	.095	.020	.078
	Anywhere	.095	.377	.521	.999	.034
Turkish						
	Initial	.287	.812	.112	.424	.033
	Final	.063	.231	.573	.004	.014
	Anywhere	.180	.649	.822	.277	.057

etymological units were stripped off, regardless of whether they had any synchronic function. Sometimes it is clear that the elements, though having no current meaning, once had a morphemic role; examples include the /d/ found at the end of so many Latin and French verbs, which may once have been a marker for the present tense (Sihler 1994:510). Others are completely obscure; these include the so-called extensions of Proto-Indo-European roots. For example, because of the hypothesis that English *grass* is related to the word *grow*, the final /as/ was identified as a formative etymon and stripped off. Finally, the sample completely excludes all words whose roots are etymologically connected with the root of some other included word.

Table 38 looks more encouraging. Most of the strong correlations are for the linker and limiter parts of speech. The correlations arise in part because there are relatively few such words in the Swadesh list, and therefore coincidences are easily magnified. Fortunately the low frequencies also mean they are not so much of an issue, because they will statistically have little effect on the computations. It may be helpful to look at one case to understand the scope of the problem. The most certain correlation reported in those columns, $p = .002$, is the case of English linking words. An inspection of the individual phoneme pair counts shows however that the program is primarily impressed with the unimpressive fact that in the word lists, few conjunctions or prepositions began with /n-/ or /s-/.

Of the remaining significant numbers, a few are readily interpretable. The strong link between Hawaiian final phonemes and verbs comes from the fact that I had just deleted possible transitive /i/ from the end of verbs; but this left most of those verbs with a final consonant, and therefore highly distinctive in a language that has no word-final consonants. Fortunately the metrics I will be developing in the next chapter are not sensitive to word-final position, and so it is probably not really necessary to now delete the vowels from the end of all words.

Other significant findings are more puzzling, although they might be clearer to one expert in the individual language history. For example, it is reported that Latin nouns begin with /k/ more often than other parts of speech. I know of no reason why this should be so. Should one go ahead and delete noun-initial /k-/ from Latin words? I think that would be a mistake. Coincidences happen. If the researcher sets about trying to correct every imbalance in a blind fashion, errors may actually be introduced in the process. If Latin initial /k/ is not really an etymological element, then removing it from nouns means one will systematically more often be comparing the wrong phonetic element against potential cognates in other languages. Moreover, one would probably simply in-

TABLE 38 Correlations Between Root Phonology and Part of Speech

Language	Position	Role				
		Verb	Adjective	Lim	Link	Noun
Albanian						
	Initial	.073	.715	.588	.349	.528
	Final	.013	.363	.635	.083	.028
	Anywhere	.250	.558	.436	.979	.705
English						
	Initial	.351	.285	.273	.002	.927
	Final	.068	.257	.452	.797	.558
	Anywhere	.118	.349	.199	.948	.452
French						
	Initial	.452	.421	.224	.669	.514
	Final	.453	.769	.247	.167	.531
	Anywhere	.109	.740	.071	.995	.395
German						
	Initial	.329	.812	.164	.005	.744
	Final	.433	.121	.184	.303	.658
	Anywhere	.124	.236	.447	.961	.216
Hawaiian						
	Initial	.395	.510	.814	.075	.365
	Final	.000	.866	.307	.100	.328
	Anywhere	.301	.175	.832	.571	.331
Latin						
	Initial	.214	.610	.296	.068	.018
	Final	.399	.170	.345	.097	.182
	Anywhere	.628	.240	.023	.999	.128
Navajo						
	Initial	.081	.953	.932	.024	.409
	Final	.477	.039	.873	.130	.051
	Anywhere	.659	.477	.738	.783	.201
Turkish						
	Initial	.166	.918	.244	.384	.076
	Final	.596	.336	.735	.070	.173
	Anywhere	.731	.556	.543	.875	.282

vert the skew, so that it would appear that nouns begin in /k/ *less* often than other parts of speech. Mathematical tools such as these significance tests should be used for finding suspicious patterns, whose precise interpretation then becomes a matter for linguistic judgment.

9.3 Stratificational Approaches

By and large I find the numbers in Table 38 reasonably satisfactory, at least for expository work, and use the trimmed roots throughout this study. But what if strong correlations had remained, and it was impossible to undo such effects? In that case there does remain one last option. If, in general, nouns tend to have similar properties to each other and verbs tend to have similar properties to each other, then one solution would be to test nouns and verbs separately. As long as the nouns have no verbs to contrast with in the same test, the fact that they have peculiar phonetic properties in common would not cause a problem. And so forth for the other parts of speech.

There are important problems that would have to be ironed out, however. One is that even such a basic concept as part of speech is not completely universal. While most if not all languages appear to have core concepts of nounhood and verbhood, other parts of speech are not so clear. In Indo-European languages, for example, adjectives are very much like nouns and can often be used as nouns; in fact, the traditional term *noun* originally referred to the classes of what we now call nouns (i.e., formerly *nouns substantive*) and adjectives (formerly *nouns adjective*) taken as a whole. In many other languages, such as Hawaiian, adjectives are very much like verbs and might well be considered the same part of speech. This suggests already that one will now need at least three tests (nouns, adjectives, and verbs). And certainly there are strong reasons for believing that grammatical words like prepositions, pronouns, and articles will be special: In comparison to lexical words, they are more likely to be short, unstressed, or clitic, if they are indeed words at all and not inflectional or derivational morphemes (Hopper and Traugott 1993, Meillet 1926). In summary, it is not clear exactly what an appropriately cross-linguistic division of parts of speech should be. Mismatches between categories may reduce one to running a separate test on every important intersection of categories.

And it is hard to imagine that one can reasonably come up with a classification scheme that will work for all languages. The goal, after all, is not to deal with part of speech for its own sake. More generally, the idea is to statistically deal with any linguistically intractable patterning along dimensions that are universally available. Languages may have not only a distinctive shape for nouns, but they may have a distinctive shape for abstract nouns, or for words denoting physical handicaps, or for verbs that denote repeated action. If it is necessary to put all of these in separate tests, one could end up with many tests indeed.

That need to run many different tests has two significant ramifications. One is the problem discussed in an earlier chapter, where it was pointed out that multiple tests often introduce false positives. Here the tests are independent, so the probabilities can be multiplied together, but that is not a great win; if two tests are run at $p \leq .05$, the overall p level if only one test is successful is .0975, which is a serious loss. More importantly, even if the lists have 200 words, splitting that list into many categories results in much shorter word lists for each test, especially after one has discarded loans, motivated words, and recurrent morphemes. As mentioned earlier, shorter word lists give the overall effect of reducing the power of the tests, in this case, intolerably so.

In the end, this technical problem appears to be intractable from a statistical point of view, especially within the context of a χ^2 test. At best, I might recommend that linguists particularly interested in grammatical morphemes should best incorporate them by running a separate test. Otherwise, the most useful recommendation appears to be that linguists have to do their best to do the hard work of undoing the morphology, through either comparative or internal reconstruction of the basic roots. If serious doubt remains, one should refrain from testing on the part of the word that has the most problems.

10

Recurrence Metrics

The χ^2 metric is a natural solution to the problem of doing statistical inference on tables of the sort advocated by Ross and Ringe. But in a couple of respects it falls short of perfection. This chapter will explore possible improvements.

10.1   Chi-Squared Versus Recurrence

The first problem is that the χ^2 metric just does not correspond to what historical linguists think of as a natural measure. It must be admitted that it has some odd properties. It is based fundamentally on a measure of how unexpected frequency counts of phoneme pairs are, relative to the frequencies of the individual phonemes. By and large, this seems intuitively correct. Assume for example that the English word list contains only 3 instances of /p/, and a German word list contains only 3 instance of /pf/, and in all three cases they team up with each other. That would surely mean a lot more than finding that very frequent phonemes like /b/ match up 3 times. Where the concept of expectation becomes more questionable is when rare phonemes match up very few times indeed. In reality, the Swadesh 100 word list has only one word with initial /p/ in English (*path*), and it does match up semantically with the only instance of /pf/ in the German list (*Pfad*). And the χ^2 metric finds that is very important; in one test I ran about 11% of the deviance is based on the one match, and this is in the face of wonderful evidence like 6 /s/:/z/ matches, 8 /f/:/f/ matches, and so on. Now possibly this seems right and proper. After all, it contributed toward a very low p value for the English–German connection, as is good, and it seems particularly clever that the test picked up on what is considered to be one of the most important phoneme correspondences of the dialect divide between Low and High Germanic languages.

Now consider another case. Once, when I tried testing the hypothesis that French and Hawaiian were related, by looking at correspondence among the first vowel of the word, I got astonishing results. The test gave me a significance level of .0004—it was 99.96% sure that the languages were historically connected. Even if one is willing to entertain the possibility that the languages may somehow be related by some common ancestor in the remote past, that degree of confidence seems inappropriate, especially if one considers that none of the other Indo-European languages (certain relatives of French) show such a strong connection with Hawaiian. What caused this finding? It turns out that the vowel /ɛ̃/ was found only once in the French word list for Swadesh 100, and the vowel /uː/ was found only once in the Hawaiian list. It also turns out that the word 'breast' is on the list, so French *sein* /sɛ̃/ corresponds to Hawaiian *ū* /uː/. This is the fact that the χ^2 metric found so important. It is statistically the exact same situation as the *path:Pfad* connection just discussed. But here the metric seems remarkably inappropriate. Surely no linguist would be willing to claim that French and Hawaiian were related—and therefore Indo-European and Austronesian—solely on the observation that the word for 'breast' is *sein* in French and *ū* in Hawaiian. This result does not turn up in the version of the test we are currently considering, but that is entirely adventitious. The reasons for the difference—the fact that *ū* was rejected as a motivated word and that we are not basing our tests solely on vowels anyway—have nothing to do with core problem, that a single nonrepeating phoneme match can give a dramatically low p value.

Fortunately, the χ^2 metric is not the only possible way of analyzing a table. With Monte Carlo tests of significance, researchers have a great deal of freedom in defining a metric. They do not have to limit themselves to the comparatively few metrics for which theoretical distributions have already been worked out by mathematicians. A tentative step in this direction was taken by Swadesh (1956) in the context of similarity tests. To get an idea of how frequently word pairs should achieve a certain similarity threshold by chance, he compared every word in Language A with every other word in Language B, counted how many of those combinations passed the threshold, and divided by the number of combinations. The idea that the data themselves could reveal the expected statistics was an important insight, but his procedure provided only an estimate of the mean, not of the distribution as a whole and therefore not of the significance of the result. A bigger step was taken by Oswalt (1970). His own method for testing long-distance relationships between languages involved counting how many pairs of phonemes across semantically-matching words share more than a certain number of features. There is of course no general statistical test for this, so Oswalt invented the shift test. He computed the base metric on the original pairing of words, and compared that with the comparable statistic obtained when words in the first language are compared with the words in the next language offset by 1 (e.g., word number 3 in Language A is compared with word number 4 in Language B; A-4 is compared with B-5, etc.); then again with the words offset by 2; and so forth to 100, that being the size of the Swadesh list he used. Then he saw whether the base metric is in the top 1% of those 100 numbers. In other words, the shift test is clearly in the class of permutation-type tests of significance. There are to be sure a few statistical problems with the specific procedure he uses. To make such a systematic procedure completely valid, one would actually need to do all possible arrangements, not just those that are offset by the same constant amount. As has already been mentioned, that is of course impossible, there being almost 10^{158} such arrangements, but one can make a completely valid approximation by taking a large enough random subset of such arrangements (the Monte Carlo approach). Oswalt's systematic procedure is dangerously nonrandom, especially if one is using the version of the Swadesh list that is arranged in semantic order (*I, thou, we, this, that, who, what,* and so forth). And looking at only 100 arrangements, in any event, results in what most people would consider a rather high degree of variance in the result. Nevertheless Oswalt's technique is essentially correct and he deserves credit for introducing this class of statistics into historical linguistics.

But Oswalt was working within the context of similarity metrics. I have already discussed in the introduction reasons for preferring measures based on phoneme recurrences in general and the traditional practices of historical linguists in particular. Unfortunately, no exact metric has ever been spelled out over two centuries of historical linguistic research. The rest of this chapter will explore various metrics with an eye both to algorithmizing traditional practice and to realizing the theoretical implications of arbitrariness in vocabulary. In the course of this program, I will present tables that illustrate the results as they apply to my data sets. The reader should be reminded again of the danger of comparing the tables too narrowly. The p values should in general be inversely correlated to the cognate level judgements. But because of the probabilistic nature of the tests and the fact that actual language relationships are unknown for most of the pairs, as well as the fact that the pairs are not independent of each other, it would be inappropriate to use small differences between these tables as solid proof of the superiority of one metric over another.

One clear fact about standard practice is that it is based fundamentally on the idea of recurrent sound correspondences. So perhaps the metric should be the total number of recurrent sound correspondences in the system. One can test whether a pair of languages has significantly more recurrent sound correspondences than they would if the words were paired by chance. It is interesting to recall that Ross's (1950) complex and unusable formula was primarily motivated by the desire to do just this—count recurrent correspondences rather than deviations as per χ^2. This turns out to be quite simple using permutation tests, provided the metric is defined carefully enough.

A first approach might be to sum the observed frequencies of the phoneme pairs, subtracting 1. That is, if a phoneme pair has frequency of 0 or 1, it would not count as all because there are no recurrences. If the frequency is 2, the metric would count one; there is 1 recurrence. A frequency of 3 would count as 2, and so forth. Those counts would be summed across all pairs of phonemes.

That metric has much to recommend it. It certainly appears to more directly represent the idea of phoneme recurrence. Nonrecurrent matches are ignored, including singletons as in the French–Hawaiian example just mentioned (when in fact the same test was performed replacing χ^2 with this metric, the p value became a nonsignificant .1078). Phoneme pairs that recur many times count more than pairs that recur fewer times. This seems intuitively reasonable.

There are however some serious issues that to my knowledge have never been definitively addressed in discussions of the comparative method. The most obvious question is whether relative distribution of frequencies among the recurrent matches should make a difference. For example, if a table shows that there are 5 different phoneme pairs that occur twice each, is that better or worse than one phone pair that occurs 6 times? In either case there are 5 recurrences, and the metric just proposed gives a total value of 5. Despite that general uncertainty, one special case is obvious. If phone X in one language corresponds to phone Y in another, then a few more $X : Y$ matches certainly counts for more than additional correspondences of X with some other phoneme Z. To take a concrete example, an initial /w/ in English corresponds to an initial /v/ in German; in the Swadesh 200 list (Table 7) there are 10 such matches out of 16 English words that begin with a /w/. That constitutes strong evidence that the languages are related. The fact that English /w/ is also matched twice with German /f/ is not held to strengthen that judgment; if anything, it detracts from the otherwise highly probative /w/:/v/ match. One way of interpreting that judgment mathematically is to say that differences in density are important to the comparative method.

	a	b	c
A	10	10	10
B	10	10	10
C	10	10	10

	a	b	c
A	20	5	5
B	5	20	5
C	5	5	20

FIGURE 2 Different distributions of N=90 in a 3 X 3 table.

Perhaps a more definitive reason for rejecting a simple count of recurrences is that the metric degenerates into powerlessness when the number of words gets large relative to the contingency table dimensions. Counting recurrences may seem satisfactory when one has a huge table filled mostly with zeroes, as is Table 7. However, consider what happens as one adds more and more pairs of words. The number of phoneme types will not increase very much at all, but more and more of the zeroes will fill in. Eventually one will get to the point where most of the cells of the table are filled with larger and larger numbers. At that point, most of the table geometries will become indistinguishable by the simple recurrence metric. Once every cell has at least two entries, any additional entry will lead to the same recurrence metric, regardless of what the table looks like. Consider for example the two tables in Figure 2. In each case the recurrence metric would count 81. But intuitively, the table on the right suggests much greater possibilities of language connectedness.

There are many possible ways to formalize the differences between an uninteresting table like the leftmost and an interesting one like the rightmost. One possible interpretation is that recurrences between X and y are uninteresting, no matter what their number, except to the degree that they exceed the background level. In the first table, even as many as ten recurrences between A and a are uninteresting, because A has as many recurrences with every possible phone, and the same for a. In the second, the 20 matches between A and a are interesting not so much because of the absolute number, but because those symbols each otherwise match many fewer phones. One way to capture that idea is to decrease the count in each cell by the row average, or the cell average, whichever is greater, then to sum up only the positive values. In a table like the left-hand one, all the 10's would reduce to 0, and the metric for the whole table would be 0. In the right-hand table, the 5's would reduce to 0's, but the 20's would only reduce to 10, summing up to 27 recurrences for the table as a whole.

Such a procedure would work, as I verified by implementing the algorithm. But a more elegant way to take into account differences in table density would be simply to square the number of recurrences (i.e., cell frequency less one) before summing them all together. In the left hand configuration of Table 2, one would count $9^2 + 9^2 + 9^2 + 9^2 + 9^2 + 9^2 + 9^2 + 9^2 + 9^2 = 729$. In the right hand table, $19^2 + 4^2 + 4^2 + 4^2 + 19^2 + 4^2 + 4^2 + 4^2 + 19^2 = 1179$. One might argue that historical linguists do not square their data. On the other hand, neither do they subtract the greater of two averages from each number. Squaring the numbers is one way of capturing the general intuition among linguists that the evidence of phoneme recurrence grows faster than linearly: 3 /p/:/f/ matches is not much, but 6 such matches seems much more than twice as good. Mathematically, squaring the inputs has the function of always rewarding differences in the distribution of N objects. For example, $5^2 + 5^2 = 50$, but $6^2 + 4^2 = 52$, $7^2 + 3^2 = 58$, and so forth. I will call this metric R^2 (recurrences squared).

Table 39 illustrates results from using of the R^2 metric. The procedure is very similar to that used in the preceding χ^2 tests on $C_{1.1}$. Except now, for each random reordering of the word lists, the test computes the R^2 metric, and sees whether that is larger or smaller than the R^2 metric for the original table.

In general, the R^2 metric gives satisfactory results. For languages known to be connected, it accurately identifies the relationship when the cognate count is about 18% or greater. Among the unrelated languages, the only false positive is that between English and Turkish.

The fact that the new R^2 metric appears to be at least as good as the χ^2 metric is reassuring. It illustrates the principle that the use of Monte Carlo permutation tests of significance opens the door to a much wider range of statistical metrics than the comparatively few metrics than can be approximated by available distributions (Good 1995).

One pleasant spin-off of the R^2 metric is that one can get some insight into what correspondences are the most important contributors to the overall p value. Through the 10,000 rearrangements, I counted how often each of the cells contributed more to the total R^2 metric than they did in the observed table. Low numbers indicate that the cell was important. For the English–German comparison, the cells with frequencies lower than 10 are $\emptyset : \emptyset$ (i.e., zero consonant, i.e., vowel-initial words), /d/:/t/, /f/:/f/, /g/:/g/, /h/:/h/, /l/:/l/, /m/:/m/, /n/:/n/, /r/:/r/, /s/:/ʃ/, /s/:/z/, /t/:/ts/, /θ/:/d/, /w/:/v/. These will be readily recognized as legitimate correspondences between the two languages. It is especially encouraging that pairs like /w/:/v/ are found, verifying that the test is really looking for correspondences, not phonetic similarity. Likewise, the program reports on the /ʃ/:/k/ and /v/:/w/ correspondences between French and Latin, and /h/:/k/ between German and Latin.

It is mostly on the basis of mathematical elegance that I will use the R^2 metric as the baseline for the rest of this chapter. But there is

TABLE 39 R-Squared Recurrence Metric

Languages		p	Cognates[1]
English	German	.000	.664
French	Latin	.000	.613
German	Latin	.000	.295
English	Latin	.010	.293
English	French	.003	.280
French	German	.000	.272
Albanian	French	.013	.191
Albanian	Latin	.004	.186
Albanian	German	.286	.157
Albanian	English	.322	.143
Albanian	Hawaiian	.121	.000
Albanian	Navajo	.796	.000
Albanian	Turkish	.098	.000
English	Hawaiian	.298	.000
English	Navajo	.806	.000
English	Turkish	.021	.000
French	Hawaiian	.585	.000
French	Navajo	.407	.000
French	Turkish	.510	.000
German	Hawaiian	.186	.000
German	Navajo	.951	.000
German	Turkish	.723	.000
Hawaiian	Latin	.418	.000
Hawaiian	Navajo	.670	.000
Hawaiian	Turkish	.585	.000
Latin	Navajo	.331	.000
Latin	Turkish	.486	.000
Navajo	Turkish	.797	.000

Note: R^2 metric computed over initial consonant of Swadesh 200 list, omitting motivated words, loans, and language-internal cognates.
[1] Closeness judgments from Table 29.

no reason to believe that this algorithm is the last word on adapting traditional methodology to the task of evaluating contingency tables of phoneme pairings. People could reasonably argue, for example, that two strong recurrences should count for more if they involve four different phonemes (i.e., are in both different rows and different columns of the table) than if they are in the same row or column. As long as one can provide an automatic algorithm that gives a higher number for better tables than for less good tables, one is free to develop a metric that incorporates any intuition one wishes.

10.2 Phonetic Features

For the most part this book has been concerned with comparing languages at the level of the individual segments, or phonemes. In part this has been a matter of convenience, because words are most commonly presented and analyzed at the phonemic level; in part it has been a matter of tradition, because that is how Ross and his followers did it. I have made only short departures from that paradigm. At one point, I considered what an analysis would look like if performed at a higher level. It turned out that an analysis based on strings of two phonemes at a time was just barely doable with commonly available amount of vocabulary data, and that was demonstrable only because the languages in question (English and German) were so closely related that they defied the tremendous loss of power that step entailed.

The only excursion to a lower level of analysis has been to consider what happens when phonemes are lumped together on the basis of shared phonetic features. It was shown that that is a reasonable approach when lack of computers make it impossible to perform Monte Carlo tests of significance, but there is a price of a lack of power, although not nearly to the same extent as in the case of bisegmental analysis. What has not been considered yet is the question of whether it is reasonable to perform an analysis at the level of phonetic features. Such an approach would be similar in principle to how most historical linguists approach their work. While it is always possible to describe correspondences between languages in terms of matching phonemes, it is always more satisfying when a feature-level generalization can be formulated, such as, "voiceless stops in Latin correspond to fricatives in Germanic".

To explore the possibility of working directly at the featural level, I performed a simple analysis on the phonemes in the eight languages, reducing them all to a set of twelve features, shown in Table 40. As will be obvious from examining the scheme, it was not meant to be a state-of-the art instantiation of featural theory. Its main design goal was to convey meaningful information on the phonetics of sound segments without using complicated geometries. Thus any segment can be represented as a set of twelve atomic feature-value pairs, which can be trivially linearized without worrying about hierarchies, default values, and all the other complications of advanced feature theory. Once linearized, a feature set can be processed very much like a string of phonemes, so that the same software can handle both levels of analysis.

Conceptually the most straightforward thing to do with the feature analyses is to take the initial phoneme of the word and reduce it to a particular feature value. This would be very much like what was done in Table 12, except using the R^2 metric and permutation tests rather than the χ^2 metric and distribution. One consequence of this difference is that it is no longer necessary to be careful to generate tables that have large expected frequencies in all of the cells. The tables can now have whatever configuration most naturally fits the feature analysis. For example, whereas Table 12 collapsed place of articulation into three categories, we are now free to work with five categories, without worrying whether a 5 × 5 table would be too sparse. Table 41 shows the results of running separate tests for each feature.

The usefulness of the various features varies quite a bit. Features that are not distributed broadly among the languages of the world understandably make bad comparanda. For example, in the eight languages under consideration, only Navajo contrasts pulmonary and ejective consonants. Therefore it is logically impossible to show any association between the air source features of the initial phonemes between two of the languages. Whatever the number of recurrences between pulmonary–ejective or ejective–ejective, one will get exactly the same number of recurrences with each rearrangement of the data, so the p value has to be 1. Similarly, vowel pitch is found only in Navajo. To avoid monotony, these two features were omitted from Table 41. Features that almost always take a particular value can be a problem as well, due to the limited number of words we have to work with. Even though almost any test is good enough to detect the connection between English and German, the lip rounding feature fails completely, for the simple reason that the sample does not happen to contain any words that begin with a rounded vowel in English. By the same token, nasality is a bad discriminator, because languages typically have only a few nasal phonemes; and so is centrality, which typically only picks out the phoneme /l/. Time and release were such poor discriminators that they were omitted from the table.

The other features—sonority, voicing, vowel height, place of articulation, and articulator—were all fairly good at finding significance levels

TABLE 40 Feature Set

Feature	Values	Examples
Air Source	Pulmonary	/t/, /a/, /m/
	Ejective	/t'/
Nasality	Oral	/t/, /a/
	Nasal	/n/
Sonority	Stop	/t/, /ts/, /n/
	Fricative	/s/
	Approximant	/r/, /l/
	Vocoid	/a/
Centrality	Central	/t/, /a/
	Lateral	/l/
Place	Labial	/p/, /f/
	Dentalveolar	/t/, /θ/
	Palatal	/ʃ/, /j/
	Velar	/k/
	Glottal	/h/
Articulator	Lips	/p/
	Teeth	/f/
	Foretongue	/t/
	Backtongue	/k/
	None	/h/
Time	Short	/a/, /t/
	Long	/aː/, /tː/
Height	Nondistinctive	/p/
	High Close	/i/, /u/
	High Open	/ɪ/, /ʊ/
	Mid Close	/e/, /o/
	Mid Open	/ɛ/, /ɔ/
	Low	/æ/, /ɑ/
Rounding	Round	/o/
	Unround	/i/
Voice	Voiced	/i/, /b/
	Unvoiced	/p/
Release	Nondistinctive	/t/
	Lateral	/tˡ/
	Affricate	/ts/
	Frontward rising diphthong	/oi/
	Backward rising diphthong	/au/
Pitch	Low	/a/
	High	/á/

TABLE 41 Single Feature Analyses

Languages		N[1]	S[2]	C[3]	P[4]	A[5]	H[6]	R[7]	V[8]	Cog[9]
English	German	.000	.000	.000	.000	.000	.000	1.00	.000	.664
French	Latin	.000	.000	.000	.000	.000	.000	.002	.000	.613
German	Latin	.004	.006	.236	.002	.003	.000	1.00	.020	.295
English	Latin	.006	.097	.251	.001	.003	.003	.108	.066	.293
English	French	.103	.011	.110	.009	.046	.002	.063	.033	.280
French	German	.011	.001	.084	.002	.037	.000	.103	.021	.272
Albanian	French	.102	.159	.052	.008	.035	.048	.188	.055	.191
Albanian	Latin	.008	.083	.000	.058	.025	.134	1.00	.000	.186
Albanian	German	.003	.010	1.00	.333	.014	.007	1.00	.417	.157
Albanian	English	.077	.543	1.00	.040	.122	.005	.252	.522	.143
Albanian	Hawaiian	.032	.163	.564	.084	.217	.128	.167	.224	.000
Albanian	Navajo	.774	.102	1.00	.061	.104	1.00	1.00	1.00	.000
Albanian	Turkish	1.00	.174	1.00	.573	.372	.714	1.00	.130	.000
English	Hawaiian	.946	.320	.135	.288	.240	.025	.021	.198	.000
English	Navajo	.795	.882	1.00	.282	.234	1.00	1.00	.851	.000
English	Turkish	1.00	.058	1.00	.830	.561	.860	.472	1.00	.000
French	Hawaiian	.220	.075	.195	.304	.435	.535	.314	.069	.000
French	Navajo	.562	.870	1.00	.400	.033	.073	1.00	.591	.000
French	Turkish	1.00	.517	1.00	.194	.270	.452	1.00	.001	.000
German	Hawaiian	.872	.590	.494	.038	.483	.088	1.00	.470	.000
German	Navajo	.368	.068	1.00	.221	.068	.169	1.00	1.00	.000
German	Turkish	1.00	.621	1.00	.819	.352	.639	1.00	.531	.000
Hawaiian	Latin	.220	.273	.046	.186	.366	.134	.393	.236	.000
Hawaiian	Navajo	.848	.346	.103	.830	.522	1.00	1.00	.168	.000
Hawaiian	Turkish	1.00	.741	1.00	.523	.872	.289	.545	.593	.000
Latin	Navajo	.864	.826	1.00	.519	.050	.327	.043	.736	.000
Latin	Turkish	1.00	.558	1.00	.232	.230	.190	.802	.011	.000
Navajo	Turkish	1.00	.942	1.00	.202	.191	.541	1.00	.258	.000

Note: R^2 metric computed over the specified feature of the initial phoneme of words in the Swadesh 200 list, omitting motivated words, loans, and language-internal cognates.
[1]Nasality, [2]Sonority, [3]Centrality, [4]Place, [5]Articulator, [6]Height
[7]Rounding, [8]Voice.
[9]Closeness judgments from Table 29.

that accord with the cognate level judgments. These have the advantage of being features that are found in all languages and with values that partition the vocabulary into well-populated sections. That gives us a fighting chance to fill the contingency table with numbers that will vary when the word lists are rearranged. Although it may be asking too much to try to compare these five tests more closely, it does appear that vowel height, place of articulation, and articulator are first among equals. Each of them fails to test significantly on only one of the related pairs of languages, and then only at cognate levels lower than 19%; and each of them has a false positive on only one or two of the unrelated language pairs. For the most part, I do not believe these results would surprise any historical linguist. Place of articulation (which term often is meant to include the articulator) is an exceptionally stable feature as languages change, as is recognized in Oswalt's (1970) methodology. The role of vowel height, on the other hand, is slightly more puzzling. Vowel height is one of the more fluid features, at least in Indo-European languages, and it does not even apply at all to consonants; they were all assigned the value *Nondistinctive* in my analysis. A certain amount of its efficacy here may be attributed to the fact that, as a side effect of sorts, it encodes the distinction between word-initial consonant and vowel; in a separate analysis (not printed here) that one bit of information (vowel vs. consonant) sufficed to find the connection between languages at cognate levels above 25%. But that still leaves some sizable effectiveness that is due to vowel height alone.

If we accept then that vowel height, place of articulation, and articulator are particularly good candidates for phonetic features, how would we weigh the combined evidence of the three tests? This is exactly the same problem discussed in Chapter 4 with respect to tests on different phonemes in the word. Within the context of χ^2 tests, the best course of action was to choose a suite of N tests in advance, and decrease one's significance level by N in compensation. Such a technique is completely applicable to the current test, and indeed to any set of statistical tests. In this case, if, for any particular pair of languages, any of the tests on the three features is significant at .017 or lower, we would accept the evidence that the languages were connected. It turns out that in the precise scenario shown in Table 41, that technique works out pretty well. Only one pair of related languages fails the test (Albanian–Latin, for which the lowest p was .025), and none of the language pairs with no known cognates passes it. In other scenarios, the data may be less felicitous. For example, if all three tests gave a p value of .02, we would technically have to conclude that there was insufficient evidence to announce that the languages were connected, a conclusion that intuitively feels overly conservative.

Fortunately permutation tests offer another option. We can simply include the three features in the same test. Recall that the problem with doing that in tests using the χ^2 distribution lay in the fact that any dependence between items would ruin the test. In Chapter 2 a polling example was given, and I pointed out that if one made a mistake such as including in the same test responses that came from interviewing some of the same women again at a later date, the test could come out artificially significant. The core of the problem is that women would agree with themselves more than they would agree with other respondents. This dependence among the data points would artificially make the sample look more homogeneous than it really is, lowering the p level. The same sort of thing would happen with multiple features. Consider for example the close connection between the features of articulator and place of articulation. Once one knows that the place of articulation is the lips, that greatly constrains the possible articulators. The features simply are not independent. The same problem obtains, perhaps less spectacularly, with almost any pair of features. In a permutation test, however, independence of the data points is not an issue as long as any potential dependencies are among items that are permuted as a unit. This is due to the fact that the test metrics are not being looked up in some table of standard distributions, but are determined by comparing the metric computed for the observed state of the world with the metric computed for all possible rearrangements of the data. Because the rearrangements keep the dependent features together (we are rearranging whole words within a language), then if that dependence in some sense inflated the R^2 value for the observed tableau, it also has the opportunity to inflate it for all of the rearrangements. That is sufficient to correct for the dependence, because the reported p value is simply the fraction of all the permutations that have R^2 values at least as large as the observed tableau.

Table 42 shows what happened when the three most promising features were combined in the same test. I used a strict alignment technique for comparing the features. That is, the place feature of the phonemes in Language X were paired with place features in Language Y; height features in Language X were paired with height features in Language Y; but place features were not paired with height features. In this tableau, all of the pairings of Indo-European languages test significant at the .05 level, which is a desirable outcome. Most of the other languages (with no known cognates) do not test significant, but there are three exceptions. That is perhaps three more than we would like (unless perhaps we were proponents of certain long-range relationships) but, as always, the numbers must be interpreted carefully because the processes are probabilistic. There is a 6% chance of finding at least three spurious positives when one runs eighteen independent tests at the .05 level, and that number can rise when the tests are not independent, which is the case here; two of the putatively false positives involve Navajo and an Indo-European language. Of course I would not want to explain away bad results and unquestioningly publish the good. The true positives are also the result of a probabilistic process and there certainly is no guarantee that any pair of related languages, even those with cognate levels of 14% or higher, will have correct results under this test. Nevertheless, the results are promising.

Despite the good results, however, some of the upcoming experiments in this book will revert to the practice of comparing languages at the phoneme level. This is not because I fear the feature-level analysis is unreliable, but because the process is a little more understandable and informative when whole segments are involved. When English and German are compared, the program can tell us, for example, that there is an exceptionally high number of matches between English /θ/ and German /d/, /s/ and /z/, /t/ and /ts/, etc. That kind of information really helps us both in verifying the correctness of the tests (as in this book, where the true correspondences are already known) and in further investigating the common history of the languages (as in using these tests in real life). At the feature level, however, a test of English and German gives rather unedifying information, such as the fact that labials tend to correspond to labials. In general, I find that by far the strongest sources of evidence in these feature-level tests are exact matches of features. That is, the tests are mostly picking up on residual similarities between the languages. Perhaps that general failure to find correspondences between dissimilar features can be referred to the fact that features rather seldom change en masse across all segments that have that feature. Usually there is some condition involved; maybe a shift in place of articulation from dental to velar applies only to stops, for example. When that sort of thing happens, feature by feature comparison becomes noisy (e.g., dentals in Language A now match dentals part of the time and velars part of the time), whereas segment by segment comparison may still be exact (/t/:/k/, /n/:/n/).

10.3 Multiple Phoneme Matches

Because permutation tests can allow one to substitute for the χ^2 metric a count that is much closer in form to the recurrence count whose legitimacy has long been accepted by historical linguists, the next step

TABLE 42 Analysis by Feature

Languages		p	Cognates[1]
English	German	.000	.664
French	Latin	.000	.613
German	Latin	.000	.295
English	Latin	.000	.293
English	French	.000	.280
French	German	.000	.272
Albanian	French	.005	.191
Albanian	Latin	.019	.186
Albanian	German	.001	.157
Albanian	English	.002	.143
Albanian	Hawaiian	.053	.000
Albanian	Navajo	.191	.000
Albanian	Turkish	.681	.000
English	Hawaiian	.032	.000
English	Navajo	.404	.000
English	Turkish	.915	.000
French	Hawaiian	.451	.000
French	Navajo	.009	.000
French	Turkish	.330	.000
German	Hawaiian	.089	.000
German	Navajo	.033	.000
German	Turkish	.670	.000
Hawaiian	Latin	.129	.000
Hawaiian	Navajo	.959	.000
Hawaiian	Turkish	.356	.000
Latin	Navajo	.172	.000
Latin	Turkish	.145	.000
Navajo	Turkish	.287	.000

Note: Applies R^2 metric to the aligned features place, articulator, and height of the initial phoneme of words in Swadesh 200, omitting motivated words, loans, and language-internal cognates.
[1] Closeness judgments from Table 29.

should be to see if those techniques can be approximated even more closely. One drawback to the statistics as they have so far been applied is that the tests only consider one phoneme out of each word. I've mostly been presenting data about the initial consonant of a word, because it appears to be the most accurate single position, and because that maintains continuity with the proposals of Ross, Villemin, and Ringe, although of course any other precisely definable element, such as the first vowel, can also be used. But it is desirable to be able to look at data from more than one part of the word. For one thing, it seems wasteful to discard information. For another thing, there is always a chance that too narrow a gaze could make one completely blind to relationships such as the ones between the aforementioned Australian languages that have dropped word-initial consonants. There is also the fact that comparative linguists generally argue that one cannot prove languages are historically connected just by looking at matches in a single phoneme in the word (Campbell 1973, Hock 1991:558, Nichols 1996). While these linguists understand that that is a purely statistical matter, there is always the danger that mathematically less sophisticated practitioners will not understand that many recurrences across a single phoneme can be as probative as a few recurrences across two phonemes, and so may naturally distrust our R^2 test as it has been so far developed.

Fortunately the flexibility of permutation tests readily accommodates the incorporation of multiple phonemes, just as it did for multiple features in the preceding section. Perhaps the most obvious approach is to adapt the technique used for features. Namely, we compare the ith phoneme of the word in the first language with the ith phoneme of the word in the second language, counting left to right. For example, using the roots for the concept 'cold', English /kol/, German /kal/ (the final consonants are not part of the root; cf. *cool*), we would compare /k/:/k/, /o/:/a/, and /l/:/l/. This can be called the Linear alignment technique. A couple of new issues do arise here that did not arise when we were uniformly comparing three features for each word. First, we have to consider the possibility that one word is longer than the other; what do we do with the final /g/ in German when comparing for the concept 'rain' the roots English /re/ and German /reːg/? Could the fact that some word pairs have many potential matches because the words are longer than in other word pairs have any detrimental effect? I will address this issue in a later section, but for now let us take the position that because we are interested in studying recurrent phoneme correspondences, not word length effects, we will try to control the latter. Solving word length mismatches by matching the leftover phonemes with a zero (e.g., /r/:/r/, /e/:/eː/, /∅/:/g/) only serves to encode the fact that the one root is

shorter than the other, so that will be avoided in favor of simple truncation. The longer word will simply be shortened to the same length as the
shorter word. Now each word within a pair will be the same length, but
the words in different pairs could be different lengths. In case that difference among the lengths of pairs should turn out to be a problem, we
can compensate in the following way. If every concept makes a constant
contribution to the overall R^2 metric, then their length will in no way
be incorporated in the analyses. Say therefore that each concept should
add one unit to the metric, as was done before when we only considered
one phoneme out of the word. Because the R^2 metric squares each cell
before adding them, we need to transform each of our contributions so
that when they are squared then added, they will sum to 1. Therefore,
each contribution should be the square root of the inverse of the number
of phonemes being compared in the word. For example, in comparing
/kol/ to /kal/, for each of the matches /k/:/k/, /o/:/a/, and /l/:/l/ we
will add to the appropriate cell of the contingency table not a 1, but
rather $\sqrt{1/3}$.

In Table 43, the column labeled *Linear* gives the results of this
multiple-phoneme alignment technique. These are sane values that give
us no reason to suspect that the statistical reasoning behind the multiple
phoneme matching is flawed. And yet, it is a bit disappointing that this
technique did not do much better. Now that we are finally looking at
virtually all of the information in the word, we might well have expected
the results to be an order of magnitude better than before. As it is,
we got better results from just comparing one feature of one phoneme
(Table 41).

In a sense, though, we are not actually making use of all the data yet.
First there is the obvious problem of the truncations. Worst, it is obvious
that we are sometimes comparing the wrong things. For example, for the
concept 'knee', an English–German match compares /ni/ with /kniː/,
which the Linear alignment method asks us to compare as /n/:/k/ and
/i/:/n/, ignoring the obvious /n/:/n/ and /i/:/iː/ matches. It is difficult
to tell the computer to just do the obvious thing, but we can make sure
that it considers every conceivable bit of information. One way of doing
this would be to use what I call the Cross Pairs method. In this method,
every phoneme in the first word is matched up with every phoneme in
the second word. To make the contributions add up to 1 after squaring and addition, each pair is entered into the contingency table with
a contribution of the square root of the inverse of the product of the
word lengths. For example, in /ni/:/kniː/, each of the matches /n/:/k/,
/n/:/n/, /n/:/iː/, /i/:/k/, /i/:/n/, and /i/:/iː/ are entered into the table, each with a value of $\sqrt{1/6}$. The idea here is that we are guaranteed

TABLE 43   R-Squared over Multiple Phonemes

Languages		Linear	Cross	Fibonacci	Cognates[1]
English	German	.000	.000	.000	.664
French	Latin	.000	.000	.000	.613
German	Latin	.000	.000	.000	.295
English	Latin	.000	.002	.000	.293
English	French	.000	.000	.000	.280
French	German	.000	.000	.000	.272
Albanian	French	.152	.176	.117	.191
Albanian	Latin	.040	.004	.004	.186
Albanian	German	.287	.186	.084	.157
Albanian	English	.135	.429	.281	.143
Albanian	Hawaiian	.336	.935	.850	.000
Albanian	Navajo	.154	.867	.758	.000
Albanian	Turkish	.670	.332	.417	.000
English	Hawaiian	.298	.877	.778	.000
English	Navajo	.960	.768	.803	.000
English	Turkish	.002	.268	.147	.000
French	Hawaiian	.322	.059	.041	.000
French	Navajo	.060	.497	.283	.000
French	Turkish	.340	.385	.221	.000
German	Hawaiian	.140	.358	.225	.000
German	Navajo	.672	.149	.201	.000
German	Turkish	.377	.008	.007	.000
Hawaiian	Latin	.603	.832	.827	.000
Hawaiian	Navajo	.436	.757	.623	.000
Hawaiian	Turkish	.521	.401	.524	.000
Latin	Navajo	.046	.272	.142	.000
Latin	Turkish	.745	.233	.158	.000
Navajo	Turkish	.608	.663	.700	.000

Note: Tests use Swadesh 200 list, omitting loans, motivated words, and language-internal cognates. Metric is R^2, evaluated over all phonemes, using three different alignment techniques.
[1] Closeness judgments from Table 29.

that if there are any correct matches, they will be counted in the table. Hopefully the other matches, because they will not recur very much, will recede into the background. The results, in column *Cross* of Table 43, are fine, though they do not suggest any real improvement over the Linear method. But the fact that it is no worse than the Linear method may be comforting when one considers that Linear alignment may work worse in some situations than it does in this set of tests. For one thing, the Indo-European languages do not have any rampant aphaereses that would invalidate tests on initial consonants and indeed completely throw off simple linear alignments. For another, we have taken advantage of the advanced state of Indo-European scholarship to strip off prefixes, suffixes, and even occasional infixes whose status as such may not be obvious from a synchronic, noncomparative standpoint. In the field, where we are more likely to unwittingly let an ancient affix slip into the data, it would be good to know that the alignment mechanism is potentially forgiving because it considers all possible pairings.

Comparing all possible pairs and correcting with the square root of the product of the word lengths is a solution that is mathematically intuitive and elegant. Some may feel however that there is something fishy about the Cross Pairs method. Linguists do not favor such undirected matchings, matching a word-initial /p/ in Language A with a word-final /b/ in Language B, and then adding in to the mix a match of word-final /p/ with word-initial /b/. That sort of correspondence does not normally happen in language. Possibly the dross from counting all the correspondences between unlikely matches drowns out the useful information there may be in the parts that are likely to match. If languages do not ordinarily change so that the same segment becomes the beginning of the word in one descendant language and the end of the word in the other descendant, then the correspondences between those two locations should be chance. Chance noise increases the p value.

A metric I have implemented to compensate for that possible weakness in the Cross Pairs technique is the Fibonacci alignment algorithm. This metric uses the R^2 summation; the new addition is a refinement in the tabulation method that Cross Pairs uses. Instead of tabulating a constant value for each match, the amount varies depending on how far apart the two phones are. For example, if both sounds are at the beginning of the word, or if both are the 3rd sound in the word, the match is assigned the highest score, 13. If however they are respectively the first and second, or the fourth and third, that is, if they are offset by 1, they receive the next higher score, 8. Pairs offset by 2 get the score 5; and so forth with the scores 3 and 2. Pairs of phonemes more than 4 positions apart are given the score 1. For example, in the 'cold' match /kol/:/kal/, the matches /k/:/k/, /o/:/a/, and /l/:/l/ are assigned the weights 13; the matches /k/:/a/, /o/:/k/, /o/:/l/, and /l/:/a/, being offset by 1, all get the weights 8; the matches /k/:/l/ and /l/:/k/, being offset by 2, get the weights 5. To make sure these will also sum to 1 at the right time, they are actually all divided by the sum of all the contributions (in this case, $3 * 13 + 4 * 8 + 2 * 5 = 81$), then their square root is taken before being entered into the contingency table. The intuition here is that we are still looking at all possible combinations of phonemes, just in case something odd has happened like a general aphaeresis or prothesis; but we are paying much closer attention to the pairings that line up next to each other, which should be by far the most usual case in a well stemmed word list. The progression of scores used, 13, 8, 5, 3, 2, 1, terms from the beginning of the Fibonacci sequence (Knuth 1973:78), was chosen almost completely arbitrarily. It simply has the desirable property of rewarding aligned matches most heavily, and falling off fairly rapidly as the phonemes get farther apart. Results of a trial run are listed in the *Fibonacci* column of Table 43. The results are not much different from the other multiple-phoneme alignment methods, but perhaps the optimist can note that the p values for related languages do seem a little lower.

To a reasonable approximation, though, it looks like the three alignment techniques give very similar results. The main point I wish to make is that the general framework of the tests I am proposing is such that one has a great deal of flexibility in tailoring the procedure to one's own preferences with respect to alignment. There is a great difference between Linear and Cross Pairs alignment, but they both are valid, they both work, and neither is particularly hard to define. And it is not hard to imagine a host of other candidates. One candidate that looks particularly promising is to adapt the algorithm of Covington (1996), who proposed a general scoring mechanism for determining the best alignment between a pair of words in the context of automated work in historical linguistics.

Perhaps the major difficulty in working out a new metric is my earlier stipulation that it is important that the alignment method be carefully chosen so as not to unwittingly encode word length. There are many ways such an encoding can come about. Imagine for example a Cross Pairs technique that does not scale for the number of phonemes in the word. Imagine too that the data are such that longer words in one language tend to match longer words in the other, and the shorter words match the shorter words. If the metric is comparing two long words, the sum of the cross pairs will be very high, the product of the two lengths. That number is much greater than when two small words are compared, or even a large word and a small word. If two languages tend to have matching word lengths, then there will be many comparisons of long words, resulting in large scores. But when the lists are randomized, a much larger number of the comparisons will be between words mismatched for size, giving metrics much smaller than in the base case. To take a concrete example, assume half the words in each language are 5 phonemes long and the other half are 1 phoneme long, and the long words match the long words and the short ones match the short ones, but the actual number of recurring phoneme matches are at chance levels. The 5-phoneme words will have $5 \times 5 = 25$ pairs of sounds to enter into the table, the short words will have 1. Since there are equal numbers of each type of word, the average pair of words will make $(25 + 1)/2 = 13$ entries. When we rearrange the words to do the permutation tests, we will on the average get an even mix of 5×5, 5×1, 1×5 and 1×1 matches, for an average of 9 entries per concept. Thus the original arrangement will be making more entries into the table than will the average rearrangement, meaning that the original will have more of an opportunity to build up an R^2 higher than that of its rearrangements. That is the very definition of a low p score, but it is due entirely to matching word lengths. The example is extreme, but the same reasoning applies to any difference in word lengths. It is also worth noting that the bias is always in one direction: Matching word lengths always make languages look more closely related.

But why is it bad when a metric takes word length into account? In principle, there is no statistical argument against it. It is not as if the p value is invalid or artificially lowered. It is just that while the metric is considering whether the number of recurrent phoneme matches is greater than chance, it is also weighing whether the matching of word lengths is greater than chance, and factoring that into the equation. There is basically nothing wrong with that, except insofar as the researcher may not wish to be confounding the different types of evidence.

From a real-world standpoint, however, the incorporation of word length into the metric may be problematic, because universally, words with the same meaning tend to have the same length. In other words, this seems to be another instance in which the pure Saussurian arbitrariness hypothesis is invalid. I am using *meaning* in a fairly broad sense here. In practice, word length is probably sensitive in a universal way to only a few peripheral aspects of a word's meaning, such as its frequency and its centrality to the culture; one can predict that the term for *fire insurance policy* will be longer than the term for *water*. The grammatical use to which a word is put can also be a predictor. Everyday experience teaches that function words (which I have been referring to as linkers and limiters) tend to be very short, often as little as one or two phonemes long, whereas lexical words tend to be longer, and indeed often are subject to grammatical requirements for minimal length. This can be verified by checking the *WORD* section in Table 44. Linking words are the shortest in all languages, and in most of the languages, limiters are the next shortest words. There may also be systematic differences between nouns and verbs as well, due to the fact that the latter are universally more likely to be marked with a wide range of morphemes indicating such things as tense, number, person, mood, and aspect, whereas nouns typically inflect for fewer categories such as case and number. This observation holds up for most of the languages in the sample (nouns are significantly shorter than verbs in English, French, German, Navajo, and Turkish; all tests in this paragraph are parametric t tests, $p \leq .05$), but not too much should be made of this, because this is not a random sample of languages and much depends on what form one chooses to cite the noun and verb in. Besides, for our purposes, the length of the whole word is not the relevant factor, because one would try to strip off at least the inflectional endings, as does Ringe (1992). The *STEM* section in the table shows that the differences become much smaller under that condition. But what we really want is to strip off all derivational affixes as well, preferably stripping the word down to a bare root. The *ROOT* section in the table gives even smaller numbers, but some trends are significant. In all the languages considered, linking words (conjunctions, prepositions, pronouns) are significantly shorter than all other words, except that in French the size difference between them and limiters is not significant. In some languages, limiters (determiners, demonstratives, quantifiers, etc.) are significantly shorter than certain other word classes: shorter than verbs in English and German; shorter than adjectives in English, French, and Turkish. In Turkish, verb roots are significantly shorter than adjective or noun roots. Thus significant differences in sizes remain after one has done one's best at stripping away all extraneous material. And differences in average lengths could potentially affect the tests even if their variances make the differences nonsignificant. For example, the fact that adjective roots are about 10% longer than verb roots in both Albanian and Hawaiian could theoretically be enough to make these languages look connected, regardless of whether a t-test finds the size difference significant.

Therefore it is desirable for metrics to not be influenced by the length of the word. One easy way to eyeball whether there is any problem with a metric in that regard is to send it data that has been randomized for content, but not for length. That is, one feeds it what is essentially normal data, but with every phoneme replaced by some randomly selected phoneme. If the metric is biased for length, numbers will tend to come out clustered. I have unintentionally developed metrics that, when

TABLE 44 Comparanda Lengths by Part of Speech

Role	Level	Alb	Eng	Fre	Ger	Haw	Lat	Nav	Tur
WORD	Link	3.00	2.17	2.64	2.50	3.07	2.50	3.14	3.31
	Lim	3.86	3.00	3.79	3.63	4.50	4.65	4.44	3.67
	Verb	4.12	4.24	3.84	4.55	4.66	5.22	5.98	7.20
	Adj	5.05	3.39	3.56	3.76	5.30	5.90	5.57	4.48
	Noun	3.84	3.39	3.34	3.93	4.56	4.97	4.71	4.14
STEM	Link	2.18	2.17	2.21	2.29	3.00	2.14	2.86	2.69
	Lim	3.57	3.00	3.26	3.21	4.50	4.40	4.00	3.56
	Verb	3.31	3.18	3.49	3.49	4.40	4.18	5.58	3.37
	Adj	4.33	3.39	3.96	3.76	5.30	5.03	5.20	4.48
	Noun	3.56	3.33	3.35	3.97	4.52	4.61	3.79	4.06
ROOT	Link	1.64	1.58	1.50	1.50	2.50	1.36	2.00	1.62
	Lim	2.64	2.15	2.00	2.11	3.35	2.40	2.78	2.44
	Verb	2.60	2.74	2.27	2.89	3.43	2.51	2.67	2.65
	Adj	2.86	2.71	2.56	2.66	3.67	2.70	2.87	3.52
	Noun	2.97	2.49	2.43	2.59	3.56	2.75	2.77	3.54

Note: Average number of phonemes in entries in the Swadesh 200 list, omitting loans and motivated words.

fed such randomized data, accurately show that the Indo-European languages are related and that the others are not, with very good p values ranging down to .0001—all based entirely on the length of the words. Conversely, a metric biased the other way will tend to give related languages very high p values when given data randomized in that fashion.

In the metrics so far considered, scaling seems to produce the desired effect. The basic idea has been that the contribution of each concept is adjusted during the tabulation phase, so that after the R^2 metric is applied to it, the overall effect in the table is constant. However, the technique of scaling has drawbacks. Mathematically, it is not even obvious what the exact meaning is of leveling out the effects of word length. The idea sounds clear until one considers concrete cases. What does it mean to say that when one is comparing /do/ and /terg/ (French and Latin stems for 'back'), one wants the answer to come out the same as if the words were the same length? Does that mean to treat /do/ as if it were as long as /terg/, or to treat /terg/ as if it were as short as /do/? Scaling the tallies by word length may be a sensible approach, but that is far short of saying that it can be mathematically proved that the overall results are the same as one would get if all the words had been of the same length, because that very idea is a bit nebulous.

And from a linguistic point of view, scaling certainly gets away from the desideratum of approaching the problem in a way that most naturally agrees with the traditional methods of historical linguists; very few would wish to say that they had counted a couple of inverse square roots of 8 /d/:/t/ pairings. More to the point, scaling has the effect of averaging the contribution of multiple phoneme matchings, rather than cumulating them. Consequently, a /d/:/t/ match within long words counts as less evidence than a /d/:/t/ match within short words.

In the final analysis, the best approach may well be to actually make the words some constant length. Unfortunately, simply padding shorter words with some null character will not do the trick, because there is no reason to expect that the matchings that the null character participates in will not in some way skew the results. It does however make sense to truncate words. When one has chosen a word length k, one needs to discard all words shorter than k, and to truncate all words longer than k. Table 45 presents the results of comparing the first two phonemes of words using the same alignment techniques that were used for the whole entry in Table 43. Again, the results do not seem very much different from previous types of multiple-phoneme tests; all have in common the fact that they are weak on the low-cognate relationships, which in this language set all involve Albanian. Because the results are so similar, one is left to theoretical arguments for deciding which test to prefer. The k-phoneme technique has the disadvantage that it throws away more information, but the advantage that one can be absolutely sure that there is no effect due to word length.

The alert proponent of multiphoneme comparisons may have noticed one important respect in which the metrics I have been proposing do not simulate the methodologies of most comparative linguists. These metrics to be sure look for matches in different parts of the word, but they do not in any way give special precedence to multiple matches in the same word. For example, for the concept 'new', the English–German pair /nu/:/nɔʏ/ is treated as evidence for a connection between the languages because of the frequent /n/:/n/ recurrence ($R = 10$). But the vowels /u/:/ɔʏ/ match only this one time in the word list ($R = 0$). Because only one phoneme of the *new:neu* takes part in a recurrence, most practitioners (e.g., Nichols 1996) would urge that the word should not be used in evidence. To simulate this idea, I developed the W (word-oriented) metric. Under this scheme, the contingency tables are built as usual, based on phoneme correspondences. But after that, the program takes a second look at the words to compute the metric. For each word pair, the program revisits the relevant phoneme pairings, retrieves their R^2 value by looking at the table (i.e., number of occurrences, minus 1,

TABLE 45   R-Squared on Initial Two-Phoneme Sequences

Languages		Linear	Cross	Fibonacci	Cognates[1]
English	German	.000	.000	.000	.664
French	Latin	.000	.000	.000	.613
German	Latin	.000	.011	.003	.295
English	Latin	.000	.010	.003	.293
English	French	.000	.000	.000	.280
French	German	.000	.000	.000	.272
Albanian	French	.337	.211	.211	.191
Albanian	Latin	.042	.138	.090	.186
Albanian	German	.629	.878	.789	.157
Albanian	English	.091	.745	.647	.143
Albanian	Hawaiian	.215	.522	.359	.000
Albanian	Navajo	.057	.025	.015	.000
Albanian	Turkish	.635	.809	.799	.000
English	Hawaiian	.253	.489	.578	.000
English	Navajo	.965	.975	.975	.000
English	Turkish	.001	.236	.113	.000
French	Hawaiian	.265	.083	.053	.000
French	Navajo	.031	.059	.058	.000
French	Turkish	.131	.231	.120	.000
German	Hawaiian	.079	.564	.398	.000
German	Navajo	.942	.648	.791	.000
German	Turkish	.517	.393	.388	.000
Hawaiian	Latin	.289	.458	.363	.000
Hawaiian	Navajo	.836	.877	.816	.000
Hawaiian	Turkish	.660	.789	.801	.000
Latin	Navajo	.256	.129	.156	.000
Latin	Turkish	.537	.539	.435	.000
Navajo	Turkish	.870	.230	.411	.000

Note: Computed over Swadesh 200 words, omitting loans, motivated words, language-internal cognates, and entries with fewer than 2 phonemes.
[1] Closeness judgments from Table 29.

squared), then multiplies those R^2 values together, across all the relevant phoneme pairings. Those products are then summed across all the word pairs. For example, assuming Linear alignment, the /nu/:/nɔʏ/ pair would have the value $10^2 * 0^2 = 0$, while 'go', /go/:/geː/, because /g/:/g/ and /o/:/eː/ each recurs twice, would have the value $2^2 * 2^2 = 16$. The metric for the entire word list would begin 0 (for 'new') + 16 (for 'go'), summing the values for each of the word pairs. Table 46 summarizes the results when the W metric is applied to my data, trying the various alignment techniques, truncating each entry to the first two phonemes.

Alas, the results do not seem any better, and in fact may be a bit of a retreat, in that there seems to be some difficulty in detecting even the connection between French and German, at a 27% cognate level. On the other hand, when the technique does work, one pleasing side effect is that, because it is word-oriented, one can easily get it to calculate which particular word pairs contribute most to the results. In effect, the program can return lists of likely cognates, as in Table 47. Attempting to do this with a metric that is not word-oriented, such as R^2, would have given a good many false cognates.

Some readers may have found this section on multiphoneme matching disappointing. To be sure, most of the techniques I have put forward reliably detect the relationship between the Italic and Germanic branches of Indo-European using modern representatives, which is really no small feat. But the results appear to be a step back from the feature level analyses of an earlier section, where even Albanian seemed to be coming into the fold. And the results seem somewhat paradoxical. The closer we approach Meillet's classic advice on how to compare languages, the less confident the methodology is that the languages are actually related. Is the methodology wrong?

It would be rash to claim that the methodology is perfect. It is easy to think up new metrics or alignments that would address perceived shortcomings in the ones I have described, or more closely approach standard techniques. The flexibility to do so is one of the strengths of the methodology. Instead of blindly matching phonemes from left to right, for example, one might try an alignment that is more likely to line up historically corresponding phonemes. Ringe's (1992) technique for extracting single phonemes based on the word's CV template may be a starting point for this, or the aforementioned algorithm of Covington (1996). It is even theoretically possible to go so far as to try to compute a system of alignments that is optimal across the entire word list, although that is certain to be computationally expensive, especially when one ponders doing so 10,000 times. One could also try accommo-

TABLE 46 W Metric on Initial Two-Phoneme Sequences

Languages		Linear	Cross	Fibonacci	Cognates[1]
English	German	.000	.000	.000	.664
French	Latin	.000	.000	.000	.613
German	Latin	.002	.033	.009	.295
English	Latin	.014	.002	.002	.293
English	French	.002	.013	.003	.280
French	German	.070	.062	.026	.272
Albanian	French	.337	.317	.299	.191
Albanian	Latin	.036	.024	.024	.186
Albanian	German	.414	.722	.708	.157
Albanian	English	.093	.258	.157	.143
Albanian	Hawaiian	.847	.494	.506	.000
Albanian	Navajo	.187	.187	.107	.000
Albanian	Turkish	.873	.905	.935	.000
English	Hawaiian	.363	.862	.801	.000
English	Navajo	.923	.965	.957	.000
English	Turkish	.068	.223	.090	.000
French	Hawaiian	.217	.290	.216	.000
French	Navajo	.084	.142	.075	.000
French	Turkish	.191	.308	.260	.000
German	Hawaiian	.065	.206	.171	.000
German	Navajo	.957	.776	.889	.000
German	Turkish	.528	.356	.355	.000
Hawaiian	Latin	.511	.763	.699	.000
Hawaiian	Navajo	.779	.811	.809	.000
Hawaiian	Turkish	.543	.829	.817	.000
Latin	Navajo	.104	.284	.196	.000
Latin	Turkish	.595	.563	.494	.000
Navajo	Turkish	.984	.741	.901	.000

Note: Computed over Swadesh 200 words, omitting loans, motivated words, language-internal cognates, and entries with fewer than 2 phonemes.
[1]Closeness judgments from Table 29.

TABLE 47 English–German Cognates, per W Metric

Concept	English	German
all	/ɔl/	/al/ *alle*
ashes	/aʃ/	/aʃ/ *Asche*
drink	/dr/	/tr/ *trinkt*
dry	/dr/	/tr/ *trocken*
fall	/fɔ/	/fa/ *fallen*
far	/fɑ/	/fe/ *fern*
fish	/fɪ/	/fi/ *Fisch*
freeze	/fr/	/fr/ *friert*
grass	/gr/	/gr/ *Gras*
hand	/ha/	/ha/ *Hand*
hold	/ho/	/ha/ *hält*
man	/ma/	/ma/ *Mann*
name	/ne/	/na:/ *Name*
old	/ol/	/al/ *alt*
say	/se/	/za:/ *sagt*
sharp	/ʃa/	/ʃa/ *scharf*
sing	/sɪ/	/zi/ *singt*
sleep	/sl/	/ʃl/ *schläft*
snow	/sn/	/ʃn/ *Schnee*
stab	/st/	/ʃt/ *sticht*
stand	/st/	/ʃt/ *steht*
star	/st/	/ʃt/ *Stern*
stick	/st/	/ʃt/ *Stock*
stone	/st/	/ʃt/ *Stein*
swell	/sw/	/ʃv/ *schwillt*
swim	/sw/	/ʃv/ *schwimmt*
three	/θr/	/dr/ *drei*
white	/wai/	/vai/ *weiß*
wide	/wai/	/vai/ *weit*
wipe	/wai/	/vi/ *wischt*

Note: Concepts whose *W* metric is significantly higher than when the German words are rearranged. Computation performed as per Table 46.

dating the fact that sound changes are often conditional on factors like position in the word. Throwing several phonemes from each word into the same contingency table obscures such correspondences; using separate bins (contingency tables) for different environments could alleviate such obscuration. Another tack would be to try various modifications to the Fibonacci weighting procedure.

But while there is manifestly room for improvement, it would also be foolish to overlook the fact that there is a strong theoretical reason why multiphoneme comparison should make it harder to demonstrate a historical relationship between languages. It will be recalled from Chapter 4 that the initial consonant of the word is demonstrably a more reliable piece of evidence than any other part of the word. Throwing the other parts of the word into the equation can only weaken the accumulated evidence. This may run counter to everyday experience, where we take every independent piece of evidence as contributing to an overall conclusion. But logically, if evidence A makes one 95% sure of a conclusion, then introducing evidence B, which by itself would make one 80% sure of a conclusion, does not increase one's overall confidence in the conclusion. If anything, it should weaken it, by the Bonferroni correction.

Why then have the great theorists of the comparative method insisted on multiphoneme comparison? Paradoxically, it is because they want to make it harder to demonstrate the connections between languages. In the absence of a mathematically grounded procedure for significance testing, linguists have traditionally had to insist on accumulating only the sorts of evidence that are very difficult to come by. It happens very infrequently that languages have strong correspondences for three consecutive phonemes in the words for a certain concept; therefore if only a few such examples are found, one can reasonably guess that one has found a significant connection (Bender 1969). By contrast, it is much harder to informally guess the significance of matches on a single consonant or feature, when the linguist may easily accumulate dozens of examples. Therefore the latter type of evidence has been avoided. But that sort of reasoning does not hold when one can do statistical significance testing on a mathematical basis. To the contrary, statistical tests generally work out better when one gathers data that is widely distributed, that is, the easy data. Holding out for data that are individually rare, such as a full three-phoneme match, simply makes it more likely that one will miss evidence that could reveal a language connection.

But being freed from some of the requirements put forth by theoretists of the traditional methodology does not resolve all the issues. Although testing on the initial phoneme of the root does seem the procedure most likely to be most probative most of the time, nothing rules out the possibility that an unusual language pair will only manifest evidence of their relationship if some other parts of the roots are tested. Several approaches to this problem are possible, and the choice between them depends largely on the goals and the temperament of the linguist. One researcher may prefer to take the best shot at finding a language connection and not look back if it fails, using a test that can uncover weak relations, but only of an ordinary sort (e.g., without rampant aphaeresis). Another may prefer a test that is more likely to catch unusual relationships but more likely to fail on weak relationships. Another may prefer to try a series of specific tests, of course scrupulously reporting the ones that fail as well as the ones that succeed. Best of all, another may take up the challenge of developing a new test that handles a broader range of conditions with power and elegance.

10.4   Subsample Averaging

One conceptual difficulty people may have with the methods being explored here is that they are thoroughly stochastic. On any given run, the computer is likely to return a different answer for the same question; yesterday English and Turkish were connected, today they are not. This is due in part to the inherently stochastic nature of Monte Carlo techniques for estimating significance, but it is compounded by fluctuating sets of comparanda. When the source word list contains words whose interrelationship is motivated, such as *that, there, they*, all but one of each such set must be discarded, and the outcome of the tests can differ depending on which word remains.

It is true that decisions could be made that would end this fluctuation. The random number generator could always be seeded with the same value, and there could be a deterministic rule for deciding which cognate to retain. But that would simply succeed in masking what is and should be a statistical process. One is not processing the entire language and returning the one true analysis. Instead, one is sampling the language, and making inferences based on that sample. That fact should not be hidden. And there are ways to take advantage of certain properties of a stochastic process. One of those is the property of large numbers of observations. The more the process is run, the closer the average observation will approach the true value. Such an average is inherently more accurate than the outcome of a single run, and is more satisfying than a large set of more approximate numbers.

The process could be helped along even more by performing the trials multiple times on different subsets of data. Although that will produce more widely varying numbers, the experimenter will be less likely to see outcomes that are due to one or two particularly deviant pieces of data. In Table 48 I present the result of an experiment with such resampling. For each pair of languages, the process was repeated five times on randomly selected subsets of 100 words. It should be noted that after removing loans, internal cognates, and the like, 100 is a good-sized subset of what is left of the Swadesh 200 list. Nevertheless it is small enough to ensure some reasonable amount of variation for most language pairs. The table gives a good idea of that variation by listing the outcomes of each of the five trials, as well as their average.

10.5 Multilateral Comparisons

It has become traditional, in any discussion of techniques in historical linguistics, to weigh in on the question of multilateral comparison. Multilateral comparison is a technique for discovering linguistic relationships by looking for similarities in morphemes across a large number of languages simultaneously. The motive for multilateral comparison, which is most prominently propounded by Greenberg (e.g., Greenberg 1987, Greenberg and Ruhlen 1992), is that relatively weak evidence, which may look like chance when it is found between two languages, can become more convincing when repeated across many languages.

Whether one accepts that conclusion depends precisely on how one models the technique. If one takes the idea to mean that one only accepts a match or similarity when it appears across all languages in a multilingual comparison, then one agrees wholeheartedly with the conclusion (Justeson and Stephens 1980). If one takes it to mean that one can accept any two-language match among many languages and count it the same as such a match in a two-language comparison, then one concludes that the idea is in error (Ringe 1992). The reasoning is simple. If one is looking for a pair of similar words for 'dog', it is going to be a lot harder to find a similarity if there is only one candidate word pair to look at (one word for each of two languages), than if there are many candidate pairs (with three languages there are three candidate pairs of words, with four languages six pairs of words, and so forth). Thus there is a much greater likelihood that one will pick up chance similarities. It is also possible to look for cutoff points in between "all" and "two". For example, Baxter and Manaster Ramer (1996) concluded that finding a match across 7 or more languages in a 15-language comparison is as probative as finding a match across 2 languages in a binary comparison.

Which is the correct model for Greenberg's work? The answer appears to be "none of the above". Although Greenberg's own calculations suggest that he is thinking of the case of all-language matches when he

TABLE 48 Subsample Averaging

Languages		Trials					Ave.	Cognates[1]
English	German	.000	.000	.000	.000	.000	.000	.664
French	Latin	.000	.000	.000	.000	.000	.000	.613
German	Latin	.000	.000	.000	.000	.000	.000	.295
English	Latin	.001	.002	.000	.000	.001	.001	.293
English	French	.001	.002	.001	.000	.002	.001	.280
French	German	.000	.000	.000	.000	.000	.000	.272
Albanian	French	.018	.006	.004	.005	.020	.011	.191
Albanian	Latin	.093	.051	.013	.051	.027	.047	.186
Albanian	German	.001	.003	.003	.003	.013	.005	.157
Albanian	English	.002	.002	.002	.002	.002	.002	.143
Albanian	Hawaiian	.244	.025	.109	.087	.046	.102	.000
Albanian	Navajo	.244	.281	.233	.225	.366	.270	.000
Albanian	Turkish	.447	.435	.654	.809	.763	.622	.000
English	Hawaiian	.038	.059	.008	.011	.028	.029	.000
English	Navajo	.672	.604	.430	.461	.296	.493	.000
English	Turkish	.988	.863	.782	.921	.944	.900	.000
French	Hawaiian	.589	.393	.358	.289	.435	.413	.000
French	Navajo	.003	.005	.047	.063	.010	.026	.000
French	Turkish	.607	.362	.341	.587	.506	.481	.000
German	Hawaiian	.107	.333	.061	.008	.445	.191	.000
German	Navajo	.007	.085	.020	.047	.141	.060	.000
German	Turkish	.905	.731	.410	.704	.717	.693	.000
Hawaiian	Latin	.353	.348	.129	.284	.560	.335	.000
Hawaiian	Navajo	.894	.856	.884	.942	.959	.907	.000
Hawaiian	Turkish	.502	.477	.332	.115	.165	.318	.000
Latin	Navajo	.763	.086	.055	.153	.273	.266	.000
Latin	Turkish	.080	.191	.344	.251	.738	.321	.000
Navajo	Turkish	.324	.640	.467	.338	.593	.472	.000

Note: Computation as per Table 42, repeated over ten random samples of 100 words.

[1] Closeness judgments from Table 29.

argues mathematically for the superiority of his multilateral approach, the evidence he publishes on the whole contradicts that idea. He does not claim to disregard similarities that are found between some but not all of the languages he is comparing, nor does he invoke any such cut-off as 7/15. Thus there is a mismatch between the mathematical justification and the praxis, and, furthermore, we simply do not have enough information about the praxis to judge whether it is mathematically justifiable (Manaster Ramer and Hitchcock 1996). One cannot work backwards from published examples, because that would constitute the fallacy of trying to convert a coincidence into a statistical study. Nor is it helpful that Greenberg and Ruhlen (1992) attempted to do just that. They supported the 'suck' etymology (presented here in the introduction) by simply giving a very rough calculation of the odds that a template like m-R-K could appear in a language, then multiplying together that number for the number of languages being compared. It is not clear why such a computation is remotely relevant. I might just as well compute the odds that the German word for 'dog' should be *Hund*, and the odds that the Hawaiian word for 'bird' should be *manu*, and so forth across a dozen languages, and multiply them together. I would get a very low number, but I would not be proving that the languages are related.

In the absence of hard facts, my own informal judgment is that in much of the work published on long-range comparisons, too many degrees of freedom are being used for the results to be truly convincing. In global and near-global etymologies such as the 'suck' example, researchers appear to be free to select words ad libitum from synonyms, near synonyms, and words that are semantically connected; they can project words from any member of a language group backwards into the protolanguage, giving them a great deal of freedom to select from virtually any language in the group; at the top level they can omit language families that simply do not work; they can define phonetic similarity in any way that makes a particular etymology work; and they present the most striking results, without any indication of how many comparisons using the same criteria fail to be convincing. These indictments all amount to the same observation: The more ways one can produce a "noteworthy" outcome, the less probative it is as evidence belying the possibility of chance. For a good deal of elaboration on this theme, see for example Campbell (1988), Hock (1993) and Matisoff (1990). This line of statistical argumentation does not of course mean that Greenberg's conclusions are necessarily wrong; in fact the communis opinio concurs with his conclusions concerning the African languages, at least. In virtually all linguistic research, the last step has always been for the linguist, in his

TABLE 49 Three Languages

Concept	Language		
	Alpha	Beta	Gamma
1	X	A	Z
2	X	A	Z
3	X	B	W
4	Y	A	W
5	X	A	Z
6	X	B	W
7	Y	A	W

or her expert but unquantified judgment, to decide whether the evidence is stronger than chance. Greenberg may well be doing this correctly in his head. The trouble is that there is no way for that ratiocination to be conveyed in print to other scholars, so that they can independently confirm it. However great Greenberg's record, linguistics cannot be reduced to appeal to authority. If the methodology and data do not convince fellow scientists, then another line of attack is needed.

As currently practiced, I believe it is true that, all things being equal, a demonstration of recurrent correspondences should be taken as more convincing than an equal amount of text devoted to expounding the results of a massive multilateral comparison based on similarities. But it is by no means obvious, despite claims such as that of Jones (1989), that multilateral comparison is in principle incapable of working. Standard statistical techniques such as those advocated in this book provide a good deal of assistance in controlling the effects of freedom of choice. In addition, the Monte Carlo test allows one to cleanly judge the significance of the number of matches one finds, without having to rely on specific statistical distributions or on empirical rules of thumb such as those of Bender (1969) or Swadesh (1956). Under these more controlled conditions, can a multilateral comparison ever be useful? In particular, can multilateral comparison be helpful to the sort of tests being considered in this book, where connections between languages are being explored by means of statistical tests of recurrent correspondences?

The narrow answer is yes, multilateral comparison can make sense in the context of these techniques. An artificial example should suffice to illustrate this contention. Consider the situation in Table 49. There are seven concepts (1, 2, etc.) and three languages Alpha, Beta, and Gamma. For each concept, I note the initial phoneme of the corresponding word in the relevant language.

	A	B
X	3	2
Y	2	0

Alpha × Beta
$p = .53$

	A	B
W	2	2
Z	3	0

Gamma × Beta
$p = .15$

	Z	W
X	3	2
Y	0	2

Alpha × Gamma
$p = .15$

FIGURE 3 Significance of the pairwise comparisons.

		Z	W
X	A	3	0
X	B	0	2
Y	A	0	2

$p = .03$

FIGURE 4 Significance of three-way comparisons of the data in Table 49.

If one applies χ^2 techniques on any of the pairwise combinations of languages, one gets the results in Figure 3. Because of small expected frequency, I use permutation techniques to estimate the p.

None of the association values approaches very close to any commonly accepted significance values. But if one collapses two of the languages, the situation is quite different. Figure 4 shows that the same data associate significantly. In less mathematical terms, the situation is obvious by inspection. In Table 49, the X-A-Z alignment pops right out at the reader, appearing three times in a seven-row table. But if one covers up one column at a time, none of the associations seem particularly compelling.

We could easily scale up our methodology to handle multilateral analyses like this. However, I will not delve into this topic in this book. It is a rather difficult problem to work with and reason about contingency tables with many dimensions. Finding that the distribution of numbers in a three-dimensional phoneme table is not likely due to chance at $p \leq .05$ is not the same as saying that all of the languages are connected to each other. All we actually find out is that *some* of the languages are historically connected. To show that there is a connection between all the languages, one needs to show that the degree of deviance (or other metric such as R^2) found in the multilanguage comparison is significantly greater than the amount that can be attributed to any subset of the languages (Agresti 1990, Agresti 1992, Wickens 1989). That is more complicated than it sounds even for three languages; when one begins to look at four or more languages at a time the task gets mind-boggling. It can also require massive computational resources, or at least some clever programming, to handle multidimensional tables with millions or billions of cells. The discussion begins to slip beyond the target audience of this book, linguists who are curious about statistical argumentation but do not necessarily wish to devote an inordinate amount of attention to its more tortuous byways.

Another reason to shy away from multilateral statistics is that in practice, one suspects that truly multilateral evidence will be fairly rare. In my informal experiments, I have found a handful of good recurrent triplets among the Indo-European languages in the data base (e.g., there are four triplets that match up initial /s/:/z/:/s/ between English, German, and Latin), but only one good example of a recurring quadruplet ('not' and 'night' with /n/ in English, German, Latin, and Albanian), and no quintuplets at all. I suspect this case will turn out to be analogous to that of the multiphone comparisons discussed earlier. While multilateral statistics sounds exciting because the individual pieces of evidence it turns up can be stunning, the very rarity of the data that makes it so stunning works against it in a statistical test. It is often pointed out, for example, that Greenberg rarely finds a comparandum that runs through all of the ten or so languages he typically compares at a time. In the end it is not wise to make ourselves dependent on finding a few pieces of killer evidence. We are much better off looking for simple, not too uncommon evidence that is widely distributed.

11

Conclusions

11.1 Recipe

I have discussed many statistical issues in this book and presented many options whose relative merits are either unknown or a matter of personal preference. In part, my goal has been precisely to show that a statistical approach to historical linguistics, while certainly imposing some strict constraints, is not a straitjacket. There is much room for individual linguists to take into account their own attitudes about risk of error and their own view about what constitutes probative evidence. On the other hand, the mass of detail may have obscured the simplicity of the basic procedure. To remedy that problem, I present here a step-by-step summary of the method I am advocating. Because this summary can serve as a recipe for those wishing to try this method for themselves, I present a few more practical tips about implementation than I have in what heretofore has been a largely theoretical discussion.

0. The very first step is to make decisions about any part of the process where there is room for options. Some of these are choices that nobody would dream of putting off until later, such as deciding what languages you will be studying. Other issues should be decided at this point in order to protect your experiment from any possible bias on your part. For example, if, when gathering data, you have not yet decided whether to use Swadesh 100 or Swadesh 200, then later you find some great-looking potential cognates in the latter list, you may suddenly find yourself subconsciously leaning toward picking the Swadesh 200 list if in fact you want to prove the languages are connected. Worse, you might be tempted to run the test both ways to see which one works better, in which case you would be honor-bound to apply a potentially debilitating Bonferroni correction. Therefore the following questions should be answered at the outset.

What two languages will I compare? This may seem too obvious to note, but there are some nontrivial issues. At the very least, you should settle on a particular dialect, so you do not end up picking a Northern word when that is a better match with the other language and a Southern word when that works better. You may also find that you are not necessarily interested in a particular language so much as in an entire group of languages, in which case some language choices are better than others. Naturally, one will prefer languages one can get data from, and languages whose history is well documented, and whose morphology is well analysed. A less obvious desideratum is information about language contacts. As Martha Ratliff pointed out (personal communication), if one wants to test the hypothesis that the Indo-European languages are related to the Uralic languages, one would not look at Finnish, with its history of many prehistoric borrowings from Common Germanic and Baltic, but rather at a language in the Samoyedic group. There is also the issue of what historic stage you should consider. All things being equal, it is best to go back to the earliest stage that is reliably documented. Unquestionably, one would never compare French with some non-Indo-European language (except in a methodological study such as this). At the very least one would use Latin, and I myself would feel comfortable using Proto-Indo-European reconstructions.

Where will I get the data? There is no single best source of information, but try to understand the biases of your source. Native speakers can be an excellent source of information about nuances, but if you elicit a translation by giving them the English word, there is a good chance that if there are several possible translations, the first one they will think of is one that sounds most like the English word, which could be a real problem if you are comparing the language to Indo-European. Dictionaries may have a similar bias, or hypercorrect by putting the most similar-looking word at the end. Word lists published by historical linguists may be biased toward selecting words that credibly correspond to words in languages they believe related. Find a source that has no bias that could affect your results, e.g., a dictionary that lists words in order of frequency, or think of a way to correct for possible biases.

How will I decide what the roots are and whether they are related to roots for other words on my list for the same language? The ideal source of information is deep etymological analysis on a comparative basis; weaker contenders, roughly in order, are internal reconstruction, morphological analysis over the whole vocabulary, native speaker intuition, and eyeballing the word list. One works with what one has, but this decision is critical, and it may be worthwhile to base the language choice in part on which language has the best etymological analyses.

Am I looking for any historical connection, or a specifically genetic relationship? These tests do not provide any inherent defense against loans, so you will be happiest if you are looking for historical connection simpliciter and remind yourself at the end that that is what you have found. If you are looking for genetic relationship, you will have to take extra-strong measures against loans and then stoically remind yourself that no human endeavor is perfect.

Which list of concepts will I use? E.g., Swadesh 100, Swadesh 200, a hand-tailored list of grammatical morphemes? At present the best choice I know of seems to be the Swadesh 100 list, although it might be best to eliminate the grammatical words on that list, especially the linker words (prepositions, conjunctions, pronouns). In comparison, the advantages of using the longer Swadesh 200 list appear at this point to be in surprisingly good balance with the disadvantages, so if you particularly like to collect twice as much data, it will probably not do any harm, particularly if you are careful about doing the necessary exclusions (loans, etc.), which are more frequent on that word list. If you decide to use another list or tailor your own, you must be careful to select concepts whose exponents do not change very frequently (through semantic shift or borrowing), and of course you must be careful not to pick or choose concepts with any thought to how their exponents compare in the two languages. This last difficulty is so pernicious that a hand-tailored list will immediately arouse suspicions of bias, so you will have to be especially careful to justify such a list.

How will I decide between multiple exponents for a single concept, and what will I do if a language does not have an exact word for a concept? Ad hoc decisions can lead to bias. What one normally wants to do is to find a word that is used as a basic, ordinary, generic term in the language: e.g., *dog*, not *mammal, canine*, or *dachshund*. Ideally one needs to think in advance of all possible mismatches between the taxonomy in your concept list (e.g., the English words in Swadesh 100) and how other cultures are likely to express terms. That is a tall order. One idea that can help is to think of a fairly specific instantiation of the concept, e.g., if it is 'foot', think 'foot of a male adult human'. Then for each language find the basic, generic term that people would use when talking about that. This will work best if you think of the most bland, everyday instantiations. 'Male adult human' is in fact a good rule of thumb for many concepts.

Which part of the roots will I compare? E.g., initial consonant, first two phonemes, entire root? The best bet appears to be the initial phoneme. Collapsing vowels into one category works best for many languages, although it is not certain that that is universal (e.g., vowels seem particularly stable in Polynesian). Comparing more of the word will probably give weaker results, but is more likely to catch certain rare problems, and may be more politically acceptable.

At what level will I inspect the phonetics? E.g., phonemes, features? The best choice appears to be features that show place of articulation, though segment-level analysis may be more politically acceptable and take greater advantage of the fact that the methodology is based on recurrences rather than similarity.

If working phonemically, where will the string be segmented? E.g., will diphthongs and affricates be treated as single or multiple units? This can be decided on the basis of whether they normally originated by fusion or by breaking in the language in question. In case of doubt I would certainly treat affricates as units.

If looking at more than a single phoneme or feature, how will I align them? E.g., Linear, Cross Pairs, Fibonacci, or some hand-tailored method? The Fibonacci technique looks best but there probably is a lot of room for improvement. One could also consider the possibility of using bins, e.g., putting phonemes in different environments into different contingency tables. My recommendation for those with no personal preferences would be to concentrate on a single phoneme.

What metric shall I use for measuring association between languages? E.g., χ^2, R^2, W, something new? The best bet at the moment appears to be R^2.

Will I be running several randomized trials and averaging the results? This is probably a good practice but is certainly not mandatory if that proves difficult or time-consuming. The reason it needs to be decided in advance is because it is easy to fool oneself and decide on doing several trials only after the first one did not give good results. Averaging several trials is only legitimate if it is done blindly.

It is also a good idea to ask yourself at any early stage what your philosophy is about significance levels. Does the conventional .05 of many social sciences seem appropriate or too weak for you? What about Ringe's .01? Deciding later will not necessarily bias your results, but you will feel uncomfortable if you end up deciding your preferred significance level on the basis of which one gives you the results you like better.

1. After having made all those decisions, the next step is gathering the words. Ideally this should be done by two different people working under identical instructions. If you gather both sets of words yourself, you will have to fight a very natural bias against picking a word in Language B that sounds much like the word you picked for the same concept in Language A. And it is just as bad to correct for such a bias by picking the most dissimilar word. On the other hand, the danger of having two different people prepare the lists (or using lists found in published sources) is that the two people may be working with different interpretations of what the concepts mean. If one person takes 'hair' to mean hair on a human head and another takes it to be body hair or animal hair, they may well come up with mismatching words. Both problems can be addressed by using a very detailed set of instructions. If one person works on both languages, then at least the two languages should not be prepared at the same time.

If after applying all the rules (using the most basic, generic word that would express a narrowly defined instantiation of the concept) you have more than one candidate, it is well to take them all, but sort them according to some predefined criterion, such as frequency, if known, or which one gets listed first in the greatest number of independent dictionaries.

A last piece of advice is that the analysis will have to be done by computer, so it is well to start building up your data files on computer from the beginning. I myself used XML for the project described here and am happy with that choice. But I was manipulating a lot more diverse data than one would do in a simple two-language experiment where all the decisions are made in advance. Your informal notes can of course be kept in any format you like. If you use a structured format like XML, a database or a spreadsheet, try to keep the concept gloss and the pronunciation in their own separate fields, because those are the two parts that will be used in later computer processing. For the pronunciation, I would recommend coding it in a transcription system like SAMPA (Wells 1999), which expresses IPA in ASCII. This is much more portable, if less glamorous, than IPA fonts, and if a colleague does not already know the system, you can point her or him to the Web page that documents it. If you will be comparing multiple phonemes in a word, it simplifies matters if you separate the phonemes with an explicit delimiter like a space.

2. After gathering all the words, delete all candidates that are loan words. Of course if you are investigating whether there is any reasonable chance that two languages had any historical connection at all, then you probably do not want to delete putative loans between those two languages. You still would want to delete loans between the languages that postdate the period under investigation, e.g., if one were comparing English and Hawaiian for remote connections between their language families, one would delete modern, post-contact loans like 'aila 'fat' and naheka 'snake'. And you would almost certainly want to delete loans from third parties; e.g., Arabic loans into English could disrupt any residual sound correspondences between Hawaiian and English. How to actually identify loans is a very difficult issue that can scarcely be given justice here. Some hints that a word may be a loan are when it names a recent cultural importation (not common in a good word list); when it is not broadly disseminated among related languages in a form that obeys otherwise normal sound correspondences; when it has unusual sounds or unusual sequences of sounds; when it is fairly long but has no morphological analysis; when historical records do not contain instances of the word before a certain date; when the word is very similar to a word in a language with which the language has had known or suspected contact. See Campbell (1999:64) for some further discussion and examples.

3. Identify and delete words that are likely to be motivated. Several words on the Swadesh list should be looked at particularly closely: words for 'bird', 'breast', 'fly', 'large', 'small' and in the Swadesh 200 list 'bad', 'belly', 'blow', 'breathe', 'cut', 'dirty', 'fall', 'father', 'fight', 'hit', 'laugh', 'mother', 'rotten', 'spit', 'split', 'squeeze', 'suck', 'wind'. One should also reject words that are ideophones or based on onomatopoeia, expressive interjections, or nursery words. Phonetically, words of unusual shape or with rare sounds should be treated with suspicion (these could also be loans); this includes reduplication in languages that do not usually employ reduplication. Words that seem to correspond to suggested sound symbolic universals, especially the frequency–size code (acute or voiceless segments for small things, grave or voiced segments for large things) invite special attention. In the end, of course, this, like loan word detection, is a matter of art. It is easy to err too far in either direction.

4. For the remaining words, consider their structure carefully. Extract the single morpheme of the word that expresses the core meaning. Ideally, that should be a root element whose meaning is modified at best slightly by attendant affixes or other reversible changes like umlauting or ablaut. If you can identify a meaning for the root that is very different from that of the word as a whole, consider discarding the word entirely. For example, if the word for 'dull' in a language is composed of morphemes for 'not' plus 'sharp', the odds that there will be a root match with another related language drop considerably, so the word would likely only add noise to a comparison. Ideally do the analysis based on the best historical comparative information, but in the absence of that, do the best that you can with an internal reconstruction of the language in question. Remove any affixes from the root. To the extent that that can be done knowledgeably and consistently, undo any sound changes and nonconcatenative morphological processes like ablaut or umlaut. This can be tough work but there is no need to overdo it. If your metric only compares the first consonants of each entry, you do not really need to worry about the rest of the entry, except insofar as they may help you better reconstruct the initial.

5. If more than one word in a list for one language reduces to the same root, omit one of them. This test does not call for the deletion of homophones; it would be a serious mistake to eliminate a word just because it sounds the same as another word. For example, Hawaiian *hou* may be selected for both 'new' and 'stab'. But if two words, once reduced to their root, seem reasonably likely to have the same origin, one must be deleted; and in case of reasonable doubt, it is better to err on the side of deleting. Which to delete? If there is any reasonable likelihood that one of the concepts is closer to the root meaning of the word, that one should be retained (see for example Sweetser 1990, Traugott 1989, Wilkins 1996 for discussion on natural directions of semantic shifts). Otherwise, if that root is the only remaining candidate for one of the concepts but the other concept(s) have multiple candidates, delete from those with multiple candidates. Otherwise, delete by some criterion you have chosen in advance. If possible, prefer the word that has been shown to be the most stable in studies like that of Swadesh (1955), Dyen et al. (1967), Kruskal et al. (1973), and Dolgopolsky (1986). Otherwise, it may be best to flip a coin.

6. At this point you have done with filtering candidates. If you have multiple candidates remaining for any of the words, delete all but

one of them. You can probably use the same last-recourse method as in the step above.

7. You will now have 0 or 1 entries for each concept. For the entries that remain, trim them down to the parts of the word that you will be analysing, if you have not already done so. If you are using a stratified analysis over several phonemes, prefix each of the phonemes with the designation for the bin that the phoneme goes into. The actual names do not matter, as long as they are unique. For example, if you want to compare $C_{1.1}$ and $C_{2.1}$ in separate categories, you might reduce English /stæb/ 'stab' to "C1.1:s C2.1:b".

8. If you are using a feature-based analysis, convert the phonemes to a feature-set description. If you are using multiple features, you will probably want to select value names that are unique across all features, so that they will go into separate bins. For example, if the phoneme /b/ is to be analysed as place of articulation "lips" and articulator "lips", that could be encoded as "P:l A:l".

9. You should now have the data transformed about as far as it makes sense to do by hand. You will now need to produce a machine-readable data file. You will probably find it easiest to make that a simple tab-delimited ASCII text file. One possible format, for example, is to have one concept per line. Start each line with the concept gloss, then tab, then the data for the word. If you are using multiple phonemes or segments, I would recommend separating each of them with a space.

10. While not strictly required, it may be a good idea at this point to run some statistical tests to see if there are any associations between the forms you have recorded and their part of speech. You do not need to be particularly rigorous about this step. It is purely an exploratory device to call to your attention any patterning you may have overlooked in your etymological analysis in Step 4. Go back and make adjustments if any meaningful patterns appear.

11. Now you can go back to Step 1 and do the same things with the second language. One advantage of doing the two languages sequentially is that you now know which concepts did not have suitable words in the first language, and so do not need to be researched in the second language. After you find the words, they can be added to your ASCII text file as a third field preceded by a tab.

12. Delete any concepts that are not found for either or both of the languages.

13. Further steps should probably be done by computer. While none of the metrics I have presented are particularly time-consuming,

the necessity of doing thousands of rearrangements of the data means that there is a certain value in seeking a reasonably efficient implementation. A few tips for implementors:

If you use a fast language like C or Fortran, as opposed to a user-friendly language like Perl, it may make the difference between a program that runs in a couple of seconds and one that runs in a couple of minutes. I personally use Perl for everything except that which necessarily takes place within the permutation loop, where control is handed off to a C program.

Most of the program involves incrementing the cells of a contingency table as indexed by a pair of phonemes (or feature values). The table cell reference can be speeded up if one does not have to look up each time which row or cell goes with a particular phoneme. Therefore it is best if the first thing the program does is tokenize all the phonemes or feature values in the words from their user-friendly strings like "P:l" to cell or row indices like 0, 1, etc. The program does not need to know that 0 is a labial place of articulation, just that it is different from the feature coded by 1, and 2, etc. A good trick is for the tokens in one language to be assigned in sequence starting from 0 and incrementing by 1 up to the highest needed number r, and for the tokens in the other language to start from 0 and increment by $r + 1$. That way, the contingency table can be implemented as an array of integers, and the index into it for any pair of phonemes can be determined simply by adding the two numbers together.

Permutation is the other intensive operation, so you should make sure you are rearranging pointers, not arbitrarily sized chunks of memory. With each rearrangement, it is only necessary to rearrange the words of one language; rearranging both languages does not make the arrangement any more random. But you do have to ensure that that one rearrangement is truly random. One way to ensure your algorithm is random is to use the Fisher-Yates shuffle (Christiansen and Torkington 1998:121).

14. If you have decided to do the test several times and average the results, then you should extract a subset of the word list. One reasonable approach would be to test on about 90% of your data. That subset needs to be randomly selected. One way to do that is to rearrange the word list and then select the first 90% of the concepts.

15. Compute the association metric for the table; we will call this the base measurement. The exact computation will depend on whether

you choose χ^2 or R^2, Linear Alignment or weighted alignments like Fibonacci, etc. The metrics I have described all proceed in two steps, however. First, zero every cell in the contingency table. Then go through each pair of words and pair up the tokens that represent the phonemes or features. For example, in the Linear feature-based method, one pairs up each of the tokens representing the features, in order, left to right. For each of those matchings, decide which cell of the contingency table represents it; if you use the tokenization technique described in Step 13, you just add the two tokens together to get the index into the contingency table array. Then increment that table cell in the appropriate way. For example, if, as recommended, all of the words have been truncated to the same length, you would simply increment by 1. If they are of different lengths, you should probably scale for the lengths of the two words, as described in the previous chapter.

16. After incrementing the contingency table, the second step in computing the metric is interpreting the tallies in the table. In most of the metrics discussed, this involves looking at each cell of the table, computing a measurement, and summing that measurement across all cells. The recommended per-cell measurement, R^2, is the easiest to compute. One takes the cell count O, subtracts one to get the number of recurrences R, then squares that: $(O-1)^2$. The χ^2 metric is a little more involved in that it requires knowing the expectation of the cell, which in turn requires knowing the sums of each row and column. Note that those sums and expectations will stay constant over each rearrangement of the data, and so can be computed just once, at the beginning. The word-based W metric is a little different in that instead of summing the R^2 across all cells of the table, one sums them across word pairs; therefore there is the added step of making a second iteration through the word pairs, looking up the R^2 metric for each token pair, multiplying those together, then summing for each word pair. Note that some of the measurements inherently return an integer metric (e.g., R^2 metrics without weighting), others give floating-point values (e.g., χ^2 and those that weight, such as Fibonacci alignment). Because you will be doing many comparisons for equality with this metric, if it is a floating-point number there is some danger that truncation errors will cause equal numbers to appear unequal. It is therefore best to convert all values to integers, perhaps after multiplying by a number like 1,000 to ensure that sufficient degrees of precision are retained.

17. After computing the base measurement, you are ready to do the Monte Carlo portion of the test. First you need to decide on how many rearrangements of the data to make. One rule of thumb might be to divide 10 by the most precise p value you feel important to report. For example, if you want to report p values to three decimal points, you should minimally do $10/.001 = 10,000$ rearrangements, as I did throughout in this book. The more rearrangements, the better.

18. Randomly rearrange the words in one of the languages.

19. Compute the metric again, in the same way you did for the base metric (Steps 15 and 16). If the new measurement is greater than or equal to the base measurement, increment a tally.

20. Go back to Step 18 until the desired number of rearrangements has been achieved.

21. Divide the tally by the number of rearrangements. That is the p value.

22. If you have decided to do several trials, go back to Step 14 until all trials have been performed. Five to ten trials will probably suffice for most purposes. Sum the p value across all trials, and in the end, divide by the number of trials.

11.2   Appraisal

Statistical techniques are potentially a useful adjunct to the work of the historical linguist. While I cannot hope or even want to reduce all comparative methodology to techniques that are tractable by rigorous experimental statistics, it often comes about that disagreements among linguists boil down to subjective disagreements about how to interpret the evidence: Is it proof, or a heap of coincidences? Post hoc probabilistic demonstrations usually turn out suspiciously supportive because of various statistical fallacies, usually having to do with making inferences by calculating the probability of a particular set of coincidences. The methodology presented in this book is an accurate and unbiased way for historical linguistics to put a reliable p value on claims that languages are historically connected.

Even more optimistically, I might hope that the material presented in this book could help bridge the divide between adherents of mass lexical comparison and its opponents. Debate about the validity of the well-publicized claims about some spectacular long-range comparisons has sometimes led to polarizing positions on linguistic methodology, often projected back into the past. At one extreme there are claims that linguists have always decided that languages were related by studying lists of words and morphemes without bothering with recurrent sound correspondences (Greenberg 1987); at the other extreme are claims that mere words (as opposed to more abstract grammatical patterns) were never considered proof, even when recurrent sound correspondences were adduced (Nichols 1996). I will resist the temptation to add to the historiographical arguments and instead focus on the future: How does my methodology resolve the essentially statistical issues in the debate? These I think are the key points:

- Statistical methodology and mathematical hypothesis testing can be applied to the task of deciding whether languages are historically connected. The contrary opinion of the eminent Meillet (1925) may have discouraged development in this field, but his reasoning suggests its own solution. He argued that the full corpus of facts gathered by historical linguists could not be statistically weighed because they were incommensurable. That is absolutely true; but one can operate on the commensurable subset, which is the arbitrary association of form and meaning in the morphemes. I believe the doctrine of the arbitrariness of the sign, as well as its preëminent usefulness for demonstrating language connections, is accepted by virtually all parties. It is true that controlled hypothesis testing does not look at all the data one can adduce, and therefore may on first consideration appear unacceptable. But there is nothing to discourage the linguist from using the methodology as a supplement to other techniques. There is nothing wrong with saying that a lexical significance test failed, while still asserting that other, less controllable evidence nevertheless seems like good evidence that two languages are connected.

- A quantitative study does not make up for lack of rigor in filtering out loan words and motivated vocabulary or in stemming words properly. Many of the critics of works like Greenberg (1987) claimed the data is rife with errors of that sort (e.g., Campbell 1988); defenders countered that any mistakes are drowned in the sea of nonmistakes. Such excuses definitely will not be tolerated in the type of tests I am advocating here. I have shown how such errors can easily make the tests invalid and throw off the results. One sometimes gets the impression that errors do not matter as much when one is analysing data statistically. I think that that idea comes from the observation that when one is analysing large amounts of data, small random errors may not matter very much because they will tend to cancel each other out. Unfortunately, in linguistics we are mostly talking about small amounts of data, and these kinds of errors are for the most part nonrandom. They largely tend to bias in favor of finding connections between languages. The other false impression is that the Swadesh 100 list is adequate guard against these errors; that appears to be the idea that caused Ringe (1992), who is otherwise very careful in his methodology, to let down his guard in examining his word lists for such problems. While it is true that the Swadesh 100 list is much better than most sets of 100 words, it unavoidably has many pitfalls that the linguist must be on guard against. Nor does it seem at all feasible that extensions of the method can ever make it possible to relax vigilance.

- Some of the dichotomies that prominently figured in the debates during the past decade or so can take some surprising turns when looked at statistically. The contrast *similarities* vs. *recurrent correspondences* ceases to be vitally important when exact definitions of similarity are employed and numerical methods are used to control for the large number of similarities that can turn up. A concrete example of that occurred when I tried comparisons at the featural level, which were some of the most accurate experiments I ran. Almost all of the recurrent correspondences were matches of identical feature values, which is expressible by a similarity metric. While I still believe finding recurrent correspondences is a good idea, mostly because of the greater time depth they are potentially useful for, it is clear that observing that someone looks for mere similarities is not in itself damning. The contrast between studying *word lists* vs. *grammar* is a bit more complicated. But once one factors out typological considerations such as word order and type of inflectional categories, which appear to be of dubious value in any methodology, the contrast really boils down to a decision of whether to use lexical (i.e., content) morphemes or grammatical (i.e., function) morphemes. While I can state definitively that it is important not to mix the categories in the same test, the decision of which to prefer is not so clear-cut. Paradigmaticity per se is not a consideration in a test like this; the only real advantage grammatical morphemes have is that they are more resistant (but by no means completely resistant) to borrowing. Their difficulties, on the other hand, are legion. Many languages have very few grammatical morphemes; it is hard to match grammatical categories across languages; they are subject to much analogical and paradigmatic shifting within a language. While the contrast, used polemically, makes it sound like a linguist using *word lists* instead of *grammar* is just thumbing through dictionaries instead of doing serious linguistic analysis, the decision to use word lists may be statistically more responsible, especially if one can confidently control for loans.

- Most of the rules of thumb that the great comparativists like Meillet (1925) have left us are designed specifically to compensate for the fact that linguists did not have tests of significance. It may seem shocking at first consideration, but the rules were designed to decrease the amount of evidence a linguist could find. If linguists had no way to statistically weigh how their evidence compared to chance levels, then it was necessary to ask them to throw away everything but pieces of evidence that were rather suggestive even by themselves. Meillet's suggestions that linguists always do multilateral comparison, or that they insist on three corresponding phonemes in a pair of words, are of this nature. After hearing such rules for enough decades, one comes to feel that they must be logical requirements for proving words cognate, and that any methodology that does not follow those practices is inherently incorrect. But in the context of statistical significance testing, at least, that is clearly not true. My tests that operated by comparing a single phoneme, or even a single feature of one phoneme, generally worked better than those that looked for multiple phonemes. In statistical terms, that makes a lot of sense. We can prove more by looking for characteristics that are widely distributed, than by looking for characteristics that are rare. Although it may seem strange that we overlook the probative value of, say, a four-phoneme match between a pair of words, insisting on using only that kind of data actually entails throwing away much more information. Worse, as Meillet himself noted, his recommendations may actually prevent one from making progress in languages that have short words and few inflections. Such languages would not pose any special problems for any of the tests I have presented in this book.

The test or, rather, family of tests presented in this book appears to be reasonably powerful and precise. It is encouraging that these tests all manage to uncover the Indo-European family, without a lot of false positives. Although from the vantage point of today's mature knowledge of Indo-European, the power to detect relationships such as those between German and Latin may seem trivial, it must be remembered that that is really quite an accomplishment. Jones's insight is not celebrated because it was trivial. The relationship between French and the other languages (besides Latin) is even less obvious, once loans are removed.

Meillet (1926) stated, and Greenberg (1993) concurred, that relationships between French and English would not be demonstrable without the evidence of Latin or ancient Germanic dialects, but this proves no problem for the statistical tests I have presented.

I ran tests over pairs like English and German simply to give a feel for how the tests work. In practice, I don't see how such tests help matters in cases where the comparative method has already been applied and gives incontrovertible results. The comparative method, after all, gives much more than a mere index of how likely language connections are. Conditional correspondences, reconstruction of the protolanguage, and subgrouping of the families are all matters I have not touched on here, but which reveal worlds of information about the language and its history. On the other hand, there are many cases where the comparative method has not already given incontrovertible results, and it is in those cases that my methodology may prove helpful. I see two major classes of applications. The first is where linguists simply have not yet gotten around to the intensive work entailed by the comparative method. My tests, to be sure, are not trivial to apply, but they are arguably a good deal easier than a full-blown comparative analysis. And when all one knows about a language is a list of words, it would be useful to quickly learn what other languages it may be connected to. Hopefully one would not treat the result of preliminary tests as gospel, because, just as in any other procedure, reliability increases as one learns more about the languages and can, for instance, better identify loans and language-internal cognates. And there can always be fairly surprising coincidences that result in false positives. But the tests could well give linguists guidance as to where to direct resources for further investigations. Perhaps more importantly, the tests are not something that is done instead of the comparative method, but are largely coextensive with its initial steps. Anyone working within the comparative method would benefit from gathering data about the Swadesh 100 concepts and from vigorously controlling for loans, motivated vocabulary, morphemic composition, and so forth. It is true that entering the words into a computer might not be needed otherwise, but in recompense one does get a list of pairs of phonemes that co-occur more often than expected, which is one of the first things one looks for when applying the comparative method.

The other major application would be in testing comparisons where the results of the comparative method, or other methods such as mass lexical comparison, are indeterminate or even negative. Languages with short roots may be an example of a situation where it is difficult to judge how to apply some of the rules of thumb of the traditional comparative method, but where my statistical tests would be fully functional.

Languages where the evidence is simply weak, perhaps because of extended time depths, may benefit from the application of objective and statistically controlled tests. Serious proposals for long-range comparisons abound, many of which have been alluded to in other parts of this book. Campbell (1999:312) gave quite a long list of proposed language groupings which he considered to be unproven. It would not be inappropriate to put these proposals to the test. A good starting point would be language groupings that many linguists accept but which perhaps require more rigorous proof, such as Afro-Asiatic, Altaic, Hokan, Khoisan, Niger-Kordofanian, Nilo-Saharan, and Penutian. The next step might be to look at more tentative groupings that are nevertheless often considered likely, such as Australian and Indo-Uralic, and various versions of Nostratic. From there one could move on to a dizzying array of broader groupings, such as Amerind, Ural-Altaic-Japanese-Korean, Eurasiatic, and on to Proto-World. It is true enough that the absence of significant findings (i.e., high p values) would by no means prove that the languages are not related. The real issue is whether the evidence being amassed is significantly greater than chance levels, and a true statistical test is a serious way to pursue that question. Theorizing about the relative merits of multilateral comparison versus binary comparison, or about similarity measures versus the comparative method with recurrent correspondences, is interesting and useful. But in the end, the question at issue is whether a particular linguistic analysis has correctly controlled for chance. None of the traditional methodologies addresses that matter in a direct and quantifiable way. Only a statistical test such as the ones elaborated in this book can give meaningful answers in specific cases.

I would not however be so chauvinistic as to assert that I have presented here the best possible statistical test. To the contrary, opportunities for improvement abound. It would, for example, be useful to work out better metrics with better alignment techniques; to add the capability of multilateral comparison; and perhaps to find better ways to incorporate evidence from different parts of the word so that less probative alignments do not drown out the better evidence. A special strength of this methodology is that these goals are easily pursuable. The method allows great flexibility in plugging in various metrics. I would hope in fact that a major use to which this book may be put is in adapting other researcher's proposals for measuring chance. In many cases researchers have had promising metrics for measuring similarities or other types of connections between languages, but attempts to determine significance have foundered because the precise mathematical distributions required for those metrics have not been available. Use of the permutation method, with all the provisos and safeguards here put forth, could help make these proposals statistically valid. Furthermore, the focusing on exact significance levels rather than rules of thumb peculiar to particular methodologies could do much to make various analyses of language connections commensurable with each other, opening up the possibility of deeper dialogue between investigators. "Four significant cells" in Ringe's test are four cells, but a p of .002 means something to researchers working with all sorts of different metrics.

As a last resort, even if researchers choose not to use my tests, or even choose not to graft their own metrics onto my general methodology, it is to be hoped that my aforementioned provisos and safeguards will be heeded. Linguists devising new tests for language connections, as well as linguists evaluating the reliability of published accounts of such tests, are invited to ponder the urgency of considering some of the problems that one can run into in data collection, data preparation, and statistical analysis. In particular, I have shown the bad effects that can be wrought by the presence of loan words, sound symbolism, onomatopoeia, and language-internal cognates; by picking words that may be biased toward or against the research hypothesis; by using metrics that, perhaps unintentionally, take into account the length of the words; and to be sure, by applying statistical tests that are flawed, or by improperly combining the results from multiple tests. None of these ideas is completely new to linguistic research, and yet vanishingly few statistical studies have taken them all into account. My hope is that from my discussing all of these issues in the same place, future practitioners and their readers will have at their disposal a checklist of pitfalls to be avoided as well as recommended practices to adapt to their own needs.

Appendix: Word Lists

This appendix lists the words that were used in this book to compare the various metrics and other approaches to the statistics. Each entry has the same structure. It is headed by the gloss of the concept, followed by part of speech, and an indication of whether the concept is found in the Swadesh 100 list or only the Swadesh 200 list. The parts of speech are as follows: *adj* (adjective), *lim* (limiter, such as demonstratives, numerals, determiners), *link* (linking words, such as conjunctions, prepositions, pronouns), *noun*, and *verb*.

Then the data are given for each of the eight languages. The block of data for a language is introduced by the first three letters of the language's English name. That is followed by the word that was chosen to be the exponent of the concept for that language. It is first given in the standard orthography of the language. That is followed by an IPA transcription of the distinctive part (root) of the word, with spaces separating each phoneme. If the word is here treated as nonarbitrary, that is followed by the word MOTIVATED; if it is treated as borrowed, it is followed by LOAN. If there is some possibility that it is cognate to the word chosen for some other concept in the same language, those concepts are listed. Those data are generally followed by free-form notes telling the etymology of the distinctive part of the word. It must be recalled that under the methodology for which these lists were developed, in cases of doubt one must err on the side of treating a word as motivated, borrowed, or cognate. Thus I do not necessarily agree with all the decisions recorded here or even think they are the most likely etymologies; I merely adjudge them as not being without merit.

Abbreviations

The following abbreviations are used for names of languages in the notes. Asterisks are not used for reconstructed forms in prehistoric languages.

Alb	Albanian.
Angl. OE	Anglian Old English.
Arm	Armenian.
Class. Lat.	Classical Latin.
Eng	English.
Fre	French.
Ger	German.
Haw	Hawaiian.
Lat	Latin.
ME	Middle English.
MHG	Middle High German.
Nav	Navajo.
OE	Old English.
OF	Old French.
OHG	Old High German.
OLat	Old Latin.
ON	Old Norse.
PA	Proto-Athapaskan.
PAlb	Proto-Albanian.
PCEMP	Proto-Central/East-Malayo-Polynesian.
PCP	Proto-Central Polynesian.
PEP	Proto-Eastern Polynesian.
PG	Proto-Germanic.
Pop. Lat.	Popular Latin.
PIE	Proto-Indo-European.
PMP	Proto-Malayo-Polynesian.
PNP	Proto-Nuclear Polynesian.
POc	Proto-Oceanic.
PPA	Pre-Proto-Athapaskan.
PPN	Proto-Polynesian.
PWG	Proto-West Germanic.
Tur	Turkish.

The notes for most of the words end with a list of bibliographical references. It should not be inferred that all the listed sources necessarily agree with the etymological information given. Because they are cited so frequently, the works in this appendix are cited in the following abbreviated forms:

Barnhart	Barnhart (1988).
Blust	Blust (1993).
Boretzky	Boretzky (1975–1976).
BW	Bloch and von Wartburg (1964).
Çabej	Çabej (1982).
Dankoff	Dankoff (1995).
Dauzat	Dauzat (1938).
Drizari	Drizari (1957).
EM	Ernout and Meillet (1979).
EP	Elbert and Pukui (1979).
Haile	Haile (1926).
Hoijer	Hoijer (1945).
Huld	Huld (1984).
Janson	Janson (1986).
Kacori	Kacori (1979).
Kiçi	Kiçi (1976).
Kluge	Kluge (1995).
Lubotsky	Lubotsky (1989).
Mann	Mann (1977).
Marck	Marck (1996).
OED	Simpson and Weiner (1989).
Pinnow	Pinnow (1974).
Pokorny	Pokorny (1959).
PE	Pukui and Elbert (1986).
Räsänen	Räsänen (1969).
Ross	Ross (1995).
Sihler	Sihler (1994).
Skeat	Skeat (1980).
Stachowski	Stachowski (1986).
Steingass	Steingass (1930).
Tucker	Tucker (1931).
Walde	Walde (1930).
Watkins	Watkins (1985).
Wehr	Wehr (1979).
YM 1980	Young and Morgan (1980).
YM 1987	Young and Morgan (1987).
YM 1992	Young and Morgan (1992).

Words

all lim (plural, e.g.: All Cretans are liars), Swadesh 100

ALB *gjithë* /ɟ i/. Meyer suggested a loan from Kurdish, but generally considered PIE *sem-k̂o-*, root *sem* 'one'. Huld 69. Mann 109. Walde 489*. Pokorny 903.

ENG *all* /ɔ l/. PG *alla-*, PIE root *al-*. Walde 90. Pokorny 25. Kluge *all*. Watkins *al-5*.

FRE *tous* /t u/. Pop. Lat. *tottus* for Classical *tōtus*, replacing the unrelated *omnis*. PIE *tēu-* 'swell'. Pokorny 1080. BW.

GER *alle* /a l/. PG *alla-*, PIE root *al-*. Walde 90, Pokorny 25, Kluge *all*, Watkins *al-5*.

HAW *apau* /p a u/. Since *a-* is a common prefix, perhaps etymologically connected with *pau* (PNP *pau*), which can share the meaning 'completely'. PE.

LAT *omnes* /o m/. Best guess is **opnis*, PIE root *op* 'work'. EM. Sihler 43. Walde 123. Pokorny 780. Watkins *op-1*.

NAV *t'áá 'áłtso* /ts o h/ DERIVED FROM 'BIG'. The *t'áá* is a particularizing particle used in dozens of expressions; the distinctive word is *'áłtso*, root *tsoh* 'big', q.v. YM 1992, p. 613 (TSOH), 932, 954. Haile 38.

TUR *bütün* /b y t/. Räsänen derives from *büt* 'finish, be complete'. Räsänen 93b.

and link, Swadesh 200

ALB *e* /e/ LOAN. Generally considered a loan from LAT, but possibly native, perhaps cognate. Mann 209, Huld 60. Not in Pokorny.

ENG *and* /a n/ COGNATE WITH 'IN'. Cognate with GER *und*. The two main theories are that these words are related to Lat. *ante*, PIE $H_2 anti$ 'facing'; and that they connect with *en* 'in'. OED. Skeat. Walde 67. Pokorny 50. Kluge *und*. Watkins *en*.

FRE *et* /e/. From LAT *et*. BW.

GER *und* /u n/ COGNATE WITH 'IN'. PIE $H_2 anti$ 'facing' or *en* 'in'. Skeat. Walde 67. Pokorny 50. Kluge *und*. Watkins *en*.

HAW *ā* /aː/. PCP *a*. PE.

LAT *et* /e t/. PIE *eti* 'beyond'. EM. Sihler 441. Pokorny 344. Watkins *eti*.

NAV *dóó* /d óː/. YM 1992, p. 938, 943.

TUR *ve* /v e/ LOAN. Loan from Arabic *wa*. Not in Räsänen. Wehr 1224.

animal noun, Swadesh 200

ALB *kafshë* /k a f ʃ/ LOAN. Also means 'thing', and is a loan from Lat. *causa*, which has no known cognates. Mann 2. Huld 79 (citing also Meyer, Haarmann). EM *causa*. Not in Pokorny.

ENG *animal* /a n ə/ LOAN. Loan from LAT. Skeat.

FRE *animal* /a n i/ LOAN. Loan from LAT (native development of *animalia* was *aumaille*). BW.

GER *Tier* /t iː/. PG *deuza-*, PIE *dhewes*, PIE *dheu-* 'fly about'. Pokorny 261. Kluge.

HAW *holoholona* /h o l o/. Root is *holo* 'run', from PPN *solo*. PE.

LAT *animal* /a n i/. Nominalization of adj. *animālis*, from *anima*, PIE $H_2 anH_1$- 'breathe'. Sihler 46. Walde 56. Pokorny 38. Watkins *anə-*.

NAV *naaldeehii* /d eː ʔ/. The root means approximately *multiple objects moving through space*. PA *dèG*, PPA *de'G*. YM 1992, p. 128, *DEE'*.

TUR *hayvan* /h j/ LOAN. Via Persian from Arabic *ḥaywān*, root 'alive'. Räsänen 153a. Wehr 257.

ashes noun, Swadesh 100

ALB *hi* /h i r/. PAlb *hin-*, possibly cognate to the LAT *cinis* as *sk̂ino-*, though there are problems. Huld 74. Walde 392.

ENG *ashes* /a ʃ/. PG $H_2 ask$- or $H_2 azg$-, velar extension of PIE $H_2 as$-

'hearth', from 'burn'. Skeat. OED. Pokorny 69. Kluge *Asche*. Watkins *as-*.

FRE *cendre* /s ā/. LAT *cinerem*. BW.
GER *Asche* /a ʃ/. PG H_2ask- or H_2azg-, as ENG. Skeat. OED. Pokorny 69. Kluge *Asche*. Watkins *as-*.
HAW *lehu* /l e h u/. PPN *refu*. Cf. POc. *qapu*. PE. Blust.
LAT *cinis* /k i n/. Best guess for PIE root is *ken* 'scratch', possibly *sk̂in* if ALB ties in. Sihler 37. Walde 392, Pokorny 560. EM. Watkins *keni-*.
NAV *łeeshch'ih* /ł e: ʒ/ DERIVED FROM 'EARTH', COGNATE WITH 'DUST'. *Łeezh* by itself means 'dirt, soil, ashes' plus *ch'ih* 'blows'. YM 1992, p. 104 (CH'I), 393.
TUR *kül* /k y l/. Räsänen 307b.

at link, Swadesh 200

ALB *në* /n ə/ COGNATE WITH 'IN'. 'In, on, at'; source perhaps PIE *endo*. Mann 207. Huld 97 (citing Meyer, Tagliavini, Jokl, Rosetti).
ENG *at* /a t/. PIE H_2ad 'to, at'. Skeat. Walde 45. Pokorny 3. OED. Sihler 441. Watkins *ad-*.
FRE *à* /a/. From LAT. BW.
GER *an* /a n/ COGNATE WITH 'NEAR'. PG *ana*, extension of *an*. Pokorny suggests a connection to *nahe* 'near'. Walde 58. Pokorny 39. Kluge. Watkins *an-1*.
HAW *ma* /m a/. PCP *ma*. PE.
LAT *ad* /a d/. PIE H_2ad. Pokorny 3. EM.
NAV *-di* /d i/. YM 1992, p. 939.
TUR *-de* /d e/

back noun, Swadesh 200

ALB *shpinë* /ʃ p i/ LOAN. Loan from Lat. *spina* 'thorn, backbone', from PIE *spei-nā-*, root *spei*. Not in Mann, Huld. Çabej 131. Janson 113. EM *spina*. Pokorny 981. Watkins *spei-*.
ENG *back* /b a k/. PG *bakom*. Skeat. OED. Not in Pokorny.
FRE *dos* /d o/. Pop. Lat. *dossum*, Class. *dorsum*. I assume *-so-* is a suffix. BW. Not in Pokorny. EM

184. Watkins [dorsum].
GER *Rücken* /r/. From PG *hrugjaz*, PIE *kreuk-*, extension of *(s)ker*, 'turn'. Kluge. Pokorny 938. Watkins *sker*.
HAW *kua* /k u a/. PPN *tuʻa*. POc word is *muri*. PE.
LAT *tergum* /t e r/. There is no certain etymology beyond Latin, but one guess is a PIE *(s)terg* 'stretch', root *(s)ter* 'stiff'. Tucker further suggests a connection with *terra*, but that's very dubious. Walde 629*. Pokorny 1023.
NAV *'anághah* /n á/. Dependent stem noun. Probably *ná* 'back' + *ghah*. YM 1992, p. 988.
TUR *arka* /a r k a/. Räsänen 26a.

bad adj, Swadesh 200

ALB *keq* /k e c/ LOAN, MOTIVATED. Generally believed to be a loan from Greek *kakos*, although this isn't clear; it could be native. Either way, it sounds like a common nursery word, /kaka/. Mann 112. Huld 79. Walde 336.
ENG *bad* /b a d/. OED, Skeat.
FRE *mauvais* /m ɔ/. Pop. Lat. *malifātius* 'ill-fated'. The first element is descendant of LAT *malus*. BW.
GER *schlecht* /ʃ l e/ COGNATE WITH 'LIVER'. PG *slihta-* 'level, smooth' (Eng. *slight*), base *slik-*. Generally referred to *(s)lei-* 'slip, slimy'. *Leip* 'fat', basis of *Leber*, *liver*, may be a *-p-* extension. Walde 391*. Pokorny 504. Kluge.
HAW *'ino* /ʔ i n o/. PNP *kino*. PE.
LAT *malus* /m a l/. FRE *mau(vais)*. PIE root *mel* 'bad', conceivably *(s)mēlo-* 'small animal'. Walde 291*, 296*. Pokorny 724. EM. Watkins *mel-5*, *mēlo-*.
NAV *doo yá'áshǫ́ǫ da* /ʒ o:/. 'It is not good', neuter imperf.; root is 'good'. PA *zhu-ny*, PPA *shu*. Haile. YM 1992, p. 795–6, *ZHQQD*.
TUR *kötü* /k ø t y/ LOAN. Not in Räsänen. Dankoff, p. 163: may be a loan from Arm. /godi/.

bark noun (of tree), Swadesh 100

ALB *shkëlbozë* /ʃ k ə l b o z/. Kiçi. Not in Drizari, Kacori.

ENG *bark* /b ɑ r k/ LOAN. From ON *bark-*, PG *barkuz*. OED. Skeat.

FRE *écorce* /e k ɔ r/ COGNATE WITH 'SHORT'. LAT *scortea*, from *scortum* 'hide', PIE *skere-t-*, from *(s)ker-* 'cut'. Dauzat. BW. Pokorny 941.

GER *Rinde* /r i n d/. PWG *rendōn*. Perhaps a PIE root *rendh* 'tear'. Pokorny 865. OED *rind*.

HAW *'ili* /ʔ i l i/ COGNATE WITH 'SKIN'. Also 'skin'. PPN *kili*. PE.

LAT *cortex* /k o r/ COGNATE WITH 'MEAT'. PIE *kr̥t* or *kort*, extension of PIE root *(s)ker* 'cut'. Walde 578*. EM. Pokorny 941. Watkins *sker-1*.

NAV *'akásht'óózh* /k á: ʔ/ DERIVED FROM 'SKIN'. *-káá'* 'surface' + *t'óózh* 'cover, bark'. YM 1992, p. 981, 310 *-KÁÁ'*.

TUR *kabuk* /k a b/. From *Kāp* 'cover'. Räsänen 234b.

because link, Swadesh 200

ALB *sepse* /p/. I.e., *se për se. Për* 'for, by, because of', PIE *per*. Pokorny 810. Mann 122, 210. Huld 110. Walde 520.

ENG *because* /k ɔ z/ LOAN. *Be-* is native, *-cause* is from French (cf. Alb. 'animal'). Skeat.

FRE *parce que* /p a r/. *Par* from Lat. *per*, PIE *per*. BW. Pokorny 810. Watkins *per-1*.

GER *weil* /v ai/. OHG *dia wīla so*, PG *hwīlō*, PIE *kʷeiH* 'be quiet'. Kluge. Pokorny 638. Watkins *kʷeiə-2*.

HAW *ā mea* /m e a/. *Mea* (PPN *me'a*) means 'thing' or 'cause' and is the core of most explicit expressions for 'because'. PE.

LAT *quod* /kw/ COGNATE WITH 'WHAT', 'WHO'. Acc. neut. of *quī*, relative developed from the interrogative/indefinites, PIE *kʷ*. Sihler 399. Pokorny 644. EM. Watkins *kʷo-*.

NAV *háálá* /h a:/ COGNATE WITH 'WHAT', 'WHO'. Because *haalá* can mean 'what', I assume this is

derived from the interrogatives, as in Latin. Haile. YM 1992, p. 943, 934.

TUR *çünkü* /tʃ/ COGNATE WITH 'IN'. From *için* 'because of', from *iç* 'inside'. Räsänen 121b.

belly noun, Swadesh 100

ALB *bark* /b a r/ COGNATE WITH 'HOLD', 'HUSBAND', 'MAN'. Generally referred to PIE *bher* 'carry'. Huld 40. Walde 154*. Pokorny 129.

ENG *belly* /b ɛ l/ COGNATE WITH 'BLOOD', 'BLOW', 'FLOWER'. OE *belg*, PG *balgiz* 'bag', from PIE *bhelĝh-* 'swell', an extension of *bhel-* 'inflate', whence also perhaps *blow*. Pokorny 121, 126. OED. Skeat. Watkins *bhelgh-*.

FRE *ventre* /v ã t r/. From LAT. Dauzat.

GER *Bauch* /b au/ MOTIVATED. OHG *būh*, PG *būka-*, PIE *bhū-* 'inflate'. Onomatopoetic for sound of blowing. Pokorny 98. Kluge.

HAW *ōpū* /ʔ o: p u:/. PPN *koopuu*. POc word is *tian*. PE. Blust.

LAT *venter* /w e n/. Possibly *wen-tri* is a tabu deformation of *udero-*. EM. Walde 191. Pokorny 1105. Watkins *udero-*.

NAV *'abid* /b i d/ MOTIVATED. PA *-wi't'*, PPA *-wit'*. *Bid* is also the onomatpoeia for 'a hollow thumping sound, as that produced by patting a dog's stomach', so this may be motivated. YM 1992, p. 969, 65.

TUR *karın* /k a r ɯ n/. Räsänen 238a.

big adj, Swadesh 100

ALB *madh* /m a ð/. PAlb *mað*, PIE *maĝyo-*, root *maĝ-*. Mann 111. Huld 88. Walde 257. Pokorny 708.

ENG *big* /b ɪ g/. Sometimes referred to ON, but this is just because it wasn't attested in pre-Viking texts. An idea that it is related to Ger. *Bauch* 'belly' (i.e., 'inflated') is not widely accepted. Skeat. OED. Pokorny 100.

FRE *grand* /g r ã d/. Lat. *grandis*, PIE

gʷrendh- 'swell'. BW. Pokorny 485.
GER *groß* /g r/. OHG *grōz*, PWG
grautaz 'coarse', PG *greuta-*
'grate', PIE root *ghrēu-* 'grate',
extension of *gher* 'scrape'. Kluge.
Pokorny 460. Watkins *ghrēu.*
HAW *nui* /n u i/ COGNATE WITH 'MANY'.
PNP *nui.* PE.
LAT *magnus* /m a g/. PIE *maĝnos*, root
maĝ-. Tucker perceives a pre-PIE
etymon *ma-* 'big' here and in many
other words such as *mare* 'sea',
maritus 'husband', but that seems
fanciful. Pokorny 708. EM. Sihler
99. Walde 258*. Watkins *meg-.*
NAV *'áníłtso* /ts o h/ COGNATE WITH
'ALL'. Neuter imperfect; optionally
also with final /h/. Adj. is from PA
kyuχ, PPA *kaxʷ.* YM 1992, p. 938,
611, TSOH.
TUR *büyük* /b y j y k/.

bird noun, Swadesh 100
ALB *zog* /z o g/ LOAN. PAlb *zog.* Many
guesses as to IE etymology
(Pokorny lists *ĝhāgʷh-*), but may
well be a loan. Mann 35. Huld 135.
Walde 531. Pokorny 409.
ENG *bird* /b ə r d/. Skeat. Pokorny 132.
OED.
FRE *oiseau* /w a/ COGNATE WITH
'EGG'. Lat. *aucellus* from
avicellus, diminutive of LAT *avis*,
PIE *H₂awis.* Dauzat. BW.
GER *Vogel* /f o:/ COGNATE WITH
'FLOW', 'FLY', 'FULL', 'MANY',
'RIVER', 'WING'. OHG *fogal*, PG
fugla-. This is often referred to *flug-*
'fly, feather', with a misbehaving
/l/ (PIE *pleuk*, from *pleu* 'flow',
itself perhaps from *pelH* 'fill'); but
could also be a PIE *pu* 'young,
small'. Pokorny 835, 843. Kluge.
OED *fowl.* Watkins *pleu-, pau-.*
HAW *manu* /m a n u/. PPN *manu*,
PMP *manuk.* PE. Blust.
LAT *avis* /a w/ COGNATE WITH 'EGG'.
PIE *H₂awi-.* Sihler 44. Pokorny 86.
Walde 32. EM. Watkins *awi-.*
NAV *tsídii* /ts í d/ MOTIVATED. Refers to
smaller birds. *Tsíd* imitates a
bird's chirping. YM 1987, p. 4; YM
1992, p. 1000.

TUR *kuş* /k u ʃ/. Räsänen 305a.

bite verb, Swadesh 100
ALB *kafshon* /k a f ʃ/. Drizari: *kafshój*
'I bite'. Not in Walde.
ENG *bites* /b ai/. OE, PG *bītan*, PIE
bheid- 'split', perhaps from *bheiH-*
'hit'. Skeat. OED. Kluge *beißen.*
Pokorny 116. Watkins *bheid-.*
FRE *mord* /m ɔ r/. Pop. Lat. *morde-,*
Class. LAT *mordē-.* BW.
GER *beißt* /b ai/. OHG *bīz(z)an*, PG
bītan, PIE *bheid-* 'split', root
perhaps *bheiH-* 'hit'. OED. Kluge
beißen. Pokorny 116.
HAW *nahu* /n a h u/. PE.
LAT *mordet* /m o r/ COGNATE WITH
'DIE'. Iterative suffix PIE *-éye/o*
on PIE stem *merd-*, possibly
extension to root *merH* 'rub raw'.
Pokorny 737. Sihler 504. Walde
279*. EM. Watkins *mer-2.*
NAV *'aháshháásh* /ɣ á: ʒ/. 'I bite
something in two', momentaneous
imperf. PA *ghàtsh'*, PPA *χa'tsh'.*
YM 1992, p. 227, *GHAZH.*
TUR *ısırıyor* /ɯ s ɯ r/. Räsänen 162a.

black adj, Swadesh 100
ALB *zi* /z e z/. Fem. *zezë.* Pre-PAlb
stem *zıd-.* PIE root approximately
gʷidh- . Mann 110. Huld 134.
Walde 696. Pokorny 485.
ENG *black* /b l/. OE *blæc*, PG *blakko-*,
presumably from *blak-no-* 'burnt',
PIE stem *bhleg-* 'burn, shine', from
bhel- 'shine'. Pokorny 124. Walde
215*. OED. Skeat. Watkins *bhel-1.*
FRE *noir* /n w a/. Lat. *niger.* Watkins
sees a root *negʷ.* Dauzat. BW. EM
niger. Not in Pokorny. Watkins
negʷ-ro-.
GER *schwarz* /ʃ v a r ts/. OHG *swarz*,
PWG *swartaz*, PIE *swordos.*
Watkins *swordo-.*
HAW *'ele'ele* /ʔ e l e/. PPN *kelekele.*
Reduplicative. PE.
LAT *ater* /a: t r/. Mostly belongs to the
written language (but Fre. *airelle*),
contrast *niger.* Perhaps PIE *ātr*
'fire' (i.e., 'burnt black'). Walde 42.
Pokorny 69. EM. Watkins *āter-.*
NAV *łizhin* /ʒ i n/. 'It is black', neuter

stem *zhin*, from PA *zhiny* from
PPA *shiny*. YM 1992, p. 782 *ZHį́į́'*.
TUR *kara* /k a r a/. Räsänen 235a.

blood noun, Swadesh 100
ALB *gjak* /ɟ a k/. PIE *sokʷos* 'juice'.
Mann 37. Huld 67. Walde 515*,
468*. Pokorny 1044.
ENG *blood* /b l/ COGNATE WITH 'BELLY',
'BLOW', 'FLOWER'. OE *blōd*, PG
blōdam. Best guess is PIE *bhel-*
'swell, flow'. Skeat. OED. Kluge
Blut. Pokorny 122.
FRE *sang* /s ã g/. LAT *sanguen*. For
/g/ cf. *sanglant*. BW.
GER *Blut* /b l/ COGNATE WITH 'BLOW',
'FLOWER', 'LEAF'. OHG *bluot*, PG
blōdam; see ENG.
HAW *koko* /k o/. PPN *toto*. /ko/ itself
isn't a legal word, but I'll treat as
reduplicative anyway, to be safe.
PE.
LAT *sanguis* /s a n gw/. EM. Sihler
291. Walde 162.
NAV *dił* /d i ɬ/. YM 1987, p. 3:
independent stem noun. YM 1992,
p. 975, 136.
TUR *kan* /k a n/. Räsänen 230a.

blow verb, Swadesh 200
ALB *fryn* /f r y/ COGNATE WITH
'BREATHE'. Same morpheme as
'breathe'. Çabej suggested
onomatopoetic. Mann 38. Huld 65.
Not in Pokorny.
ENG *blows* /b l/ COGNATE WITH
'BELLY', 'BLOOD', 'FLOWER'. OE
blāwan, PG *blǣwan*, PIE *bhleH₁*,
prob. from root *bhel* 'blow up,
swell'. Pokorny 121. OED. Kluge
blasen, blähen, Blatt. Walde 177*.
Skeat. Watkins *bhlē-2*.
FRE *vente* /v ã/ DERIVED FROM 'WIND'.
12th c. derivation from *vent* 'wind'.
BW.
GER *bläst* /b l/ COGNATE WITH
'BLOOD', 'FLOWER', 'LEAF'. OHG
blāsan, PG *blǣsan*, PIE *bhleH*,
from *bhel* 'blow up, swell'. Pokorny
121. OED. Kluge *blasen, blähen,
Blatt*. Walde 177*. Skeat.
HAW *puhi* /p u h i/ MOTIVATED. PPN
pusi, POc *pusi*. This sounds like

onomatopoeia. PE. Ross.
LAT *flat* /f l/ COGNATE WITH 'FLOW',
'FLOWER', 'LEAF', 'RIVER'. PIE
bhlā-, variant of *bhleH₂-*, probably
extension of root *bhel-*. Perhaps
there is a connection with the
homonymous root 'bloom'.
Onomatopoeia? Tucker. Sihler 529.
Walde 179*. EM. Pokorny 121.
Watkins *bhlē-*.
NAV *nich'i* /tʃ i/ COGNATE WITH
'WIND'. 3d person momentaneous
perfective: '(a breeze) is blowing'
(lit., 'has arrived'). PA
tshʷɨy/tshʷi from PPA *kʷɨy*. YM
1992, p. 104, *CH'I*.
TUR *esiyor* /e s/. Räsänen 49b.

bone noun, Swadesh 100
ALB *kockë* /k o ts k/. Ringe 1993:
/kockə/. Drizari: *koskë, eshtra*.
ENG *bone* /b o n/. OE *bān*, PG *bainam*.
Skeat. OED. Kluge *Bein*. EM *os*.
FRE *os* /ɔ s/. LAT *ossum*, a less
standard form than *os*. BW.
GER *Knochen* /k n o x/ MOTIVATED,
COGNATE WITH 'KNEE'. MHG
knoche. No clear etymology;
guesses have been 'knee' (qv) and a
PIE *gen-* 'press together'
(Pokorny). Perhaps (Kluge)
onomatopoeia for cracking one's
joints (cf. Ger. *knacken* 'crack',
Eng. *knock*). Walde 582. Kluge.
Pokorny 370.
HAW *iwi* /i w i/. PPN *iwi*.
LAT *os* /o s s/. PIE *ost-*. Tucker. Sihler
319. Walde 186. Pokorny 783.
Watkins *ost-*.
NAV *ts'in* /ts' i n/ COGNATE WITH 'HIT'.
PA *ts'in*. Independent stem noun.
YM 1992, p. 1003, 637.
TUR *kemik* /k e m i k/. Räsänen 251a.

breast noun (woman's), Swadesh
100
ALB *gji* /ɟ i r/. Mann compares to Lat.
sinus. Not in Walde. Mann 92.
ENG *breast* /b r/ COGNATE WITH
'BREATHE', 'BURN'. OE *brēost*, PG
breustam, brusts, PIE *bhreus*
'swell', perhaps ext. of *bhreu* 'boil'.
Some connect to *breathe, burn*,

qq.v. Pokorny 171. Skeat. Kluge *Brust*. Watkins *breus-1*.

FRE *sein* /s ɛ̃/. Lat. *sinus*, orig. a fold in the garment where mothers carried their babies. No further etymology, although Mann compares to ALB. Dauzat. BW. EM *sinus*.

GER *Brust* /b r/ COGNATE WITH 'BURN'. OHG *brust*, PG *brusts*, see ENG.

HAW *ū* /u/ MOTIVATED. PPN *huhu*, POc, PMP *susu*. Perhaps a nursery word. PE. Blust.

LAT *mamma* /m a/ MOTIVATED. Earlier generations attempted to derive from some such root as *mad* + *-ma*, but it seems clear now that this is purely nursery talk, also means 'mother', and is same stem as in *mater*, hypocoristic gemination. Tucker. Sihler 224. Walde 221*, 232*. EM. Pokorny 694.

NAV *'abe'* /b e ʔ/ MOTIVATED. YM 1987, p. 4: *-be'* 'breast, milk', dependent stem noun. YM 1992, p. 969, 63.

TUR *meme* /m e/ MOTIVATED. 'Teat, nipple, udder, breast'. This seems a nursery word. Räsänen 333b.

breathe verb, Swadesh 200

ALB *marr frymë* /f r y/ DERIVED FROM 'BLOW'. 'He takes breath.' Distinctive part is *frymë* 'breath'. Some have attempted to connect to LAT.

ENG *breathes* /b r i/ COGNATE WITH 'BREAST', 'BURN'. OE *bræþ* 'odour', PG *bræþaz*. Pokorny has a PIE *bher-*, which has extension *bhreus* 'swell', as in *breast*. Kluge, Watkins prefer PIE *gʷhreH* 'smell', extension of *gʷher* 'heat'? Kluge *Brodem*. OE. Pokorny 132. Skeat. Watkins *gʷhrē-*.

FRE *respire* /s p i r/ LOAN. Loan from LAT *respiro*. BW.

GER *atmet* /a: t/. Noun *Atem*, from OHG *ātum*, PWG *æð(u)ma-*, PIE *ētmó-* 'breath', root *ēt*. Kluge. Pokorny 345. Watkins *ētmen-*.

HAW *hanu* /h a n u/. PPN *fangu*. PE.

LAT *spirat* /s p i: r/. Onomatopoeia? Pokorny sees a PIE *(s)peis* 'blow';

possibly connected to ALB. EM. Walde 11*. Pokorny 796. Watkins *[spīrāre]*.

NAV *ńdísdzih* /j i: h/. 'I breathe'. Repetitive durative, stem *dzih*. Root (perfective) *dzíí'*, probably from obsolescent *yih*, PA *ghyiky*, PPA *xik'*. YM 1992, p. 172, 702. Pinnow 23.

TUR *nefes alıyor* /n f s/ LOAN. 'To take (*almak*) breath (*nefes*)', the latter Arabic /nafas/, root /nfs/ 'precious'. Stachowski, v. 3, p. 18. Wehr 1155.

burn verb (intransitive), Swadesh 100

ALB *digjet* /d i ɟ/. *Digjet* is reflexive; base transitive is *djeg* (both numbers). PA *djeg-*, earlier *deg-*, PIE *dhegʷh-* 'burn'. Drizari. Kacori 256. Huld 53. Mann 154. Pokorny 240, 7.

ENG *burns* /b ə r/ COGNATE WITH 'BREAST', 'BREATHE'. OE *brinnan*, PG *brinnan*. Probably nasal extension of *gʷher* 'heat' (Kluge). But possibly tied in with PIE *bher* 'boil'. Pokorny 144, 132, 170. Walde 167*. Skeat. Kluge *brennen*. Watkins *gʷher*.

FRE *brûle* /b r/ LOAN, COGNATE WITH 'FOG'. Appears to be a crossing of OF *usler* (Lat. *ustulare*, from *uro*, PIE *eus*) with OF *bruir* (Frankish *brōjan*, close to Ger. *brühen* 'brew', MHG 'burn', PIE ultimately *bher* 'boil'). Dauzat. BW. Pokorny 348. EM *uro*.

GER *brennt* /b r/ COGNATE WITH 'BREAST'. OHG *brennan*, PG *brinnan*. See ENG.

HAW *'ā* /ʔ a:/. PNP *kaa*, from PPN *kakaha*. Replaces POc *tunu*. This sounds rather like a cry of pain; motivated? PE. Blust.

LAT *ardet* /a: r/. Extension of *āreo* 'be dry'. EM *areo*. Tucker. Pokorny 68. Lubotsky, p. 59. Watkins *as-*.

NAV *diltłi'* /tˡ á: d/. 'It burns', neuter. PA *tłaxd* from PPA *tłixd*. YM 1992, p. 561, *TŁAH*. Pinnow 19.

TUR *yanıyor* /j a/. Root is *ya*, cf. *jak*

'kindle'. Räsänen 184b.

child noun, Swadesh 200

ALB *fëmiljë* /f ə/ LOAN. Loan from Lat. *familia*, from *famulus* 'domestic servant', derived from a *fam-* not elsewhere attested, but perhaps from *dheH* 'put'. EM *famulus*. Huld 10. Janson 116. Pokorny 238.

ENG *child* /tʃ ai l/ COGNATE WITH 'CLAW', 'CLOUD'. OE *cild*, PG *kilþam*. Uncertain, one theory has PIE *gelt-* 'pregnant belly' from *gel* 'ball'. Pokorny 358. OED. Walde 614. Skeat.

FRE *enfant* /f ã/. From Lat. *infantem*, *in-* 'not' + *fa* 'speak' (PIE *bhaH*) + *nt* pres. part. BW. EM *infans*, for. Pokorny 105. Watkins *bhā-2*.

GER *Kind* /k i n/. OHG *kind*, PG *kinþa-*. Probably *to*-participle of PIE *ĝenH* 'be born'. Kluge. Pokorny 373. Walde 576. Watkins *genə*.

HAW *keiki* /k e/ COGNATE WITH 'HUSBAND', 'HUMAN', 'MAN'. PCP *taiti*, from *ta(ma)* 'child' + *ʔiti* 'small'. Is the *ta* at some point the same element as in *kāne* and *kanaka*, i.e., 'person'? PE. Marck.

LAT *puer* /p u/ COGNATE WITH 'FEW', 'SMALL'. PIE *pu-, pau-* 'small'. Tucker. Walde 76*. Pokorny 843. Watkins *pau-*.

NAV *'ashkii* /k iː/. Stem + prefix noun. YM 1992, p. 967.

TUR *çocuk* /tʃ o dʒ u k/. Räsänen 113a.

claw noun, Swadesh 100

ALB *thua* /θ o n/. 'Finger-, toe-nail, claw'. PAlb *θua, θoNi̦*, earlier *θonC-*. Approx. PIE *ḱeH₁n(T)s*. Huld 120. Mann 44.

ENG *claw* /k l/ COGNATE WITH 'CHILD', 'CLOUD'. OE *clawu* (influenced by the verb), PG *klǣwō*. Perhaps PIE *gel*. Skeat. Kluge *Klaue*. Walde 618. Pokorny 361. Watkins *gel-1e* .

FRE *griffe* /g r i f/ LOAN. Most direct interpretation is that it is a loan from post-OHG-cons-shift Frankish **grif*, PG *grip-*, PIE *ghreib-* 'grip'. BW. Watkins *ghreib-*.

GER *Klaue* /k l/. OHG *klāwa*, PG *klǣwō*, see ENG. Kluge *Klaue*. Walde 618. Pokorny 361.

HAW *miki'ao* /m i k i ʔ a o/. PCP *mitikao*. PE.

LAT *unguis* /u n gw/. PIE *H₃nogʷh-* 'nail, hoof'. Sihler 86, 97. Walde 180. Pokorny 780. Watkins *nogh-*.

NAV *'akéshgaan* /k éː ʔ/ DERIVED FROM 'FOOT', COGNATE WITH 'ROOT'. *-késhgaan* 'toenail, talon, hoof, claw' from *ká*, combining form of *kee'* 'foot' + *-gaan* (root *GAN*) 'be dry'. YM 1992, p. 982, 319.

TUR *tırnak* /t ɯ r/ COGNATE WITH 'SCRATCH'. 'Fingernail, toenail, claw, hoof'. Räsänen 479a.

cloud noun, Swadesh 100

ALB *re* /r e/. PAlb *ren*, pl. *ran*. Meyer equated with Ger. *Rauch* 'smoke' (PIE *reug*), Huld with Lat. *ros* 'dew'. Huld 107. Walde 357*. Pokorny 872.

ENG *cloud* /k l/ COGNATE WITH 'CHILD', 'CLAW'. OE *clūd* 'rock', PG *klūdaz*, PIE *glūto-*, root *gel* 'ball'. OED. Pokorny 357. Skeat. Walde 618. Watkins *gel-*.

FRE *nuage* /n y/. Archaic *nue*, Pop. Lat. *nūba*, for Class. *nūbes*. Dauzat. BW.

GER *Wolke* /v o l k/. OHG *wolcan*, PG *wulkna-*, PIE *wl̥g*, root *welg*. Kluge *Wolke*. Pokorny 1145.

HAW *ao* /a o/. PPN *'ao*, POc *qaRoq*. PE. Ross.

LAT *nubes* /n u: b/ COGNATE WITH 'SWIM', 'FOG'. PIE *sneudh* 'fog'; some have connected to *snaH* 'flow' (Lat. *no* 'swim'). Pokorny 978. Walde 697*. EM. Watkins *sneudh-*.

NAV *k'os* /k' o s/. Independent stem noun. YM 1987, p. 3. YM 1992, p. 984, 355. Hoijer 10.

TUR *bulut* /b u l/. From *bulǧa* 'to cloud, darken'. Räsänen 88b, 88a.

cold adj, Swadesh 100

ALB *ftohët* /t o h/. PAlb *ftohëtë*, adj. form of *ftoh* 'I chill'. *F-* is privative, root is PIE *tep* of Lat. *tepeo*. Mann 160. Pokorny 1070. Huld 65. Walde

719. Watkins *tep-*.

ENG *cold* /k o l/. Angl. OE *cald*, PG
kaldaz, *-to-* part. to PIE root *gel*.
OED. Walde 622. Pokorny 365.
Skeat. Watkins *gel-3*.

FRE *froid* /f r w a/. Pop. Lat. *frĭgidus*,
Class. LAT *frigidus*. BW.

GER *kalt* /k a l/. OHG *kalt*, PG *kaldaz*,
see ENG. Kluge *kalt*.

HAW *anu* /a n u/. PCP *anu*.

LAT *frigidus* /f r i: g/. *Frīgeo* 'be cold',
PIE stative stem *srigeH₁*, root *srīg*.
Pokorny 1004. EM *frigus*. Sihler
497, 214. Walde 705*. Watkins
srig-.

NAV *sik'az* /k' a z/. 'It is cold', neuter
perf. PA *K'áts'*, PPA *K'ats'*. YM
1992, p. 338, *K'AAZ*.

TUR *soğuk* /s o ɣ u/. From *soğu*,
'become cold'. Räsänen 425a.

come verb, Swadesh 100

ALB *vjen* /v j e n/ LOAN. Loan from
Lat. *venit*. Mann 163, Huld 129.

ENG *comes* /k ə m/. OE *cuman*, PG
kuman, PIE root *gʷem*. Skeat.
Walde 675. OED. Pokorny 463.
Watkins *gʷā-*.

FRE *vient* /v j ɛ̃/. LAT. BW.

GER *kommt* /k o m/. OHG *kuman*, PG
kuman, see ENG. Kluge *kommen*.

HAW *hele mai* /h e l e/ COGNATE WITH
'GO'. Same as 'go', but with
'hither' particle (POc *ma(R)i*,
PMP *um-aRi* 'come') added. PCP
sele, PPN *sa'ele*. PE. Blust.

LAT *venit* /w e n/. PIE *gʷm̥-ye/o-*, see
ENG. Sihler 96, 503. EM *venio*.

NAV *yíghááh* /y á/ COGNATE WITH 'GO'.
'He comes, goes'. Momentaneous
imperfect stem. PA *ya*, PPA *haw*.
The *gh* is an old classifier. YM
1992, p. 668, *YÁ*. Pinnow 15.

TUR *geliyor* /g e l/. Räsänen 248b.

count verb, Swadesh 200

ALB *numëron* /n u m/ LOAN. Not in
Mann, Huld, Janson. I assume
from Lat. *numerare*, viz., noun
numër. Kiçi.

ENG *counts* // LOAN. OF *conter*, see
FRE. Skeat. OED.

FRE *compte* //. Spelling variant of

conte. Lat. *computo*, prefix *com-*
'with, together' + Lat. root *put*
'think', orig. 'prune', PIE *pu-to-*,
participle of *pēu* 'strike'. Note no
part of the original root *pu-*
survives. OED. BW. Pokorny 827.
EM *puto*. Watkins *peu-*.

GER *zählt* /ts a: l/. OHG *zellen*, PG
talja, PIE *del* '(re)count'. Orig.
vowel visible in *Zahl* 'number'.
Pokorny 193. Kluge *zählen*.
Watkins *del-2*.

HAW *helu* /h e l u/. PE.

LAT *numerat* /n u m/. *Numerus*
'number', PIE *nomeso-*, root *nem*
'assign, take'. Sihler 43, 70. Walde
331. Pokorny 763. EM *numerus*.
Watkins *nem-*.

NAV *'íínishta'* /t a ʔ/. 'I count'.
Repetitive durative. PA *táK'*,
perhaps from PPA *taK'*. YM 1992,
p. 472, *TA'*.

TUR *sayıyor* /s a j/. Räsänen 390a.

cut verb, Swadesh 200

ALB *pres* /p r/. PIE *per* 'hit'. Huld 105.
Pokorny 819. Walde 42*. Watkins
per-5.

ENG *cuts* /k ə t/ LOAN. Prob. ME
borrowing from Scand., cf. dial.
Swed. *kuta*, *kata* 'cut'. Skeat. OED.
Barnhart.

FRE *coupe* /k u/ LOAN, COGNATE WITH
'MANY'. Noun *coup*, Pop. Lat.
colpus, *colaphus* 'punch', Greek
kólaphos 'a blow', PIE *kolH₁bho-*,
root *kelH* 'strike'. Dauzat. BW.
Pokorny 545. Watkins *kel-1*.

GER *schneidet* /ʃ n ai d/. OHG *snīdan*,
PG *sneiþa*, PIE root *sneit* 'cut'.
Pokorny 974. Kluge *schneiden*.
Walde 695*. Watkins *sneit-*.

HAW *'oki* /ʔ o k/. Possibly a 'transitive'
-i. PPN *koti*. PE.

LAT *secat* /s e k/ COGNATE WITH
'KNOW'. PIE *skei-* 'cut', root *sek*
'cut'. EM *seco*. Pokorny 895. Walde
474. BW *couper*. Watkins *skei-*.

NAV *'aháshgéésh* /g i ʒ/ COGNATE WITH
'SPLIT'. *I cut (sth) apart*.
Momentaneous imperfect. PA
Gidzh. YM 1992, p. 210, *GIZH*.

TUR *kesiyor* /k e s/ COGNATE WITH

'SHARP'. Räsänen 257b.

day noun, Swadesh 200

ALB *ditë* /d i/. Possibly PIE *dei* 'shine', but more likely a PIE *diHti* 'period of time', root *diH* 'divide'. Pokorny 175. Mann 77. Huld 52. Janson 182. Walde 774*. Watkins *deiw-, dā-*.

ENG *day* /d e/. OE *dæg*, PG *dagaz*. Not connected to Lat. *dies*. PIE *dhogʷh* 'burn', root *dhegʷh-*, crossed with PIE *aĝher* 'day', root *aĝh*. Skeat. Walde 849. OED. Pokorny 7, 240. Watkins *dhegʷh-. agh-1.*

FRE *jour* /ʒ/. Earlier *jorn*, Pop. Lat. *diurnus*, adj. (infl. by *nocturnus*) of Class. LAT *dies*. BW. EM *dies*.

GER *Tag* /t aː g/. OHG *tag*, PG *dagaz*, see ENG. Kluge *Tag*.

HAW *lā* /l aː/ COGNATE WITH 'SUN'. 'Sun, day'. PPN *la'aa*. PE.

LAT *dies* /d i/. PIE *dyews* (crossed with *divus*, etc.), root *dyeu* 'shine bright'. EM *dies*. Pokorny 184. Sihler 338. Walde 773. Watkins *deiw-*.

NAV *jį́* /dʒ í/ COGNATE WITH 'MOON'. Independent stem noun. PA *dzhʷen*. YM 1992, p. 981, 274.

TUR *gün* /g y n/ COGNATE WITH 'SUN'. Räsänen 309a.

die verb, Swadesh 100

ALB *vdes* /d e s/. Huld relates to *ndjek* 'chase', and derives from a $H_2aw(o)-tokʷ-ey-$, root *tekʷ* 'run'. Huld 124. Mann 165. Kacori 247.

ENG *dies* /d ai/ LOAN. ME *degen*; its absence in OE suggests it is a loan from ON *deyja*, PIE root *dheu*. Skeat. OED. Barnhart. Pokorny 260. Watkins *dheu-3*.

FRE *meurt* /m œ r/. Pop. Lat. *morire*, Class. LAT *mori*. Dauzat. BW.

GER *stirbt* /ʃ t e r/. OHG *sterban*, PG *sterban*, PIE root *(s)ter* 'stiff'. Kluge *sterben*. Pokorny 1022. Walde 632*.

HAW *make loa* /m a k e/. *Make* by itself also means 'die'; *loa* is intensive. PPN, POc *mate*, PMP *matay*. PE. Blust.

LAT *moritur* /m o r/ COGNATE WITH

'BITE'. PIE *mr̥ye* (durative thematic), root *mer* 'die'. Perhaps derived from *merH* 'be rubbed out'. Pokorny 735. EM *morior*. Sihler 503. Walde 276.

NAV *daaztsą́* /ts a/. 'He died'. PA *tsany*, PPA *tsa*. YM 1992, p. 594, *TSĄ́*.

TUR *ölüyor* /ø l/ COGNATE WITH 'KILL'. Räsänen 371a.

dig verb, Swadesh 200

ALB *gërmon* /g ə r m/. Kiçi & Aliko (1969). Not in Mann, Huld, Janson.

ENG *digs* /d ɪ g/ LOAN. Likely a loan ultimately from Dutch, perhaps via Fre. *digue* 'trench'. PIE root *dheiHgʷ* 'stick'. Skeat. OED. Barnhart. Pokorny 244. Watkins *dhīgʷ-*.

FRE *creuse* /k r ø z/ LOAN. Noun *creux*, *crosus, unknown, prob. Celtic. BW. Dauzat.

GER *gräbt* /g r aː b/. OHG *graban*, PG *graban*, PIE *ghrebh* 'dig, scratch', perhaps from *ghrebh* 'seize'. Pokorny 455. Kluge *graben*. Watkins *ghrebh-2*.

HAW *'eli* /ʔ e l/. PPN, PCEMP *keli*, PMP *kali*. PE. Blust.

LAT *fodit* /f o d/ COGNATE WITH 'STAB'. Trans. used as 'stab'. PIE *bhodh-ye* (for expected *bhedhye*), root *bhedh* 'stab'. Pokorny 113. EM *fodio*. Sihler 121, 536. Watkins *bhedh-*.

NAV *hahashgééd* /g eː d/ COGNATE WITH 'STAB'. 'I dig a hole'. Momentaneous imperfect stem. PA *Ged*, from PPA *Gʷe'd*. YM 1992, p. 202, *GEED*. Pinnow 21.

TUR *kazıyor* /k a z/. Räsänen 243a.

dirty adj, Swadesh 200

ALB *ndyrë* /d y/. Jokl derives *ndynj* 'I am dirty' from PIE *dhūgnyō*, with prefix *n(ə)-*. Janson 86. Not in Mann, Huld.

ENG *dirty* /d ə r/ LOAN. From *dirt*, earlier *drit*. Absence in OE suggests loan from ON *drit* 'excrement', verb *drītan*, PG *drītanan*, PIE *dhreid*, ext. of root *dher*. Barnhart. Pokorny 56.

FRE *sale* /s a l/ LOAN. Borrowed from

Frankish *salo*, PG *salwa*, PIE *sal*
'dirty grey'. Dauzat. BW. Watkins
sal-2.
GER *schmutzig* /ʃ m u/. Verb *smutzen*
'stain', PIE root *meu* 'moist'.
Kluge *Schmutz*. Pokorny *(1)meu*.
Walde 251*.
HAW *lepo* /l e p o/ COGNATE WITH
'EARTH'. 'Dirt, earth; dirty'. PEP
lepo. PE.
LAT *sordidus* /s o r d/. *Sordes* 'dirt',
PIE *sword*. EM *sordes*. Pokorny
1052. Walde 535*. Watkins
swordo-.
NAV *baa'ih* /ʔ i h/ MOTIVATED. *T'óó
baa'ih* 'it is dirty', neuter
imperfective 3rd person only; based
on interjection of revulsion. YM
1992, p. 241, *'IH*.
TUR *pis* /p i s/ LOAN. Loan from Persian
pīs. Räsänen 385b. Steingass *pys*.

dog noun, Swadesh 100
ALB *qen* /c e n/ LOAN. Loan from LAT.
Mann 98. Huld 107.
ENG *dog* /d ɑ g/. Found once in OE
(*docga*), but no further etymology
known. Apparently native. OED.
Barngart.
FRE *chien* /ʃ j ɛ n/. Fem. is /ʃjɛn/.
LAT. Dauzat.
GER *Hund* /h u n/. OHG *hunt*, PG
hundaz, PIE *k̂wn̥tos*, extension of
PIE *k̂won*. Kluge *Hund*. Pokorny
633. Walde 466. Watkins *kwon*.
HAW *'īlio* /ʔ i: l i o/. PPN *kulii*,
Tuamotu *kurio*. PE.
LAT *canis* /k a n/. PIE root *k̂won*,
vowel unclear (as if from *k̂an-i*);
perhaps crossed with *cano* 'sing'.
EM *canes*. Pokorny 633. Walde
351, 466. Watkins *kwon*.
NAV *łééchąą'í* /ɬ í: ʔ/. *łéé-* combining
form of *łįį́'* (PA *łiny* 'dog')
'domesticated animal' + *-chąą'*
'excrement' + suffix *-í*. YM 1992,
p. 986, 394.
TUR *köpek* /k ø p e k/. Räsänen 291b.

drink verb, Swadesh 100
ALB *pi* /p i/. PAlb *pi*, PIE zero-grade
piH₃, root *poH₃i*. Pokorny 840.
Mann 170. Huld 103. Walde 71*.

Watkins *pō(i)-*.
ENG *drinks* /d r ɪ k/. OED *drincan*, PG
drinkan. Possibly from PIE *dhreĝ*
'pull'. Barnhart. Walde 874. Skeat.
Pokorny 273, 1089. Kluge *trinken*.
Watkins *dhreg-*.
FRE *bois* /b/. LAT. Dauzat. BW.
GER *trinkt* /t r i k/. OHG *trincan*, PG
drinkan, see ENG.
HAW *inu* /i n u/. Passive stem is *inuh-*.
PPN *inu*, POc, PMP *inum*. PE.
EP, p. 84. Blust.
LAT *bibit* /b/. Assim. form of **pibo*,
redupl. PIE stem *pi-pH₃-*, root
poH₃(i). Sihler 496. Pokorny 840.
EM *bibo*. Walde 71*. Watkins
pō(i)-.
NAV *'adlą́* /d¹ ą̄:/. 'He is drinking (an
unspecified liquid)'. Durative
imperfect. PA *di-nany*. YM 1992,
p. 153, *DL.Ą́Ą́'*. Pinnow 27.
TUR *içiyor* /i tʃ/. Räsänen 168b.

dry adj, Swadesh 100
ALB *thatë* /θ a/. PAlb. *θatë*, adj. suffix
-të to *θa* (*thaj* 'I dry'), prob. PIE
sausni, root *saus* 'dry'. Mann 75.
Pokorny 881. Huld 117. Walde
447*.
ENG *dry* /d r/. OE *drȳge*, PG *drūgiz*.
Possibly PIE *dhreugh* 'tremble',
root *dher* 'hold'. Pokorny 255, 275.
Skeat. Walde 860. OED. Barnhart.
Kluge *trocken*. Watkins *[dreug-]*.
FRE *sec* /s ɛ ʃ/. LAT. BW. Dauzat.
GER *trocken* /t r/. OHG *truckan*, PG
drūgiz, see ENG. Kluge *trocken*.
HAW *malo'o* /l o ʔ o/. *Ma-* could be
stative prefix here. Cf PMP
ma-Raŋaw. PE. Blust.
LAT *siccus* /s i k:/. Perhaps a PIE root
seikʷ 'pour out'. Another school of
thought derives from *sitis*
'drought', as **sitikos*. Pokorny 893,
889. EM *siccus*. Sihler 200. Walde
506. Watkins *seikʷ*.
NAV *yíttseii* /ts e i:/. 'It is dry'. Neuter
perfective. Stem is *tsaii* or *tseii*,
possibly from PA *tsághy*. YM 1992,
p. 590, *TSAII*. Pinnow 36.
TUR *kuru* /k u r/. Cf. *kurak* 'dry'.
Räsänen 302a.

dull adj, Swadesh 200

ALB *pa mprehtë* /p r/ DERIVED FROM 'SHARP'. *Pa* 'without, not' from PIE *apo*; *mprehtë* 'sharp', qv. Huld 156. Mann. Pokorny 54.

ENG *dull* /d ə/ COGNATE WITH 'DUST'. Difficult vowel, perhaps a borrowing from another Low German language. PG root *dwel* 'be foolish', PIE *dhwel*, root *dheu* 'fly about'. OED. Skeat. Pokorny 261. Watkins *dheu-1*.

FRE *émoussé* /m u s/. Noun *mousse*, Pop. Lat. *muttius, root *mutt*. Dauzat. BW.

GER *stumpf* /ʃ t u pf/. OHG *stumpf*, PIE root *step*. Pokorny 1012. Kluge *stumpf*.

HAW *kūmūmū* /m u:/. PE. *Kū-* may be a stative prefix.

LAT *hebes* /h e b/. EM *hebeo*. Walde 349.

NAV *doo deení da* /n í/ DERIVED FROM 'SHARP'. Don't know a word for 'dull', so I constructed this phrase, which is supposed to mean 'it is not sharp'. PA *yen*. YM 1992, p. 431.

TUR *kör* /k ø r/ LOAN. 'Blind, dim, blunt'; loan from Persian *kūr* 'blind'. Räsänen 292a. Steingass *kvr*.

dust noun, Swadesh 200

ALB *pluhur* /p l u h/. Mann derives from PIE *pleus* 'pluck', as Lat. *plūma* (see Fre. 'feather'). No better guesses. Mann 50. Pokorny 838. Not in Janson, Huld.

ENG *dust* /d ə/ COGNATE WITH 'DULL'. OE *dūst*, PG *dunstaz*, PIE *dhwens*, from *dhwes*, root *dheu* 'fly about'. Skeat. OED. Barnhart. Pokorny 268. Watkins *dheu-1*.

FRE *poussière* /p u/ COGNATE WITH 'PUSH'. Extension of *pous*, from Pop. Lat. *pulvus*, Class. LAT *pulvis*, PIE root *pel-* 'dust, flour'. May be same *pel* as in *pousser* 'push': PIE *pel*. Dauzat. BW. Pokorny 801. Watkins *pel-1*.

GER *Staub* /ʃ t au b/. OHG *stoub*. Kluge *Staub*.

HAW *'ehu* /e h u/. Absence of /ʔ/ in derivatives shows that /ehu/ is the more basic form. PPN *efu*. POc, PMP is *qapuk* PE. Blust.

LAT *pulvis* /p u l/. *-w-* ext. of *pel* 'dust'. Pokorny 802. EM *puluis*. Walde 60*. Watkins *pel-1*.

NAV *ɬeezh* /ɬ e: ʒ/ COGNATE WITH 'ASHES', 'EARTH'. Same word as 'earth', q.v. So glossed at YM 1992, p. 986, #1113.

TUR *toz* /t o/ COGNATE WITH 'EARTH'. Same root as *toprak* 'earth'. Räsänen 492a.

ear noun, Swadesh 100

ALB *vesh* /v e ʃ/. PAlb *veš*, earlier *as*, PIE *oH₃us* 'hear'. Mann 28, 81. Huld 127. Walde 18. Watkins *ous-*.

ENG *ear* /ɪ r/ COGNATE WITH 'EYE'. OED *ēare*, PG *auzan*, PIE root *oH₃us*. Barnhart. Pokorny 785. OED. Skeat. Watkins *ous-*.

FRE *oreille* /ɔ r/. LAT. *auricula*, from *auris*. Dauzat. BW.

GER *Ohr* /o: r/ COGNATE WITH 'EYE'. OHG *ōra*, PG *auzan*, see ENG. Kluge *Ohr*.

HAW *pepeiao* /p e p e i a o/. Replaced POc *taliŋa*. Blust.

LAT *auris* /au s/ COGNATE WITH 'HEAR'. Earlier dual *ausi, root *ōus*. Possibly related to *audio* 'hear'. Pokorny 785. EM *auris*. Walde 18. Watkins *ous-*.

NAV *'ajaa'* /dʒ a: ʔ/. Dependent stem noun. YM 1987, p. 4. YM 1992, p. 980, #801, p. 255.

TUR *kulak* /k u l a k/. Räsänen 298b.

earth noun, Swadesh 100

ALB *dhe* /ð e/. PAlb *ðee*, earlier *ðeCo*. Usu. referred to PIE *deĝhom* 'earth', but Huld prefers *dheiĝh* 'daub with mud'. Mann 84. Huld 57. Walde 662. Pokorny.

ENG *earth* /ə r/. OE *eorþe*, PG *erþō*, PIE root *er* 'earth'. Pokorny 332. Skeat. OED. Barnhart. Watkins *er-2*.

FRE *terre* /t ɛ r/. LAT. Dauzat. BW.

GER *Erde* /e: r/. OHG *erda*, PG *erþō*, PIE root *er* 'earth'. Pokorny 332. Kluge *Erde*. OED. Barnhart.

Watkins *er-2*.

HAW *lepo* /l e p o/ COGNATE WITH
'DIRTY'. PE. PEP *lepo*.

LAT *terra* /t e r:/. From **tersā*, PIE
root *ters* 'dry'. EM *terra*. Pokorny
1078. Sihler 504. Walde 738.
Watkins *ters-*.

NAV *łeezh* /ɬ e: ʒ/ COGNATE WITH
'ASHES', 'DUST'. Independent stem
noun. Same as 'dust'. YM 1987,
p. 3. YM 1992, p. 393, 986.

TUR *toprak* /t o/ COGNATE WITH
'DUST'. Same root as *toz* 'dust'.
Räsänen 492a.

eat verb, Swadesh 100

ALB *ha* /h a/. PAlb *haa*, earlier *haC*.
There are many guesses (Skt.
khādati, ghas-, Lat. *avere*), but
probably not *ed-*. Kacori 242.
Mann 169. Huld 72. Walde 119.

ENG *eats* /i t/ COGNATE WITH 'TOOTH'.
OE *etan*, PG *etan*, PIE H_1ed.
Skeat. Pokorny 287. OED. Watkins
ed-.

FRE *mange* /m ã/. Lat. *manducare*,
from *mandere* 'chew', PIE root
mendh. Dauzat. BW. Pokorny 732.
Watkins *mendh-1*.

GER *ißt* /e s/ COGNATE WITH 'TOOTH'.
OHG *ezzan*, PG *etan*, PIE H_1ed.
Kluge *essen*. Pokorny 287. Watkins
ed-.

HAW *'ai* /ʔ a/. Possibly a transitive *-i*.
PPN *kai*, POc *kani*, PCEMP *kan*,
PMP *kaen*. Blust.

LAT *edit* /e d/ COGNATE WITH 'TOOTH'.
PIE H_1ed 'eat'. EM *edo*. Pokorny
287. Sihler 148. Walde 118.
Watkins *ed-*.

NAV *'ayą́* /j ã:/. 'He is eating
(unspecified obj.)'. PA *hany, yany*,
PPA *hany*. YM 1992, p. 691, YÁ̧Á̧.
Hoijer 31. Pinnow 29.

TUR *yiyor* /j e/. Räsänen 194a.

egg noun, Swadesh 100

ALB *vezë* /v e/. Generally *ve*. PAlb.
voe, vee, from *veo*, from *o(y)om*,
PIE perhaps H_3oH_2wyom. Details
are unclear, but probably same ult.
root as Lat., Eng., Ger. Huld 125.
Mann 29.

ENG *egg* /ɛ g/ LOAN. Borrowed from ON
egg, PG *ajjam*, PIE oH_2wyom, see
LAT. Barnhart. Skeat. OED.
Watkins *awi-*.

FRE *œuf* /œ f/ COGNATE WITH 'BIRD'.
Pop. Lat. *ŏvum*, Class. LAT *ōvom*.
Dauzat. BW.

GER *Ei* /ai/. OHG *ei*, PG *ajjam*, see
ENG.

HAW *hua* /h u a/ COGNATE WITH
'FRUIT'. Also means 'fruit'. PPN
fua. Replaces POc *qatoluR*. Blust.
PE.

LAT *ovom* /o: w/ COGNATE WITH
'BIRD'. PIE oH_2wyom, based on
H_2awi- 'bird'. Pokorny 784. EM
auis. Sihler 44. Walde 21.

NAV *'ayęęzhii* /j ẽ: ʒ/. Dependent stem
noun. YM 1992, p. 1005, #2206;
p. 698.

TUR *yumurta* /j u m u r t a/. Räsänen
211a.

eye noun, Swadesh 100

ALB *sy* /s/. Not clear, but most try
some sort of derivation from PIE
$H_3okʷ$. Mann 97. Huld 11. Walde
170. Pokorny 776.

ENG *eye* /ai/ COGNATE WITH 'EAR'. OE
ēage, PG *augōn*, from PIE $H_3okʷ$
'see', possibly influenced by
Germanic *auzan* 'ear'. Pokorny
776. OED. Watkins *okʷ-*.

FRE *œil* /œ/. LAT. Dauzat. BW.

GER *Auge* /au g/ COGNATE WITH 'EAR'.
OHG *ouga*, PG *augōn*, see ENG.
Kluge *Auge*.

HAW *maka* /m a k a/ COGNATE WITH
'YEAR'. PPN, PMP *mata*. Blust.

LAT *oculus* /o k/. PIE *okʷ-olos*, root
$H_3okʷ$- 'eye'. Pokorny 776. EM
oculus. Sihler 42. Walde 170.
Watkins *okʷ-*.

NAV *'anáá'* /n á: ʔ/. Dependent stem
noun. PA *neẋ*. YM 1992, p. 989,
#1294; p. 420.

TUR *göz* /g ø z/. Räsänen 295a.

fall verb, Swadesh 200

ALB *bie* /b i e/ COGNATE WITH 'RUB',
'SNOW'. 'Hit, fall', to be
distinguished from 'bring'. PAlb
bie, PIE *bher-* 'bore, strike'.

Pokorny 134. Mann 171. Huld 43.

ENG *falls* /f ɔ l/. OED *feallan*, PG *fallan*, PIE *polno*, root *pol*. Barnhart. Skeat. OED. Pokorny 851. Watkins *p(h)ol-*.

FRE *tombe* /t ɔ̃ b/ LOAN. Earlier 'fall backwards', OF *tumer* 'dance, tumble', borrowed from Frankish **tūmon*. BW. Dauzat. Barnhart *tumble*.

GER *fällt* /f a l/. OHG *fallan*, PG *fallan*, see ENG. Kluge *fallen*.

HAW *hina* /h i n a/. PPN *singa*. PE.

LAT *cadit* /k a d/. PIE *ḱad* 'fall'. Pokorny 516. EM *cado*. Walde 339. Watkins *kad-*.

NAV *naashtɬíísh* /tˡ íː ʒ/ MOTIVATED. 'I fall'. Momentative imperfective. YM 1992, p. 570 *TɫIZH*. Note that *tɬiizh* is onomatopoeia for the crashing sound of falling timber.

TUR *düşüyor* /d y ʃ/ COGNATE WITH 'THINK'. Räsänen 507b.

far lim, Swadesh 200

ALB *larg* /l a r/ LOAN. From Late Latin *large*. *Largus* 'generous' has no clear etym.; Pokorny connects with *lardum* to a root *lai* 'fat'. Mann, p. 199. Pokorny 652. EM.

ENG *far* /f ɑ r/. OE *feor(r)*, PG *fer(e)ra*, comparative of PIE *per*. Pokorny 810. OED. Skeat. Barnhart. Watkins *per-1*.

FRE *loin* /l/ COGNATE WITH 'LIE', 'LONG'. Lat. *longe*. PIE *del*. Dauzat. BW.

GER *fern* /f e r/ COGNATE WITH 'WOMAN'. OHG *ferrana*, orig. 'from afar', from *ferro*, PG *fer(e)ra*, see ENG. Kluge *fern*.

HAW *mamao* /m a o/. PPN *mamaʻo*. Stative or reduplicative *ma-*? PE.

LAT *procul* /p r o/ COGNATE WITH 'NEAR'. First element is PIE *pro* 'forwards', root *per*. Second element is obscure; cf. *simul*. Pokorny 815. EM *procul*. Walde 517. Watkins *per-1*.

NAV *nízahgóó* /z aː d/. Neuter comparative: *nízah* 'it is far'. PA *zàd*. YM 1992, p. 735, *ZAAD*.

TUR *uzak* /u z/ COGNATE WITH 'LONG'.

For root, cf. *uzun*, 'long'. Räsänen 518a.

father noun, Swadesh 200

ALB *babë* /b a/ LOAN, MOTIVATED. Also *baba*. Boretzky derives from TUR. Boretzky 19.

ENG *father* /f ɑ/ MOTIVATED. OE *fœder*, PG *fader*, PIE *pəₐ tér*, see LAT. The root is probably a nursery word. Skeat. OED.

FRE *père* /p ɛ/ MOTIVATED. LAT. BW.

GER *Vater* /f aː/ MOTIVATED. OHG *fater*, PG *fader*, see ENG. Kluge *Vater*.

HAW *makua kāne* /m a k u a/ COGNATE WITH 'MOTHER'. *Makua* means any relative of the parents' generation. Replaces POc *tama*. PPN *matuʔa*. Marck. Blust.

LAT *pater* /p a/ MOTIVATED. PIE *pəₐ tér*. Suffix H_2 *ter* 'kinsman' on nursery word *pa*. Pokorny 829. EM *pater*. Sihler 99, 118. Walde 4*. Watkins *pəter-*.

NAV *ʼataaʼ* /t aː ʔ/ MOTIVATED. YM 1992, p. 994, #1540, p. 485. Haile 114.

TUR *baba* /b a/ MOTIVATED. Nursery word. Räsänen 53a.

feather noun, Swadesh 100

ALB *pendë* /p e n d/. Not in any of the etym. dictionaries. LAT loan?

ENG *feather* /f ɛ ð/. OE *feðer*, PG *feðrō*, PIE *petrā*, root *pet* 'fly'. OED. Pokorny 826. Watkins *pet-*.

FRE *plume* /p l y/. Lat. *pluma* 'down feathers', PIE *plusma-*, root *pleus* 'pluck'. Pokorny 838. Dauzat. BW. Watkins *pleus-*.

GER *Feder* /f eː d/. OHG *fedara*, PG *feðrō*, PIE *petrā*, see ENG. Kluge *Feder*.

HAW *hulu* /h u l u/. PPN *fulu*, POc *pulu*, PMP *bulu*.

LAT *penna* /p e n/. Earlier *pesna*, **petsnaH₂*, PIE root *pet* 'fly, fall'. Pokorny 826. EM *penna*. Sihler 209. Walde 21*. Watkins *pet-*.

NAV *atʼaʼ* /tʼ a ʔ/ DERIVED FROM 'FLY', COGNATE WITH 'WING'. 'Wing, feather', from root 'fly', PA

t'aG. YM 1992, p. 532, *T'A '.*
TUR *tüy* /t y/. Räsänen 503a.

few lim, Swadesh 200
ALB *pak* /p a/ LOAN. From Lat. *pauci*,
PIE *pau.* Mann 200. Watkins *pau-.*
ENG *few* /f j u/. OE *fēawe*, PG *fawaz*,
PIE *pau* 'small', like LAT *pauci.*
Pokorny 843. OED. Skeat.
FRE *peu de* /p ø/. Pop. Lat. *paucum*,
sing. of *pauci*, see LAT. BW.
GER *wenige* /v e:/ MOTIVATED. OHG
wēnag, PG *wainōn* 'cry', PG
interjection *wai* 'woe!'. Kluge
wenig. Pokorny 1111.
HAW *kaka'ikahi* /k a h i/ DERIVED FROM
'ONE', COGNATE WITH 'SOME'.
Possibly this is derived from 'a
single row'. Kernel idea would be
kahi 'one', q.v.
LAT *pauci* /p au/ COGNATE WITH
'CHILD', 'SMALL'. PIE *pau* 'small'.
Pokorny 853. EM *pau-.* Walde 75*.
NAV *t'áá díkwíí* /k w í:/ COGNATE WITH
'HERE'. Based on *díí* 'this', *kwíí*
'here'? YM 1992, p. 932.
TUR *az* /a z/. Räsänen 32b.

fight verb, Swadesh 200
ALB *lufton* /l u f/ LOAN. From Low Lat.
lucta, from *luctor*, perhaps PIE
lugto-, root *leug-* 'bend'. Pokorny
685. Mann 67. EM *luctor*. Watkins
leug-1.
ENG *fights* /f ai/. OE *feohtan*, PG
fehtan, usually referred to PIE *pek̑*
'pluck (hair)'. Pokorny 828, 797.
OED. Barnhart. Skeat. Watkins
pek-2.
FRE *bat* /b a t/. *Se battre.* Pop. Lat.
battere, from *battuere*, of unknown
origin, perhaps a loan from Celtic
or Germanic. Dauzat. EM *battuo.*
Watkins *[battuere].*
GER *kämpft* /k a m pf/ LOAN. From
OHG *kampf*, loan from Lat.
campus 'field'. Perhaps PIE *kamp*
'bend', hence 'corner'. Kluge
Kampf. EM *campus*. Pokorny 525.
Watkins *kamp-.*
HAW *hakakā* /k a:/. Derivative of *kā* 'hit,
strike, murder' (PPN *taa*). PE.
LAT *pugnat* /p u g/ MOTIVATED.

Pugnus 'fist', PIE *puĝ-no-*, root
p(e)uĝ 'prick, box'. Could this be
an ext. of a natural expressive
sound, *pu*, like Eng. *pow!?* Pokorny
828. EM *pugnus, pungo*. Sihler 512.
Walde 15*. Watkins *peuk-.*
NAV *'ahishgą́* /ɣ ą́/. 'I fight with',
momentaneous. PA *ghan*, PPA *ɣan*.
YM 1992, p. 236, *GHÁ̧Á̧ '.* Haile
118.
TUR *dövüşüyor* /d ø v/. Also *döğ-.*
Reciprocal voice of *dövmek* 'beat'.
Räsänen 492a.

fire noun, Swadesh 100
ALB *zjarr* /z j a r/. PAlb *zjar̃* alt. with
zjarm, PIE *gʷhermo-* 'warm', root
gʷher 'hot, burn'. Mann 14, 35, 53.
Huld 135. Walde 687. Pokorny 494.
ENG *fire* /f ai/ MOTIVATED. OED *fȳr*,
PWG *fūri*, PIE *puH₂r*, root *paH₂w*
(*r/n* stem) possibly based on an
onomatopoetic *pu* 'blow'. Pokorny
828. OED. Kluge *Feuer*. Watkins
pūr-. Sihler 299.
FRE *feu* /f ø/. Lat. *focus* 'fireplace',
PIE *bhok*, an unusual root shape.
Dauzat. BW. Pokorny 162. EM
focus.
GER *Feuer* /f oi/ MOTIVATED. OHG *fiur*,
PWG *fūri*, see ENG. Pokorny 828.
Kluge *Feuer*. Watkins *pūr-.* Sihler
299.
HAW *ahi* /a h i/. PPN *afi*, POc *api*,
PMP *hapuy*. PE. Blust.
LAT *ignis* /i g n/. PIE *egnis*. Pokorny
293. EM *ignis*. Walde 323. Watkins
egni-.
NAV *kǫ'* /k ō ?/. Independent stem
noun. PA *kʷin'*, *kʉn'*. No obvious
connection to *-kǫh* 'smooth'. YM
1987, p. 3. YM 1992, p. 983, p. 330.
Hoijer 10.
TUR *ateş* /a t e ʃ/ LOAN. Loan from
Persian *ātaš*. Räsänen 31b.
Steingass 13a.

fish noun, Swadesh 100
ALB *peshk* /p e ʃ k/ LOAN. Most believe
this to be a borrowing from Lat.
piscis. Mann 56. Huld 103.
ENG *fish* /f ɪ ʃ/. OE *fisc*, PG *fiskaz*, PIE
piskos, root *peisk*. Pokorny 796.

OED. Skeat. Watkins *peisk-*.
FRE *poisson* /p w a s/. Extension of
pois from LAT *piscis*. BW. Dauzat.
GER *Fisch* /f i ʃ/. OHG *fisc*, PG *fiskaz*,
see ENG. Pokorny 796. OED.
Kluge.
HAW *iʻa* /i ʔ a/. PPN *ika*, POc *ikan*,
PMP *hikan*. PE. Blust.
LAT *piscis* /p i s k/. PIE root *peisk*,
though the *k* may be an extension.
Pokorny 796. EM *piscis*. Walde
11*. Kluge *Fisch*.
NAV *łóóʼ* /ł óː ʔ/. Independent stem
noun. YM 1987, p. 3. YM 1992,
p. 987, 395.
TUR *balık* /b a l ɯ k/. Räsänen 61b.

five lim, Swadesh 200
ALB *pesë* /p e s ə/. PIE *penkʷe*, possibly
crossed with *pn̥kʷti-*. Huld 102.
Pokorny 808. Mann 32. Walde 25*.
ENG *five* /f ai v/. OED *fīf*, PG *fimf(e)*,
pre-Germanic *pempe*, assim. from
PIE *penkʷe*. OED. Pokorny 808.
FRE *cinq* /s ɛ̃ k/. Pop. Lat. *cīnque*,
dissim. from Class. LAT *quīnque*.
GER *fünf* /f y n f/. OHG *fimf*, PG
fimf(e), see ENG. Pokorny 808.
Kluge *fünf*.
HAW *lima* /l i m a/ COGNATE WITH
'HAND'. Same word as 'hand'.
Seems safest to assume that it's no
coincidence: five fingers. PE.
LAT *quinque* /kw iː n kw e/.
Proto-Italic *kʷenkʷe* (the /iː/ from
quīntus 'fifth'), Italo-Celtic assim.
from PIE *penkʷe*. Pokorny 808. EM
quīnque. Sihler 413. Walde 25*.
NAV *ʼashdla'* /ʔ a ʃ dˡ a ʔ/. YM 1992,
p. 932, 10. Haile 124.
TUR *beş* /b e ʃ/. Räsänen 71b.

flow verb, Swadesh 200
ALB *rrjedh* /r e ð/. PAlb *r̃eð*, earlier
Crɛð, PIE prob. *enregh-*, root *regh*.
Huld 110. Mann 143. Walde 365*.
Pokorny 857.
ENG *flows* /f l/ COGNATE WITH 'FLY',
'FULL'. OE *flōwan*, PG *flō-*, PIE
plō, root *pleu*, root *pel(H)* which
either has several meanings or
several homonyms. Pokorny ties
directly to Lat. *pluvia* 'rain', Eng

and Ger 'fly' and Ger. *Vogel* 'bird',
and slightly less certainly to the
word meaning 'fill'. OED. Pokorny
835, 799. Barnhart. Watkins *pleu-*.
FRE *coule* /k u/. Lat. *cōlare* 'sift', from
cōlum 'sieve', perhaps PIE **kogh-*,
possibly a variant of PIE *kagh*
'wicker'. BW. EM *colum*. Pokorny
518. Watkins *kagh-*.
GER *fließt* /f l/ COGNATE WITH 'BIRD',
'FLY', 'FULL', 'MANY', 'RIVER',
'WING'. OHG *fliozan*, PG *fleutan*,
see ENG. Kluge *fließen*. Pokorny
835, 799. Barnhart.
HAW *kahe* /k a h e/. PPN *tafe*.
LAT *fluit* /f/ COGNATE WITH 'BLOW',
'FLOWER', 'LEAF', 'RIVER'. Source
of *flumen* 'river', Fre. *fleuve* 'river'.
Because there is a PIE root for
'flow' (*sreu*) that would give Lat.
*fruit, some see a crossing with
some such word as *pluit* 'rain', so I
delete the /l/ from root. Or it
could be a direct outcome of a
bhleugʷ 'inflate', root *bhleu-*.
Pokorny 159, 121. EM *fluo*. Sihler
214. Walde 214*. Watkins *bhleu-*.
NAV *yígeeh* /g o ʔ/. 'It flows',
momentaneous imperfect. PA *Gʉk'*
/ *Gʉtsh ʷ*. YM 1992, p. 217, *GO*'.
TUR *akıyor* /a k/. Räsänen 12b.

flower noun, Swadesh 200
ALB *lule* /l u l/ LOAN. Almost certainly
a borrowing, perhaps Lat. *lilium*.
Mann 73. Huld 88.
ENG *flower* /f l/ LOAN, COGNATE WITH
'BELLY', 'BLOOD', 'BLOW'.
Borrowed from OF *flour*, see FRE.
Skeat. OED.
FRE *fleur* /f l œ/ COGNATE WITH
'LEAF', 'RIVER', 'SWELL'. LAT
flōrem, acc. of *flōs*.
GER *Blume* /b l/ COGNATE WITH
'BLOOD', 'BLOW', 'LEAF'. OHG
bluoma, PG *blōmōn*, from *blōan*
'bloom', PIE *bhel*, see LAT. Sihler
529. Walde 179*. EM. Kluge
Blume. Pokorny 121. Watkins
bhel-3.
HAW *pua* /p u a/. PPN *pua*, POc *puŋa*,
PMP *buŋa*. PE. Blust.
LAT *flos* /f l/ COGNATE WITH 'BLOW',

'FLOW', 'LEAF', 'RIVER', 'YELLOW'.
PIE *bhlōs*, root *bhel* 'bloom'.
Possibly connected to *bhel* 'inflate,
blow', which may be connected to
Lat. *fluo* 'flow'. Pokorny 122. EM
flos. Sihler 310. Walde 177*.

NAV *ch'ilátah hózhóón* /tʃ' i l/. *Ch'il*
'plant' (from verb 'curl', PA
tshʷi'tł', PPA *kʷitł'*) + *-látah* 'tip'
+ *hózhóón* 'the beautiful one'. YM
1992, p. 972, #383; p. 108, *CH'IL*.
Haile 126.

TUR *çiçek* /tʃ i tʃ e k/. Räsänen 162a.

fly verb, Swadesh 100

ALB *fluturon* /f l/ MOTIVATED,
COGNATE WITH 'WING'. Çabej
considers this a spontaneous,
nonarbitrary creation (cf. unrelated
fly, flutter, flit in Eng.). Çabej 196.

ENG *flies* /f l/ COGNATE WITH 'FLOW',
'FULL'. OE *flēogan*, PG *fleugan*,
PIE *pleuk-*, root *pleu-* 'flow',
possibly from root *pel(H)* 'fill,
pour'. Skeat. OED. Barnhart.
Pokorny 835. Watkins *pleu-*.

FRE *vole* /v ɔ l/. LAT. Dauzat. BW.

GER *fliegt* /f l/ COGNATE WITH 'BIRD',
'FLOW', 'FULL', 'MANY', 'RIVER',
'WING'. OHG *fliogan*, PG *fleugan*,
see ENG. Pokorny 835. Kluge
fliegen. Watkins *pleu-*.

HAW *lele* /l e/. PPN *lele*. Replaces POc
Ropok. Blust.

LAT *volat* /w o l/. PIE *gʷel*. Tucker. EM
uolo. Watkins *gʷel-*.

NAV *yit'ah* /t' a ʔ/ COGNATE WITH
'FEATHER', 'WING'. Cursive: 'it is
flying along'. PA *t'aG*. YM 1992,
p. 522, *T'A'*.

TUR *uçuyor* /u tʃ/. Räsänen 509a.

fog noun, Swadesh 200

ALB *mjegull* /m j e/. Also exists in
by-forms *mjergull, njegull*. Closest
to PIE *merHgʷ* 'dark', from root
merH 'flicker'; probably crossed
with orig. *H₃mighlaH₂* 'cloud'.
Pokorny 734, 712. Huld 93. Mann
50. Walde 275*.

ENG *fog* /f ɑ g/ LOAN. Likely borrowed
from some Scandinavian source (cf.
Danish *fog* 'shower'). OED.

Barnhart.

FRE *brouillard* /b r/ LOAN, COGNATE
WITH 'BURN'. From *brouillas*, from
brou, OHG *brod*, PG *bruþan*, from
breu-, PIE *bhreu*, root *bher* 'boil'.
Duazat. BW. Barnhart. Pokorny
145.

GER *Nebel* /n e: b/. OHG *nebul*, PG
nebulaz, PIE root *nebh*. Pokorny
315. EM *nebula*. Sihler 147. Walde
131. Kluge *Nebel*. Watkins *nebh-*.

HAW *'ohu* /ʔ o h u/. PNP *kofu*, POc
kaput, PMP *kabut*. PE. Blust.

LAT *nebula* /n e b/ COGNATE WITH
'CLOUD'. PIE *nebh-elā*, root *nebh*.
EM suggest the possibility of an
irregular connection with *nubes*
'cloud'. Pokorny 315. EM *nebula*.
Sihler 147. Walde 131. Watkins
nebh-.

NAV *áhí* /á h/. Independent stem noun;
also *'ááh*. YM 1992, p. 966, #33,
p. 27 *'ÁÁH*. Haile 126. Hoijer 10.

TUR *sis* /s i s/. Räsänen 423b.

foot noun, Swadesh 100

ALB *këmbë* /k ə m b/ LOAN, COGNATE
WITH 'STAND'. Apparently from
VLat *camba, gamba* 'leg', from
Greek *kampḗ*, PIE root *kamp*
'bend'. Mann 59. Huld 80. Pokorny
525. EM *gamba*. Watkins *kamp-*.

ENG *foot* /f u t/. OE *fōt*, PG *fōtum*,
PIE *pōd*, root *ped*. Skeat. OED.
Pokorny 790. Watkins *ped-1*.

FRE *pied* /p j e t/. Lat. *pedem*, acc. of
pes. For French final, consider
pied-à-terre. Dauzat. BW. Watkins
ped-1.

GER *Fuß* /f u: s/. OHG *fuoz*, PG *fōtum*,
see ENG. Pokorny 790. Kluge *Fuß*.
Watkins *ped-1*.

HAW *wāwae* /w a e/. PCP *wae(wae)*,
PPN *wa'e*, POc *waqe*, PCEMP
waqay, PMP *qaqay*. PE. Blust.

LAT *pes* /p e d/. PIE *ped*. Pokorny 790.
EM *pes*. Sihler 117. Walde 23*.
Watkins *ped-1*.

NAV *'akee'* /k e: ʔ/ COGNATE WITH
'CLAW', 'ROOT'. Dependent stem
noun. YM 1987, p. 4. YM 1992,
p. 982, #938. Haile 127.

TUR *ayak* /a j a k/. Räsänen 5a.

four lim, Swadesh 200

ALB *katër* /k a t ə r/. Possibly borrowed from Latin, but no other numeral is; perhaps a PIE form *kʷtwor*. Huld 79. Mann 27.

ENG *four* /f ɔ r/. OE *fēower*, PWG *fewar*, PG *fedwōr*, PIE *kʷetwor*. OED, Pokorny 643. Skeat. Watkins *kʷetwer-*. Kluge *vier*.

FRE *quatre* /k a t r/. Pop. Lat. *quattor*, Class. LAT *quattuor*. BW.

GER *vier* /f iː r/. OHG *fior*, PWG *fewar*, see ENG.

HAW *hā* /h aː/. PPN *faa*, POc *pat*, PMP *epat*. Blust.

LAT *quattuor* /kw a t: u o r/. PIE *kʷetwor*. Pokorny 643. EM *quattuor*. Sihler 411. Walde 512. Watkins *kʷetwer-*.

NAV *díį'* /d ĩː ʔ/. YM 1992, p. 932.

TUR *dört* /d ø r t/. Räsänen 495a.

freeze verb, Swadesh 200

ALB *ngrin* /g r i/. Prefix *en-*. Mann makes base *krūs*, as in Greek *cryos*, PIE root *kreus*, perhaps from *kreu* 'thick blood'. Mann 185. Pokorny 621. Watkins *kreus-*.

ENG *freezes* /f r i z/. OE *frēosan*, PG *freusan*, PIE *preus*. OED. Skeat. Pokorny 846. Watkins *preus-*.

FRE *gèle* /ʒ ɛ l/ COGNATE WITH 'ICE'. LAT. BW.

GER *friert* /f r iː r/. OHG *friosan*, PG *freusan*, see ENG. Kluge *frieren*.

HAW *ho'opa'a i ka hau* /p a ʔ a/ DERIVED FROM 'HOLD'. I.e., 'make solid into the ice'. *Pa'a* means 'firm, solid, frozen'. PE.

LAT *gelat* /g e l/ COGNATE WITH 'ICE'. PIE *gel* 'cold, freeze'. Pokorny 365. EM *gelu*. Walde 622. Watkins *gel-3*.

NAV *yishtin* /t i n/ COGNATE WITH 'ICE'. 'I freeze': conclusive mediopassive. PA *tin*. YM 1992, p. 509, *TIN*. Haile 130.

TUR *donuyor* /d o n/. Räsänen 488.

fruit noun, Swadesh 200

ALB *pemë* /p e m/ LOAN. Probably loan from Lat. *pomum*. Janson 24. Not in Mann or Huld.

ENG *fruit* /f r u/ LOAN. Borrowed from

OF *fruit*, see FRE. OED. Skeat. Watkins *bhrūg-*.

FRE *fruit* /f r ɥ/. Lat. *frūctus*, earlier 'produce', PIE *bhrūg*; the root is possibly *bhreu* 'cut'. BW. Skeat. Pokorny 173.

GER *Frucht* /f r u/ LOAN. OHG *fruht*, borrowed from Lat. *frūctus*, see FRE. Pokorny 173. Kluge *Frucht*.

HAW *hua* /h u a/ COGNATE WITH 'EGG'. PE: 'Fruit, tuber, egg, seed'. PPN *fua*, POc *puaq*, PMP *buaq*. Blust.

LAT *pomum* /p oː m/. Tucker suggests a connection with *poH₃* 'drink', the root of *bibo*. EM *pomus*.

NAV *bineest'ą'* /t' ã/. From verb 'ripen', PA *t'any*. YM 1992, p. 1066, #3361. Root *T'Ą́2*. Haile 133 has short *e*.

TUR *meyva* /m e j v/ LOAN. Loan from Persian *maiva*. Räsänen 332b. Steingass 1365a.

full adj, Swadesh 100

ALB *plotë* /p l o/ COGNATE WITH 'WOODS'. Adj. suffix *-të*, root from PIE *pleH₁* 'pour, fill'. Mann 9. Huld 105. Pokorny 799. Walde 64*.

ENG *full* /f u l/ COGNATE WITH 'FLOW', 'FLY'. OE *full*, PG *fullaz*, PIE *pl̥H₁nos*. Root *pelH₁*, possibly connected to *flow*, q.v. Skeat. OED. Watkins *pelə-1*.

FRE *plein* /p l/ COGNATE WITH 'RAIN'. LAT. BW. Dauzat.

GER *voll* /f o l/ COGNATE WITH 'BIRD', 'FLOW', 'FLY', 'MANY', 'RIVER', 'WING'. OHG *fol*, PG *fullaz*, see ENG. Kluge *voll*.

HAW *piha* /p i h a/. PE.

LAT *plenus* /p l/ COGNATE WITH 'RAIN'. PIE root *pleH₁* 'fill, pour', perhaps from *pel* 'flow'. Pokorny 799. EM *plē*. Sihler 126. Walde 64*.

NAV *hadeezbin* /b i n/. 'It is full of them'. Neuter, PA *win*. YM 1992, p. 68, *BĮĮD*. Haile 133. Pinnow 30.

TUR *dolu* /d o l/. Räsänen 486a.

give verb, Swadesh 100

ALB *jep* /j e p/. PAlb *jap*, earlier *ɛp*, PIE *H₁ep* 'give'. Kacori 242. Mann 164, 171. Huld 77.

ENG *gives* /g ɪ v/ LOAN. OE *giefan*
under the influence of ON forms in
/g-/, PG *geban*, PIE *ghebh*. The
root vowel in Germanic may be
influenced by *neman* 'take'. Skeat.
Pokorny 407. Barnhart. OED.
Watkins *ghabh-*.

FRE *donne* /d ɔ/. LAT *dōnāre* 'give a
gift', from *dōnum* 'gift', from
dō-no-, PIE root *doH₃*. Dauzat.
BW. EM *do*. Watkins *dō-*.

GER *gibt* /g eː b/. OHG *geban*, PG
geban, see ENG. Kluge *geben*.

HAW *hā'awi* /ʔ a w/. *Hā-* is often
causative; *-i* is often transitive.

LAT *dat* /d a/. PIE *dH₃-*, root *doH₃*.
EM *do*. Pokorny 224. Sihler 544.
Walde 815. Watkins *dō-*.

NAV *nish'aah* /ʔ aː/. 'I bring a small
roundish object, hand it over'. SRO
is chosen here because that is the
most semantically unmarked class.
PA *'a-ny*, PPA *'a*. YM 1992, p. 16.
Pinnow 32.

TUR *veriyor* /v e r/. Räsänen 70b.

go verb, Swadesh 100

ALB *shkon* /ʃ k/ COGNATE WITH 'PATH'.
Major theories are: a denominative
formation from PIE *stoighos* 'path'
(root *steigh-* 'walk'); or a
borrowing from Germanic, e.g.,
Gothic *skēwjan* 'walk'. Pokorny
1017. Mann 146. Huld 114.

ENG *goes* /g o/. OE *gān*, PIE *ĝheH₁*
'leave'. OED. Skeat. Pokorny 418.
Barnhart. Watkins *ghē-*.

FRE *va* /v a/. Lat. *vadere*, PIE *waH₃*
with *de/o* present. Dauzat. BW.
EM *uado*. Pokorny 1109.

GER *geht* /g eː/. OHG *gēn*, PG *gǣ-*,
PIE *ĝheH₁* 'leave'. Pokorny 418.
Kluge *gehen*. Watkins *ghē-*.

HAW *hele* /h e l e/ COGNATE WITH
'COME'. PCP *sele*, PPN *sa'ele*. PE.

LAT *it* /iː/. PIE root **ey-*, which is
perhaps evident in the alternation
eo (**eyo*), *īs* (**eis*). Pokorny 295.
EM *eo*. Sihler 456. Walde 102.

NAV *yíghááh* /y á/ COGNATE WITH
'COME'. 'He comes, goes'.
Momentaneous imperfect stem.
The *gh* is an old classifier. YM

1992, p. 668, *YÁ*. Pinnow 15.

TUR *yürüyor* /j y r/ COGNATE WITH
'HEART'. Räsänen 213b.

good adj, Swadesh 100

ALB *mirë* /m i/ LOAN. Possibly PIE
meiH like Lat. *mītis* 'mild' and
Russian *mir* 'peace', but most
often considered a loan from Lat.
mīrus 'wonderful' (PIE *(s)mei-*).
Mann 109. Huld 92. Walde 244*.
Pokorny 712, 967. EM *mirus*.
Sihler 214. Watkins *smei-*.

ENG *good* /g ʊ d/. OE *gōd*, PG *gōdaz*,
PIE *ghōdh-*, root *ghedh* 'unite'.
Skeat. OED. Barnhart. Pokorny
423. Watkins *ghedh-*.

FRE *bon* /b ɔ/. LAT. Dauzat. BW.

GER *gut* /g uː t/ COGNATE WITH
'HUSBAND', 'WIFE'. OHG *guot*, PG
gōdaz, see ENG. Kluge *gut*.

HAW *maika'i* /i k a ʔ i/. PPN *ma'itaki*.
ma- can be a stative prefix.

LAT *bonus* /b o/. OLat. *duenos*, PIE
dw-enos, root *deu* 'revere'. Pokorny
218. EM *bonus*. Sihler 41. Walde
778. Watkins *deu-2*.

NAV *yá'át'ééh* /t' é: h/. 'It is good',
neuter. YM 1992, p. 544 *T'ÉÉH*.

TUR *iyi* /i j i/. Räsänen 36a.

grass noun, Swadesh 200

ALB *bar* /b a r/. PIE root *bhar-* or
bhor-. There's a slight chance there
is some connection with *bher*
'carry'. Pokorny 130. Huld 150.
Mann 91. Janson 27. Walde 134*.

ENG *grass* /g r/ COGNATE WITH
'GREEN'. OE *græs*, PG *grasam*,
PIE *ghrəso-*, root *ghrē* 'grow'.
Skeat. OED. Barnhart. Pokorny
454. Watkins *ghrē*.

FRE *herbe* /ε r/. Lat. *herba* 'grass'.
Possibly a PIE *gherdhā*, root *ghrē*
'grow'. BW. EM *herba*. Pokorny
454.

GER *Gras* /g r/ COGNATE WITH
'GREEN'. OHG *grās*, PG *grasam*,
see ENG. Kluge *Gras*.

HAW *mau'u* /m a u ʔ u/. PEP *mauku*,
PPN *mohuku*. PE.

LAT *gramen* /g r aː/. **Grasmen*, from
PIE *gras* 'graze'. Possibly a

desiderative of g^werH_1 'swallow'.
Pokorny 404, 474. EM *gramen.*
Sihler 153. Walde 645. Watkins
gras-.
NAV *tł'oh* /t^{l}' o h/. Independent stem
noun. YM 1992, p. 586.
TUR *ot* /o t/. Suspiciously close to Arm.
/xot/. Old Turkic borrowing?
Dankoff, p. 160. Räsänen 366b.

grease noun, Swadesh 100
ALB *dhjamë* /ð j a/ LOAN. Possibly a
borrowing from some form of
Greek *dēmós*, possibly from PIE *dā*
'flow'. Mann 55, 97. Huld 59.
Pokorny 175.
ENG *grease* /g r/ LOAN. Loan from OF
graisse, see FRE. Skeat. OED.
Pokorny 584.
FRE *graisse* /g r/. Lat. *crassia* from
crassus 'fat', earlier 'thick',
possibly from PIE *kert* 'turn',
extension of *(s)ker* 'turn'. Skeat.
Pokorny 584. Dauzat.
GER *Fett* /f e/ LOAN. 14th c. loan from
Low German. PIE root *pei* 'be fat'.
Kluge *fett*. Watkins *peiə-.*
HAW *'aila* /ʔ a i l a/ LOAN. A loan from
Eng. *oil*, OFre *oile*, Lat. *oleum*,
Aegean *elaywā* 'olive'. PE.
Watkins *[elaia].* Sihler 41.
LAT *adeps* /a d e p/. EM *adeps.*
NAV *'ak'ah* /k' a h/. YM 1992, p. 336
K'AII. Pinnow 36.
TUR *yağ* /j a ɣ/. Suspiciously close to
Arm. dial. /yeɣ/. Old Turkic
borrowing? Dankoff, p. 160.
Räsänen 177b.

green adj, Swadesh 100
ALB *gjelbër* /ɟ e l/ LOAN. Loan from
Lat. *galbinus* 'pale green' ('yellow'
in the Romance languages), from
galbus. That itself appears late,
may be a loanword, in which case
it may be from PIE *ĝhel-* 'yellow,
green'. Pokorny 429. EM *galbus.*
Huld 84.
ENG *green* /g r/ COGNATE WITH
'GRASS'. OE *grēne*, PG *grōnjaz*,
root *ghreH* 'grow'. Skeat. Pokorny
454. OED. Watkins *ghrē.*
FRE *vert* /v ε r/. LAT. Dauzat. BW.

GER *grün* /g r/ COGNATE WITH 'GRASS'.
OHG *gruoni*, PG *grōnjaz*, see
ENG. Kluge *grün.*
HAW *'ōma'oma'o* /m a ʔ o/. *'Ō-* is a
prefix meaning 'like'. Base *ma'o* is
also used to mean 'green'. PE.
LAT *viridis* /w i r/. Adj. in *-id-* from
vireo 'be green', possibly PIE *weis*
'sprout'. Pokorny 1133. EM *uireo.*
Walde 242. Watkins *[virēre].*
NAV *dootł'izh* /t^{l}' i: ʒ/. 'It is green',
neuter imperf. YM 1992, p. 583:
root *TŁ'IIZH.*
TUR *yeşil* /j e ʃ/. Cf. *yaş* 'fresh'.
Räsänen 192a.

guts noun, Swadesh 200
ALB *zorrë* /z o r:/ COGNATE WITH
'MOUTH'. PIE $g^werH_1naH_2$, root
g^werH_1 'swallow'. Janson 34. Mann
14. Huld 147. Pokorny 474.
Watkins $g^werə$-4.
ENG *guts* /g ə/. OE *guttas*, PIE
ĝhudnu-, root *ĝheu* 'pour'. Pokorny
447. Skeat. OED. Watkins *gheu-.*
FRE *intestins* /ɛ̃/ LOAN, COGNATE WITH
'IN'. Loan from LAT *intestina*, q.v.
BW. Dauzat.
GER *Därme* /d a r/ COGNATE WITH
'SQUEEZE'. OHG *darm*, PG
þarmaz, PIE *torHma-*, root *terH*
'turn'. Kluge *Darm.* Watkins
terə-1.
HAW *na'au* /n a ʔ a u/. PPN *ngaakau.*
LAT *intestina* /i n/ DERIVED FROM 'IN'.
From *intus*, PIE *entos*, from *en.*
EM *in.* Pokorny 313. Sihler 246.
Walde 127. Watkins *en.*
NAV *'ach'íí* /tʃ í: ʔ/. Dependent stem
noun. YM 1980, p. 10. YM 1992,
p. 973, 110.
TUR *barsaklar* /b a r/. From *bar* (*bağır*)
'belly'. Räsänen 55b.

hair noun (of head), Swadesh 100
ALB *flok* /f l o k/ LOAN. Loan from Lat.
floccus 'flock of wool', itself with
no certain etymology; Pokorny has
as earlier *flōcos*, from a PIE *bhlŏk-.*
Huld 88. EM *floccus.* Pokorny 161.
ENG *hair* /h ε r/ LOAN. Crossing of OF
haire 'hair-cloth' (ult. from PG
hǣram) with OE *hǣr*, same PG

source (*hǣram*); possibly a PIE *k̂er* 'bristle'.
FRE *cheveux* /ʃ ə v/. LAT. Dauzat. BW.
GER *Haare* /h aː r/. OHG *hār*, PG *hǣram*, see ENG.
HAW *lauoho* /l a u/ DERIVED FROM 'LEAF'. PCP *lauoso* or *lauofo*. *Lau* 'leaf' (q.v.) + *oho* 'head'. PE.
LAT *capillus* /k a p/ COGNATE WITH 'HEAD'. Looks very much like it has a relationship to *caput* 'head', but no satisfactory explanation of exactly how. EM *capillus*. Pokorny 530. Tucker 45. Walde 347.
NAV *'atsiighá* /ts iː ʔ/ DERIVED FROM 'HEAD'. From -*tsii'* 'head, hair' + -*ghá* 'wool'. YM 1992, p. 1001, 610.
TUR *saç* /s a tʃ/. Räsänen 391b.

hand noun, Swadesh 100
ALB *dorë* /d o r/. Probably something like PIE *ĝhḗsr*, similar to Greek *kheir*. Pokorny 203, 447. Mann 86, 99. Huld 54. Walde 603.
ENG *hand* /h a n d/. OE *hand*, PG *handuz*. Skeat. OED. Barnhart.
FRE *main* /m a/. LAT. Also in *maintenant* 'now'. For root vowel consider *manier*, *manette*. Dauzat. BW.
GER *Hand* /h a n d/. OHG *hant*, PG *handuz*. Kluge *Hand*.
HAW *lima* /l i m a/ COGNATE WITH 'FIVE'. PMP *lima*. PE. Blust.
LAT *manus* /m a/. PIE *r/n* stem *mH*. EM *manus*. Pokorny 740. Walde 272*. Watkins *man-2*.
NAV *'ála'* /l a ʔ/. Dependent stem noun. YM 1987, p. 4. YM 1992, p. 366.
TUR *el* /e l/. Räsänen 39a.

he link, Swadesh 200
ALB *ai* /a/ MOTIVATED, COGNATE WITH 'THAT', 'THERE', 'THEY'. Also *ay*. *A-* 'that' + PIE *so* 'that'. The *a*-for designating remote objects is unexplained; possibly expressive? Huld 37. Pokorny 979.
ENG *he* /h/ COGNATE WITH 'HERE'. PG demonstrative stem *hi-*, PIE *k̂i-*, variant of *k̂o* 'this'. Pokorny 609. Barnhart. OED. Watkins *ko-*.

FRE *il* /i l/ COGNATE WITH 'THERE', 'THEY'. Pop. Lat. **illī* (influenced by *quī*?), Class. Lat. *ille*, from PIE root *al* 'beyond', but vocalism influenced by *is*. BW. Pokorny 24. Watkins *al-1*.
GER *er* /e/. OHG *er*, PG *eiz*, PIE *eis*, pronominal root *e* (Lat, *ei* 'they'). Pokorny 281. Kluge *er*.
HAW *ia* /i a/. PPN *ia*, PMP *si-ia*. PE. Blust.
LAT *is* /i/ COGNATE WITH 'ONE', 'THAT', 'THERE'. PIE pronominal root *i* (alternating with the root *e*). Pokorny 281. EM *is*. Sihler 391, 405. Walde 97. Watkins *i-*, *oi-no-*.
NAV *bí* /b í/ COGNATE WITH 'THEY'. YM 1992, p. 930.
TUR *o* /o/ COGNATE WITH 'THAT', 'THERE', 'THEY'. Demonstrative morpheme. Räsänen 356a.

head noun, Swadesh 100
ALB *kokë* /k o k/. Not in Huld, Janson, Pokorny, Walde. Çabej lists a *kokë* as a dialectal form of *kokërr* 'grain, anything small and round'.
ENG *head* /h ε/. OE *hēafod*, PG *haubud-*, PIE *kapot*, root *kap*. Pokorny 530. Skeat. OED. Watkins *kaput*.
FRE *tête* /t ε/. Lat. *testa*, orig. 'clay pot', PIE root *tek̂s* 'weave'. Dauzat. BW. Pokorny 396, 1058. Watkins *teks-*.
GER *Kopf* /k o/ LOAN. OHG *kopf* 'cup', loan from Lat. *cuppa*, ext. of zero grade of PIE *keu* 'bend'. EM *cuppa*. Pokorny 396, 591. Watkins *keu-2*.
HAW *po'o* /p o ʔ o/. PEP *'upoko*. PE.
LAT *caput* /k a p/ COGNATE WITH 'HAIR'. PIE *kap-ot* 'head', root *kap*. Pokorny 529. EM *caput*. Walde 346. Watkins *kaput*.
NAV *'atsii'* /ts iː ʔ/ COGNATE WITH 'HAIR'. Dependent stem noun, 'Head, hair'. YM 1992, p. 1001, 610, YM 1987, p. 4.
TUR *baş* /b a ʃ/ COGNATE WITH 'OTHER'. Räsänen 64b.

hear verb, Swadesh 100

ALB *dëgjon* /d ə ɟ/. Drizari.

ENG *hears* /h ɪ/. Angl. OE *hēran*,
Proto-Northwest Germanic
haurjan, PG *hauzjan*, PIE H_2kous-,
root H_2keu 'perceive'. Pokorny
587. Skeat. OED. Watkins *keu-1*.

FRE *entend* /t ã/ COGNATE WITH
'HOLD', 'NOW'. Earlier
'understand', from *in-* + Lat.
tendere 'extend', extension of PIE
ten 'stretch'. BW. Dauzat. Pokorny
1066. Watkins *ten-*.

GER *hört* /h ø:/. OHG *hōren*,
Proto-Northwest Germanic
haurjan, see ENG. Kluge *hören*.

HAW *lohe* /l o h e/. PE.

LAT *audit* /au d/ COGNATE WITH 'EAR'.
PIE *awizdh-*, root *aw* 'perceive'.
Pokorny 78. EM *audio*. Walde 17.
Watkins *au-5*.

NAV *diits'a'* /ts' ã: ʔ/. 'He hears',
neuter imperfect stem. PPA *ts'any*.
YM 1992, p. 625 *TSʼÁ́Á́ʼ*. Hoijer
14.

TUR *işitiyor* /i ʃ/. Räsänen 51a.

heart noun, Swadesh 100

ALB *zemër* /z e/. One guess is PIE
gʷhenmer, root *gʷhen* 'swell'.
Pokorny 491. Huld 132. Walde 679.

ENG *heart* /h ɑ r t/. OE *heorte*, PG
hertōn, PIE *kerd-en-*, root *k̂erd*.
Skeat. OED. Pokorny 579. Watkins
kerd-1.

FRE *cœur* /k œ r/. Pop. Lat. *cor-*,
Class. LAT *cord-*. Dauzat. BW.

GER *Herz* /h e r ts/. OHG *herza*, PG
hertōn, see ENG.

HAW *puʻuwai* /p u ʔ u/. *Puʻu* 'internal
organ' + *wai* 'blood'.

LAT *cor* /k o r d/. PIE *k̂r̥d*, root *k̂erd*.
EM *cor*. Pokorny 579. Sihler 94.
Walde 423. Watkins *kerd-1*.

NAV *ʼajéídíshjool* /dʒ é í/. From *-jéí*
'thoracic organs, heart' + 'round'.
YM 1992, p. 261.

TUR *yürek* /j y r/ COGNATE WITH 'GO'.
From *yür* 'be in motion'. Räsänen
213b.

heavy adj, Swadesh 200

ALB *rëndë* /r ə/. PAlb *randë*, where the
ndë may be a pres. participle. But

the root is obscure. Janson 33.
Huld 108. Not in Mann, Pokorny.

ENG *heavy* /h ɛ v/. Related to *heave*.
OE *hefig*, PG *habīgaz*, PIE root
kap 'grasp'. Skeat. OED. Barnhart.
Pokorny 527. Watkins *kap-*.

FRE *lourd* /l u r/. Pop. Lat. *lūrdus*,
Class. *lūridus* 'jaundiced'; cf. *luror*.
BW. Dauzat. EM *luror*.

GER *schwer* /ʃ v e: r/. OHG *swāri*, PG
swǣraz, PIE root *swer*. Pokorny
1150. Kluge *schwer*. Watkins
swer-5.

HAW *kaumaha* /h a/. PCP *taumafa*,
PPN *mamafa*, POc *mapat*,
PCEMP *ma-bəRat*, PMP
ma-beReqat. Blust.

LAT *gravis* /g r a/. PIE root *gʷerH₂*.
EM *grauis*. Pokorny 476. Sihler 95.
Walde 685.

NAV *ndaaz* /d a: z/. 'It is heavy', neuter
absolute. PA *dàz*, PPA *da's*. YM
1980, p. xvii. YM 1992, p. 122
DÁ́ÁZ.

TUR *ağır* /a ɣ ɯ r/. Räsänen 8a.

here lim, Swadesh 200

ALB *këtu* /k/ COGNATE WITH 'HERE',
'THIS', 'WHAT', 'WHO'.
Demonstrative particle of nearness
kə plus *tu*, as in Sanskrit. The
former most likely the PIE *kʷ-*
interrogative/indefinite root. Mann
114, 203. Huld 80. Pokorny 644.
Walde 369.

ENG *here* /h ɪ/ COGNATE WITH 'HE'.
Same demonstrative base as *he*.
OE *hēr*, PG *hēr*, r-locative to PIE
k̂i-, variant of *k̂o* 'this'. Pokorny
609. Skeat. OED. Watkins *ko-*.

FRE *ici* /i/. *I-* may be from *īlicō* 'on the
spot', from *in* + *loco*; *-ci* is a
proximate suffix, from *ecce* 'behold'
(*-ce* a proximate particle in *hic*,
etc.) + *hīc* 'here' (see LAT).
Dauzat. BW. EM *ilico*.

GER *hier* /h i:/. OHG *hia(r)*, PG *hēr*,
see ENG. Kluge.

HAW *nei* /n e i/. PNP *nei*, PPN *ni*. PE.

LAT *hic* /h i:/ COGNATE WITH 'THIS'.
PIE *gheyk̂e*, root *gh*. Pokorny 417.
EM *hic*. Sihler 260.

NAV *kweʼé* /kw/ COGNATE WITH 'FEW'.

Cf. *kwii* 'here', *kǫ́ǫ́* 'here', etc. YM 1992, p. 936. Hoijer 10.

TUR *burada* /b/ DERIVED FROM 'THIS', COGNATE WITH 'I', 'WE'. Locative of *bura* 'this place', from stem *bu* 'this'. Räsänen 85a.

hit verb, Swadesh 200

ALB *qëllon* /c ə ł/. Note *qëllim* 'aim'. Not in Huld, Janson, Mann.

ENG *hits* /h ɪ t/ LOAN. Late OE *hyttan*, apparently from ON *hitta* 'hit upon', PG *hitjan*, PIE root *keid* 'fall'. Barnhart. Pokorny 542. Skeat. OED.

FRE *frappe* /f r a p/ LOAN, MOTIVATED. Dauzat says it is a borrowing from Frankish, BW considers it onomatopoeia. Dauzat. BW.

GER *schlägt* /ʃ l a: g/. OHG *slahan*, PG *slahan*, PIE *slak*. Pokorny 959. Kluge *schlagen*. Watkins *slak-*.

HAW *ku'i* /k u ʔ/. PPN *tuki*. Possibly transitive *-i*. PE.

LAT *ferit* /f e r/ COGNATE WITH 'RUB'. PIE *bher* 'apply a sharp tool to'. Pokorny 134. Walde 160*.

NAV *ńdiists'in* /ts' i n/ COGNATE WITH 'BONE'. 'I hit someone one time with my fist'. YM 1992, p. 638 *TS'IN*.

TUR *vuruyor* /v u r/. Räsänen 515a.

hold verb, Swadesh 200

ALB *mban* /b a/ COGNATE WITH 'BELLY', 'HUSBAND', 'MAN'. Prefix *en-*, but root very obscure. Several subscribe to a derivation from *bher* 'carry', with iterative *ey* suffix. Pokorny 130. Janson 165. Huld 90. Mann 170. Walde 154*.

ENG *holds* /h o l/. Angl. OE *haldan*, PG *haldan*, perhaps PIE root *kel* 'drive'. Pokorny 548. Barnhart. Skeat. OED. Watkins *kel-3*.

FRE *tient* /t ə n/ COGNATE WITH 'HEAR', 'NOW'. Pop. Lat. *tenīre*, Class. LAT *tenēre*. BW.

GER *hält* /h a l/. OHG *haltan*, PG *haldan*, see ENG. Pokorny 548.

HAW *pa'a* /p a ʔ a/ COGNATE WITH 'FREEZE'. *Pa'a* means 'firm, solid, frozen'. Passive stem is *pa'ah-*. PE.

LAT *tenet* /t e n/ COGNATE WITH 'THIN'. Stative suffix *ē* on PIE root *ten* 'stretch'. EM *teneo*. Pokorny 1066. Walde 723. Watkins *ten-*.

NAV *yótą'* /t ã ʔ/. PA *tim'*. YM 1992, p. 483 *TĄ'*.

TUR *tutuyor* /t u t/. Räsänen 502a.

horn noun, Swadesh 100

ALB *bri* /b r i r/. Pokorny groups with PIE *bhrentos* 'deer'. Pokorny 168. Mann 92. Walde 205*.

ENG *horn* /h ɔ r/. OE *horn*, PG *hurnaz*, PIE *kṛn-*, *n*-extension of zero grade of PIE root *ker* 'top of body'. Pokorny 576. OED. Skeat. Watkins *ker-1*.

FRE *corne* /k ɔ r/. Pop. Lat. *corna*, from Class Lat. *cornua*, orig. pl. of LAT *cornu*. Dauzat. BW.

GER *Horn* /h o r/. OHG *horn*, PG *hurnaz*, see ENG. Kluge *Horn*.

HAW *kiwi* /k i w i/. PE.

LAT *cornu* /k o r/. PIE *kṛn-*, see ENG. EM *cornu*. Sihler 323. Walde 406.

NAV *'adee'* /d e: ʔ/. YM 1992, p. 974. Hoijer 34.

TUR *boynuz* /b o j n u z/. Räsänen 347.

hot adj, Swadesh 100

ALB *nxehtë* /dz e h/. *-të* adj. to *nxeh* 'I heat'. There are by-forms without the *n-*; root must be *xeh*.

ENG *hot* /h ɑ/. OE *hāt*, PG *haitaz*, PIE *kaid-*, root *kai* 'heat'. Pokorny 519. OED. Skeat. Watkins *kai-*.

FRE *chaud* /ʃ o/. Lat. *caldus*, sync. from Class. LAT *calidus*. Dauzat. BW.

GER *heiß* /h ai/. OHG *heiz*, PG *haitaz*, see ENG. Pokorny 519. Kluge *heiß*.

HAW *wela* /w e l a/. PPN *wela*.

LAT *calidus* /k a l/. Deverbative in *-id-* from stative *caleo* 'be hot', PIE root *k̂elH₁* 'warm'. Pokorny 551. EM *caleo*. Sihler 629. Walde 430. Watkins *kelə-1*.

NAV *sido* /d o:/. 'It is hot' (an object like a skillet, as opposed to an area like a room). Neuter perfect. Perhaps related to *-do* 'amorphous substance moves rapidly', PA *dugh*. YM 1992, p. 145 *DOII*. Pinnow 37.

TUR *sıcak* /s ɯ/. Cf. *sıtma* 'ague'. Räsänen 173b.

human noun, Swadesh 100

ALB *njerí* /ɲ e r/. PAlb *neri*, earlier *nɛr-in-*, PIE H_2ner. Mann 85. Huld 100. Walde 32*. Pokorny 765.

ENG *human* /h j u m/ LOAN. Loan from OF *humain*, see FRE. OED. Skeat. Watkins *dhghem-*.

FRE *humain* /y m/ LOAN, COGNATE WITH 'MAN'. Borrowed from Lat. *hūmānus*, adj. ending on stem of *homo*, see LAT. OED. Dauzat. BW.

GER *Mensch* /m a n/ DERIVED FROM 'MAN'. OHG *mennisco*, PG *manniskaz* 'human', PIE *manus* 'man', sometimes associated with root *men* 'think'. Watkins *man-1*.

HAW *kanaka* /k a/ COGNATE WITH 'CHILD', 'HUSBAND', 'MAN'. PPN *tangata*, POc, PMP *tau(mataq)*. PE. Blust.

LAT *homo* /h o m/. OLat *hemo*, PIE *dĝhmon* or *dĝhomon-*, root *deĝhom* 'earth'. The adj. form is *hūmānus*, with unexplained stem vowel (*/oi/). EM *homo*. Pokorny 415. Sihler 96, 49. Walde 663. Watkins *dhghem-*.

NAV *diné* /n é/. YM 1987, p. 4: *diné* (old stem + prefix noun). YM 1992, p. 422: 'man, person, Navajo'.

TUR *adam* /a d m/ LOAN. Loan from Arabic *ādam*. Stachowski, v. 1, p. 13. Wehr 12b.

hunt verb, Swadesh 200

ALB *gjuan* /ʒ/. Cf. the noun *gjah*, 'a hunt'. Huld derives from *yeH₂ĝh* plus *-ni-*. Mann 190. Huld 70. Pokorny 492.

ENG *hunts* /h ə n t/. OE *huntian*, PG *huntōjan*, deriv. of *hentan* 'seize', suggesting a PIE *kend*. Barnhart. Skeat. OED.

FRE *chasse* /ʃ a/. Pop. Lat. *captiare*, Class. *captare* 'seek to capture', PIE root *kap* 'grasp'. Dauzat. BW. Watkins *kap-*.

GER *jagt* /j a: g/. OHG *jagōn*, PIE root *yeH₂gh*. Pokorny 502. Kluge *jagen*.

HAW *hahai* /h a/. Base *hai* has same meaning. PCP *fai* or *sai*. Possibly transitive *-i*. PE.

LAT *venatur* /w e: n/. PIE *wēnā-*, root *wen* 'strive'. Pokorny 1147. EM *venor*. Walde 230. Watkins *wen-1*.

NAV *haalzheeh* /ʒ e: ʔ/. 'He is out hunting', durative imperf. PA *zhèG*, PPA *she'G*. YM 1992, p. 769 *ZHEE'*. Kari 50.

TUR *avlıyor* /a v/. Cf. *av* 'the hunt'. Räsänen 1a.

husband noun, Swadesh 200

ALB *burrë* /b u r:/ COGNATE WITH 'BELLY', 'HOLD', 'MAN'. Same word as 'man'. PIE *bhr̥nos* 'boy', perhaps PIE *bher* 'carry', like 'belly'. Mann 45. Huld 46. Pokorny 130, 148. Walde 141, 154.

ENG *husband* /b ə/ LOAN. Late OE formation, borrowed or adapted from ON *húsbóndi*, where the head is the ON *bóndi* 'freeholder', from a Germanic root *bōwan* 'dwell', PIE *bhōw-*, root *bheuH* 'grow'. Pokorny 146. Barnhart. Skeat. OED. Watkins *bheuə-*.

FRE *mari* /m a r i/. LAT. BW.

GER *Gatte* /g a t/ COGNATE WITH 'GOOD', 'WIFE'. OHG *gegate*, PWG *gagadōn*, PIE root *ghedh* 'unite'. Pokorny 423. Kluge.

HAW *kāne* /k a:/ COGNATE WITH 'CHILD', 'HUMAN', 'MAN'. Same word as 'man'. PE.

LAT *maritus* /m a r i:/. *-to-* suffix ('provided with') on PIE *meri*, *mari* 'young woman'. Pokorny 739. EM *maritus*. Walde 281*. Watkins *mari-*.

NAV *hastiin* /t i h/ COGNATE WITH 'MAN'. '(Mature) man, husband'. Perhaps from root *tih* 'old'. YM 1992, p. 978.

TUR *koca* /k o dʒ a/ LOAN. 'Old man, husband'. Loan from Persian *xoja* 'eunuch'. Räsänen 274a. Steingass 482a.

I link, Swadesh 100

ALB *unë* /u/. The *-në* is an optional extension. The *u* remains unexplained, but could conceivably be related to PIE *eĝo*. Mann 114. Huld 122.

ENG *I* /ai/. OE *ic*, PG *ik*, PIE *eĝ*. Skeat. Pokorny 291. OED. Watkins *eg.*

FRE *je* /ʒ ə/. Pop. Lat. **eo*, Class. LAT *ego.*

GER *ich* /i x/. OHG *ih*, PG *ik*, see ENG. Pokorny 291. Kluge.

HAW *au* /a u/. Also spelt *wau*. PPN *au*, POc *aku*, PMP *i-aku*. PE. Blust.

LAT *ego* /e g/. PIE *eĝoH*, extension of *eĝ*. EM *ego*. Pokorny 291. Sihler 369. Walde 115.

NAV *shí* /ʃ í/. Emphatic independent, subjective case. YM 1992, p. 930.

TUR *ben* /b/ COGNATE WITH 'HERE', 'THIS', 'WE'. Perhaps the /b/ is shared with *bu* 'this'. Räsänen 66a, 333b.

ice noun, Swadesh 200

ALB *akull* /a k u ł/. Huld 37. Pokorny 524.

ENG *ice* /ai s/. OE *īs*, PG *īsa-*, PIE *eis*, perhaps a root *ei*. Skeat. OED. Pokorny 301. Watkins *eis-2.*

FRE *glace* /g l/ COGNATE WITH 'FREEZE'. Pop. Lat. *glacia*, Class. LAT *glacies*. BW.

GER *Eis* /ai s/. OHG *īs*, PG *īsa-*, see ENG. Kluge *Eis.*

HAW *hau* /h a u/ COGNATE WITH 'SNOW'. PPN *sau*, POc *sau* 'breeze'. PE. Ross.

LAT *glacies* /g l/ COGNATE WITH 'FREEZE'. PIE *gel* 'cold, freeze'. Pokorny 365. EM *glacies*. Walde 622. Watkins *gel-3.*

NAV *tin* /t i n/ COGNATE WITH 'FREEZE'. Verb stem *tin* 'freeze'. Hoijer 26. YM 1992, p. 508.

TUR *buz* /b u z/. Räsänen 91a.

if link, Swadesh 200

ALB *në qoftë se* /c o f/. The head word *qoftë* is not in Huld, Janson or Mann.

ENG *if* /ɪ/ COGNATE WITH 'ONE'. OE *gif*, cogn. with Ger. *ob*. Possibly PIE pronom. stem *e-/i-* with an emphatic *ba* particle. Skeat.

Barnhart. Pokorny 284. OED. Watkins *i-*.

FRE *si* /s i/. LAT.

GER *wenn* /v/ COGNATE WITH 'WHAT', 'WHO'. Variant of *wann* 'when', OHG *hwanne*, PG *hwan-*, PIE root *kʷ*. Kluge *wenn*. Pokorny 810. Watkins *kʷo-*.

HAW *inā* /i n aː/. *I* can be used in the same sense; is it the base? PE.

LAT *si* /s/. OLat *sei*, PIE root *se*. EM *si*. Pokorny 884. Sihler 395.

NAV *-ḍąą* /d ãː/. Haile 98.

TUR *eğer* /e ɣ e r/ LOAN. Loan from Persian *agar*. Räsänen 37a. Steingass 90b.

in link, Swadesh 200

ALB *në* /n ə/ COGNATE WITH 'AT'. Same word as 'at'. Source would seem to be PIE *endo*, from *en*. Mann 207. Huld 97.

ENG *in* /ɪ n/ COGNATE WITH 'AND'. OE *in*, PG *in*, PIE *en*. Skeat. OED. Pokorny 311. Watkins *en*.

FRE *dans* /ã/ COGNATE WITH 'GUTS'. Lat. *de-intus*. *Intus* is from PIE *en*. Dauzat. BW. Pokorny 311.

GER *in* /i n/ COGNATE WITH 'AND'. OHG *in*, PG *in*, PIE *en*.

HAW *i loko* /l o k o/ COGNATE WITH 'LAKE'. PPN *loto*, POc *ralom*, PMP *dalem*. PE.

LAT *in* /i n/ COGNATE WITH 'GUTS'. OLat *en*, PIE *en*. EM *in*. Pokorny 311. Sihler 439. Walde 126. Watkins *en*.

NAV *bii'* /i: ʔ/. Stem postposition. YM 1992, p. 243 *II'*.

TUR *içinde* /i tʃ/ COGNATE WITH 'BECAUSE'. Räsänen 121b.

kill verb, Swadesh 100

ALB *vret* /r/. Mann suggests a compound of *apo-* and a root like *reut*, perhaps 'fall'. Mann 165.

ENG *kills* /k ɪ l/. Possibly a reworking or a zero-grade of OE *cwellan*, PIE root *gʷel* 'pierce'. Barnhart. Pokorny 471. Skeat. OED. Watkins *gʷel-1.*

FRE *tue* /t y/. Earlier 'extinguish', from Pop. Lat. *tutare*, Class. Lat. *tutari* 'protect', *tueri*, PIE *teu* 'turn one's attention to'. Dauzat. BW. Pokorny 1079.

GER *tötet* /t ø:/. Adj. *tot* 'dead', OHG *tot*, PG *daudaz*, PIE root *dheu* 'vanish'. Pokorny 260. Kluge *tot*. Watkins *dheu-3*.

HAW *pepehi ā make* /p e h/. Intensive of *pepehi*, which alone means 'beat' or 'kill'. PNP *pepesi*. Reduplicative? *Pehi* alone means 'throw (at)' (PNP *peesi*). Possibly transitive *-i*. PE.

LAT *interficit* /f a/. Base is weakened form of *facio* 'do, put', PIE *dhə₁kyo*, root *dheH₁* 'place'. EM *facio*. Pokorny 736. Watkins *dhē-1*.

NAV *'iishhé* /ɣ é:/. 'I kill a single unspecified thing', conclusive imperfect. PA *gheny*, PPA *ɣe*. YM 1992, p. 709 *YĮ́*. Pinnow 27.

TUR *öldürüyor* /ø l/ DERIVED FROM 'DIE'. Causative of *ölmek* 'die'. Räsänen 371a.

knee noun, Swadesh 100

ALB *gju* /ɟ u r/. PAlb. *gʎun*. This can be related to the other IE langs here if assume a dissimilation of /n/ to /l/. Otherwise, a remote possibility of PIE *gleu*, extension of *gel* 'ball'. Mann 93. Janson 28. Huld 70. Pokorny 362.

ENG *knee* /n i/. OE *cnēo*, PG *knewam*, PIE *ĝnewo-*, root *ĝenu* 'knee'. Pokorny 381. Skeat. OED. Watkins *genu-1*.

FRE *genou* /ʒ ə n u/. Backformed from plural *genous*, Pop. Lat. *genuclum*, Class. *geniculum*, from LAT *genu*, see LAT. BW. Pokorny 381. Dauzat.

GER *Knie* /k n i:/ COGNATE WITH 'BONE'. OHG *knio*, PG *knewam*, see ENG. Pokorny 381. Kluge *Knie*.

HAW *kuli* /k u l i/. PPN *turi*.

LAT *genu* /g e n u:/. PIE *ĝenu* 'knee'. Possibly connected to *ĝenH* 'be born'. EM *genu*. Pokorny 381. Sihler 67, 320. Walde 586. Watkins *genu-1*.

NAV *'agod* /g o d/. Dependent stem noun. YM 1992, p. 978, 213.

TUR *diz* /d i z/. Räsänen 482a.

knife noun, Swadesh 200

ALB *thikë* /θ i k/. Mann suggests PIE *k̂uk-*. Not in Janson, Huld, Pokorny. Mann 66.

ENG *knife* /n/. OE *cnīf*, PG *knībaz*, ext. of PIE *gen* 'compress'. Pokorny 370. OED. Barnhart. Skeat. Watkins *gen-*.

FRE *couteau* /k u/. LAT *cultellus*, from *culter*, see LAT. BW. Dauzat.

GER *Messer* /r/. OHG *mezzisahs*, compound of *mezzi* (PWG *matiz* 'food') and *sahs* 'sword' (PG *sahsam*, PIE *sokso-*, root *sek* 'cut'). The only remnant of *sahs* is its initial, rhotacized in final *-r*. Pokorny 895. Kluge *Messer*. Watkins *sek-*.

HAW *pahi* /p a h i/. PCP *pafi* or *pasi*. PE.

LAT *culter* /k u l/. *kel-tro-*, PIE root *(s)kel* 'cut'. Is there some connection with *(s)ker* 'cut'? Some posit a dissimilation. EM *culter*. Pokorny 924. Walde 592*. Watkins *skel-1*.

NAV *béésh* /b é: ʃ/. Independent stem noun. PA *wesh*. YM 1980, p. 2.

TUR *bıçak* /b ɯ tʃ/. From *bıç* 'cut'. Räsänen 73a.

know verb (facts), Swadesh 100

ALB *di* /d i/. PIE *dhiH*. Huld 152. Mann 169.

ENG *knows* /n o/. OE *cnāwan*, PG *knē-*, PIE *gnē-*, root *gnoH*. Skeat. OED. Pokorny 376. Watkins *gnō-*.

FRE *sait* /s a v/. Pop. Lat. **sapēre*, Class. *sapēre*, from 'understand', PIE root *sap* or *sep* 'taste'. Dauzat. BW. Pokorny 880. Watkins *sep-1*.

GER *weiß* /v ai s/. PG *wait*, PIE *woida*, root *weid* 'see'. Pokorny 1125. Kluge *wissen*.

HAW *'ike* /ʔ i k e/ COGNATE WITH 'SEE'. Same word as 'see', which is original sense (POc 'know' is *taqu*). PPN *kite*, PMP *kita*. Blust.

LAT *scit* /s k/ COGNATE WITH 'CUT'. PIE *skei* 'cut', extension of *sek* (Lat. *seco* 'cut'). EM *scio*. Pokorny

919. Walde 542*.
NAV *bééhózin* /z i n/ COGNATE WITH
'THINK'. Neuter, impersonal. PA
zin, PPA *sin*. YM 1992, p. 753 *Zį́į́'*.
TUR *biliyor* /b i l/. Räsänen 75b.

lake noun, Swadesh 200
ALB *liqen* /l i c/ LOAN. Explained as a
loan from Lat. **lacōnis*, der. of
lacus (see LAT); or from Greek
lekánē or *lekánion* 'hollow, basin'.
Mann 93. Janson 205. Pokorny 308,
653. Not in Huld, Pokorny.
ENG *lake* /l e k/ LOAN. Borrowed from
OF *lac*, see FRE; perhaps crossed
with the native word *lake* 'stream'.
Skeat. OED. Watkins *laku-*.
FRE *lac* /l a k/ LOAN. 12th cent. loan
from LAT *lacus*. OED *lake*. BW.
GER *See* /z e:/ COGNATE WITH 'SEA'.
OHG *sēo*, PG *saiwiz*. Same word
as *See* 'sea', differentiated by
gender since 16th c. Possibly PIE
sāi 'pain'. Kluge *See*. Pokorny 877.
HAW *loko* /l o k o/ DERIVED FROM 'IN'.
Perhaps this means an 'interior'
body of water? PPN *loto*, replacing
POc *rano*. PE. Blust.
LAT *lacus* /l a k/. PIE *laku*. Pokorny
653. EM *lacus*. Walde 380*.
Watkins *laku-*.
NAV *tooh siyínígíí* /t o: h/ DERIVED
FROM 'RIVER', COGNATE WITH
'SEA', 'WATER'. *Tooh* 'body of
water' (cf. *tó* 'water') *siyínígíí*
'standing'. YM 1992, p. 717, 520,
518.
TUR *göl* /g ø l/. Räsänen 288b.

laugh verb, Swadesh 200
ALB *qesh* /c e ʃ/. Not in Janson, Huld,
Mann, Pokorny.
ENG *laughs* /l/ MOTIVATED. Angl. OE
hlœhhan, PG *hlahjan*, PIE *klak*,
probably echoic. Or this could tie
in more directly with Lat. *clango*
as *kleg*, an extension of *kel* 'call',
itself described as onomatopoetic.
OED. Pokorny 599. Watkins *kleg-*.
FRE *rit* /r i/ COGNATE WITH 'WORM'.
Pop. Lat. **rīdĕre*, Class. *rīdēre*, see
LAT. PIE root possibly *wer* 'turn'.
BW.

GER *lacht* /l/ MOTIVATED. OHG
hlahhan, PG *hlahjan*, see ENG.
Kluge *lachen*.
HAW *'aka* /ʔ a k a/ MOTIVATED. PPN
kata. Onomatopoeia? Replaces
POc *malip*. PE. Blust.
LAT *ridet* /r/ COGNATE WITH 'WORM'.
Perhaps PIE *wrizd*, -d- present of
s-extension of *wrei*, an extension of
wer 'turn'. Pokorny 1158. EM
rideo. Walde 277.
NAV *yidloh* /dˡ o ʔ/. *He is laughing*,
cursive. PA *dlə̀K'*, PPA *di-łiKʷ*.
YM 1992, p. 156. YM 1980, p.
xxvii. Pinnow 23.
TUR *gülüyor* /g y l/. Räsänen 307b.

leaf noun, Swadesh 100
ALB *gjethe* /ɟ e/. PAlb *geθ-*, perhaps
earlier *gaθi*, PIE *gʷozdo-*, ext. of
gʷes 'foliage'. Huld 69. Walde 644.
Pokorny 480.
ENG *leaf* /l i/. OED *lēaf*, PG *laubaz*,
PIE *leup*, prob. ext of *leu* 'cut off'.
Skeat. OED. Pokorny 690. Watkins
leup.
FRE *feuille* /f œ j/ COGNATE WITH
'FLOWER', 'RIVER', 'SWELL'. Pop.
LAT *folia*, reanalysed from plural
of Class. *folium*, q.v. BW. Dauzat.
GER *Blatt* /b l/ COGNATE WITH
'BLOOD', 'BLOW', 'FLOWER'. OHG
blat, PG *bladaz*, PIE *bhel* 'bloom'.
May be related to *bhel* 'blow, swell'
and *bhel* 'flow'. Sihler 529. Walde
179*. EM. Kluge *Blatt*. Pokorny
121. Watkins *bhel-3*.
HAW *lau* /l a u/ COGNATE WITH 'HAIR',
'STICK', 'TREE'. Is this POc
ndraun, PMP *dahun*? PE. Blust.
LAT *folium* /f o l/ COGNATE WITH
'BLOW', 'FLOW', 'FLOWER',
'RIVER', 'YELLOW'. PIE *bholyom*,
root *bhel* 'bloom'. See GER.
Pokorny 122. EM *folium, flos*.
Sihler 42. Walde 176*.
NAV *'at'ąą'* /t' ã: ʔ/ COGNATE WITH
'THIN'. PA *t'an'*. YM 1992, p. 995,
534.
TUR *yaprak* /j a p r a k/. Räsänen 188a.

left lim (hand), Swadesh 200
ALB *majtë* /m a j/. Not in Janson,

Mann, Huld.

ENG *left* /l ɛ f/. Kentish OE *left*, PIE *lupt*, root *(s)leup* 'hang limply'. Skeat. Barnhart. Pokorny 964.

FRE *gauche* /g o ʃ/ LOAN. Verb *gauchir*, prob. from *guenchir* 'detour', from Frankish **wenkjan*, PIE *weng* 'bend', perhaps n-infix on a root *weg*; crossed with Frankish *walkan*, PIE *wolg*, ext. of *welH* 'turn'. BW. Dauzat. Pokorny 1148, 1144.

GER *link* /l i k/. OHG *lenca*, originally 'clumsy', probably PIE *(s)lēg* 'limp'. Pokorny 959. Kluge *link*.

HAW *hema* /h e m a/. PPN *sema*, orig. meaning 'awkward'. PE. Blust.

LAT *sinister* /s i n/. Double comparative, base would be *sen* or *sin*, perhaps connected to *sen-* 'old' as in Provençal *ma sanega*. Pokorny 1147. EM *sinister*. Sihler 40.

NAV *nishtł'a* /tˡ a h/. Verb 'be disadvantaged'. PA *tł'ighixy*. YM 1992, p. 575.

TUR *sol* /s o l/. Räsänen 426b.

lie verb (stative), Swadesh 100

ALB *(rri) shtrirë* /ʃ t r/. *Shtrirë* 'stretched out', from *shtrij* 'I stretch', PAlb *štriñ*, PIE *strni-*, root *ster* 'spread'. Huld 115. Pokorny 1030. Mann 195. Walde 639*.

ENG *lies* /l ai/. OE *licgan*, PG *ligjan*, PIE *legyo-*, root *legh* 'lie'. Skeat. OED. Pokorny 658. Watkins *legh-*.

FRE *(est) allongé* /l/ DERIVED FROM 'LONG', COGNATE WITH 'FAR'. Root *long* 'long', PIE *dloH₃nghos*, root *del*. BW. Dauzat.

GER *liegt* /l i: g/. OHG *liggen*, PG *ligjan*, see ENG. Kluge *liegen*.

HAW *moe* /m o e/ COGNATE WITH 'SLEEP'. Same word as 'sleep'. PPN *mohe*. PE.

LAT *iacet* /j a/ DERIVED FROM 'THROW'. *ē*-stative of *iacio* 'throw', i.e. 'be thrown down', PIE *yak-*, root *yeH₁*. EM *iaceo*. Tucker. Walde 199. Watkins *yē-*.

NAV *sitį* /t e/. 'He lies', neuter perfect. PA *teny*, PPA *te*. YM 1992, p. 509

Tį̄.

TUR *yatıyor* /j a t/. Räsänen 192b.

liver noun, Swadesh 100

ALB *mëlçi* /m ə l tʃ i/.

ENG *liver* /l ɪ/. OE *lifer*, PG *librōn*. Apparently not connected to Lat. *iecur*, but to PIE *leip* 'fat' (i.e., way of cooking liver), perhaps extension of *lei* 'slimy'. OED. Skeat. Pokorny 670, 504. Watkins *leip-*.

FRE *foie* /f w a/ LOAN. Cross between *ficātum* '(cooked) with figs', from *ficus* 'fig', presumably from some native Mediterranean language; and *sécotom*, Greek *sykōtón* '(cooked) with figs', from same unknown source. BW. Dauzat. EM *ficus*.

GER *Leber* /l e:/ COGNATE WITH 'BAD'. OHG *lebara*, PG *librōn*, see ENG. Pokorny 670, 504. Kluge *Leber*.

HAW *ake* /a k e/. PPN *'ate*, POc *qate*, PCEMP, PMP *qatay*. Blust.

LAT *iecur* /j e k/. PIE r/n stem, nom. *yekʷr̥(t)s*, root *yekʷ*. EM *iecur*. Pokorny 504. Sihler 790, 300. Walde 205.

NAV *'azid* /z i d/. YM 1987, p. 4. YM 1992, p. 1006. Hoijer 12.

TUR *ciğer* /dʒ i ɣ e r/ LOAN. Loan from Persian *jigar*. Räsänen 126a. Steingass 366b.

long adj, Swadesh 100

ALB *gjatë* /ɟ/. PAlb. *gẋatë*, PIE *dlH₃gh* plus adj. *-të*, from root *del* 'long'. Huld 68. Mann 38. Walde 813. Pokorny 197.

ENG *long* /l/. OE *long*, PG *langaz*, PIE *dloH₃nghos*, root *del* 'long'. Skeat. OED. Pokorny 196. Watkins *del-1*.

FRE *long* /l/ COGNATE WITH 'FAR', 'LIE'. LAT *longus*. Dauzat. BW.

GER *lang* /l/. OHG *lang*, PG *langaz*, see ENG. Pokorny 196. Kluge *lang*. Watkins *del-1*.

HAW *loa* /l o a/. This is the same *loa* as in 'die'. PPN *loa*.

LAT *longus* /l/. A PIE form like *dloH₃nghos* would tie LAT, ENG and GER together with ALB. Root

would be *del*. Pokorny 197. EM
longus. Sihler 210. Walde 813.
Watkins *del-1*.
NAV *nineez* /n eː z/. 'It is long', neuter.
-nééz 'long, tall' (adjectival enclitic
or verb stem); low or high tone
root: *neez* or *nééz*. PA *nyèz*, PPA
nye's. YM 1992, p. 938, 430.
TUR *uzun* /u z/ COGNATE WITH 'FAR'.
For stem, cf. *uzak* 'far'. Räsänen
518a.

louse noun, Swadesh 100
ALB *morr* /m o rː/. One guess is
mōrkos 'itcher', from *merH* 'rub
raw'. Walde 278*, Pokorny 737.
ENG *louse* /l au s/. OE *lūs*, PG *lūs*, PIE
lūs. Skeat. OED. Pokorny 692.
Watkins *lūs-*.
FRE *pou* /p/. Backformed from plural
of Pop. LAT **pēduclus*, Class.
pēdiculus. BW. Dauzat.
GER *Laus* /l au z/. OHG *lūs*, PG *lūs*,
see ENG. Pokorny 692. Kluge *Laus*.
HAW *'uku* /ʔ u k u/. PPN, PMP *kutu*.
PE. Blust.
LAT *pediculus* /p eː d/. Diminutive of
pēdis. Perhaps PIE *pezd*. EM *pedis*.
Pokorny 829. Walde 68*.
NAV *yaa'* /j aː ʔ/. Independent stem
noun. PA *ya'*. YM 1987, p. 3. YM
1992, p. 689.
TUR *bit* /b i t/. Räsänen 76b.

man noun (male), Swadesh 100
ALB *burrë* /b u rː/ COGNATE WITH
'BELLY', 'HOLD', 'HUSBAND'. Same
word as 'husband'. PIE *bhr̥nos*
'boy', perhaps PIE *bher* 'carry',
like 'belly'. Mann 45. Huld 46.
Pokorny 130, 148. Walde 141, 154.
ENG *man* /m a n/. OE *mann*, PG
mannaz from *manwaz*, PIE *manus*
or *monus* 'man', possibly from root
men 'think'. Barnhart. Skeat.
OED. Pokorny 700. Watkins
man-1.
FRE *homme* /ɔ m/ COGNATE WITH
'HUMAN'. LAT. *hominem*, acc. of
homo 'person'. Dauzat.
GER *Mann* /m a n/ COGNATE WITH
'HUMAN'. OHG *man*, PG *mannaz*,
see ENG. Pokorny 700. Kluge

Mann.
HAW *kāne* /k aː/ COGNATE WITH
'CHILD', 'HUMAN', 'HUSBAND'.
'Male, husband'. PPN *ta'ane*,
replacing POc *maRuqane*. Do we
therefore have *ka-ane*, with same
initial morpheme as *ka-naka*
'person', perhaps even *keiki*
'child'? Blust. PE.
LAT *vir* /w i/. PIE *wīros*, root perhaps
weiH 'vital force'. EM *uir*. Pokorny
1128. Sihler 70. Walde 315.
Watkins *wī-ro-*.
NAV *hastiin* /t iː n/ COGNATE WITH
'HUSBAND'. '(Mature) man,
husband'. Perhaps from root *tih*
'old'. YM 1992, p. 978.
TUR *erkek* /e r/. Cf. *er* 'man'. Räsänen
46a.

many lim (pl.), Swadesh 100
ALB *shumë* /ʃ u m/ LOAN. Loan from
Lat. *summa* (**supma*, from PIE
(s)up(er) 'up from below'). Mann
201. Huld 166. Pokorny 1107. EM
sub. Watkins *uper*.
ENG *many* /m ɛ n i/. OE *manig*, PG
managaz, PIE *monogho-*, root
men(e)gh 'abundant'. Pokorny 730.
Skeat. OED. Watkins *menegh-*.
FRE *beaucoup de* /k u/ DERIVED FROM
'CUT'. Root is *coup* 'blow', see
under 'cut'. BW. Dauzat.
GER *viele* /f iː l/ COGNATE WITH 'BIRD',
'FLOW', 'FLY', 'FULL', 'RIVER',
'WING'. OE *filu*, PIE root *pelH*
'pour, fill'. Pokorny 799.
HAW *nui* /n u i/ COGNATE WITH 'BIG'.
Same word as 'big'.
LAT *multi* /m u l/. PIE *ml̥t-*, root *mel*
'big'. EM *multus*. Pokorny 720.
Walde 292*. Watkins *mel-4*.
NAV *łą'í* /l ã:/. From verb *łąą* 'be
many'. PA *lany*, PPA *łany*. YM
1992, p. 951, 369.
TUR *çok* /tʃ o k/ LOAN. Possibly
loanword from Arm. /jok/ 'crowd'.
Dankoff, p. 168. Räsänen 113b.

meat noun, Swadesh 100
ALB *mish* /m i ʃ/. PAlb *mis*, PIE
memsom. Pokorny 725. Mann 49.
Huld 92.

ENG *meat* /m i t/. OE *mete*, PG *matiz*, PIE root *mad* 'wet'. Pokorny 694. OED. Skeat. Watkins *mad-*.

FRE *viande* /v j/. Orig. 'food', Pop. Lat. *vīvanda*, Class. *vīvenda*, verb *vivo* 'live', PIE *gʷiHwo-*, root *gʷeyH*. Dauzat. BW. Watkins *gʷei-*.

GER *Fleisch* /f l ai/. OHG *fleisc*, PWG *flaiskjan*, perhaps PIE *ploiksk-*, root *pleH₁k̂*, perhaps extension of *plē* 'split off'. Pokorny 835. Kluge *Fleisch*. Watkins *plēk-*.

HAW *'i'o* /ʔ i ʔ o/. PPN *kiko*. PE.

LAT *caro* /k a r/ COGNATE WITH 'BARK'. Originally 'a slice of meat', n-extension of PIE root *(s)ker-* 'cut'. EM *caro*. Pokorny 939. Sihler 295. Walde 575*. Watkins *sker-1*.

NAV *'atsį'* /ts ĩ ʔ/. Dependent stem noun. YM 1992, p. 1001, 610.

TUR *et* /e t/. Räsänen 52a.

moon noun, Swadesh 100

ALB *hënë* /h ə n/. PAlb *haNë*, PIE *skandnā*, root *(s)kand* 'shine'. Huld 74. Pokorny 526. Walde 352.

ENG *moon* /m u/. OE *mōna*, PG *mænōn-*, PIE base *mēn-*, root *meH₁* 'measure'. OED. Pokorny 731. Skeat. Watkins *mē-2*.

FRE *lune* /l y/. LAT. BW. Dauzat.

GER *Mond* /m oː/. OHG *māno*, PG *mænōn-*, see ENG. Pokorny 731. Kluge *Mond*.

HAW *mahina* /h i n a/. PPN *maasina*. Possibly the base is *hina*; note forms *mahinahina* 'pale moonlight' and *hina* 'grey'. Cf. POc *pulan*. PE. Blust.

LAT *luna* /l uː/. PIE *louksnaH₂-*, root *leuk* 'shine'. Pokorny 687. EM *luc-*. Sihler 56. Walde 409*. Watkins *leuk-*.

NAV *'ooljéé'* /dʒ éː/ COGNATE WITH 'DAY'. Root appears to be *jéé'*, which is cognate to *jį́* 'day'. YM 1992, p. 261, 1069.

TUR *ay* /a j/. Räsänen 10a.

mother noun, Swadesh 200

ALB *nënë* /n ə/ MOTIVATED. PAlb *naNë*. Nursery origin is obvious. Huld 98. Janson 98. Pokorny 754.

ENG *mother* /m ə/ MOTIVATED. OE *mōdor*, PG *mōðar-*, PIE *maH₂ter*; kinship suffix added to nursery word *ma* 'mother, breast'. Watkins *māter-*.

FRE *mère* /m ɛ/ MOTIVATED. LAT. BW.

GER *Mutter* /m u/ MOTIVATED. OHG *muoter*, PG *mōðar-*, see ENG.

HAW *makuahine* /m a k u a/ COGNATE WITH 'FATHER'. *Makua* 'parent' (PPN *matuʔa*) + -*hine* 'female'. Replaces POc *tina*. PE. Marck. Blust.

LAT *mater* /m aː/ MOTIVATED. PIE *maH₂ter*, see ENG. Pokorny 700, 694. EM *mater*. Sihler 50, 293. Walde 229*. Watkins *māter-*.

NAV *'amá* /m á/ MOTIVATED. PA is *han*. Hoijer 11. YM 1992, p. 987, 395.

TUR *anne* /a n n e/ MOTIVATED. Not widespread in Turkic, probably nursery word. Räsänen 19b.

mountain noun, Swadesh 100

ALB *mal* /m a l/. PIE *ml̥H₃dho*. Pokorny has *molno*, from PIE *mel* 'emerge'. Mann 35. Huld 89. Walde 95*. Pokorny 721.

ENG *mountain* /m au n/ LOAN, COGNATE WITH 'MOUTH'. Borrowed from OF *montaigne*, see FRE. OED. Skeat. Watkins *men-2*.

FRE *montagne* /m ɔ̃/. Pop. Lat. *montānea*, Class. *montānus*, from *mons, montis*; see LAT. OED *mountain*. Dauzat. BW.

GER *Berg* /b e r/. OHG *berg*, PG *bergaz*, PIE root *bherĝh* 'high', perhaps ext. of *bher* 'carry'. Watkins *bhergh-2*.

HAW *mauna* /m a u n a/. PPN *ma'unga*. PE.

LAT *mons* /m o n/. PIE *monti-*, root *men* 'be prominent'. EM *mons*. Pokorny 726. Sihler 316. Walde 263*. Watkins *men-2*.

NAV *dził* /dz i ɬ/. YM 1987, p. 3: *dził*, independent stem noun. YM 1992, p. 171.

TUR *dağ* /d a ɣ/. Räsänen 454a.

mouth noun, Swadesh 100

ALB *gojë* /g o j/ LOAN, COGNATE WITH

'GUTS'. PAlb *goχë*, loan from Lat. *gula* 'throat'; this would be *gel*, from *gʷel*, possibly connected to *gʷerH₁* in *zorrë* 'guts'. Pokorny 365. EM 284. Mann 72. Huld 66.

ENG *mouth* /m au/ COGNATE WITH 'MOUNTAIN'. OE *mūþ*, PG *munþaz*, PIE *mn̥tos*, root likely *men* 'project'. Pokorny 732. OED. Skeat. Watkins *men-2*.

FRE *bouche* /b u/ MOTIVATED. Lat. *bucca* 'puffed cheek', ext. of PIE root *beu* 'inflate'. Dauzat. BW. Pokorny 100. EM *bucca*. Watkins *beu-1*.

GER *Mund* /m u n/. OHG *mund*, PG *munþaz*, see ENG. Pokorny 732.

HAW *waha* /h a/ MOTIVATED. PCP *wafa*, PEP *fafa*, POc *papaq*, PCEMP *babaq*, PMP *baqbaq*. I can imagine a symbolic origin of this, because of the labial. PE. Blust.

LAT *os* /o: s/. PIE *oH₁s*. EM *os*. Pokorny 784. Sihler 49. Walde 168. Watkins *ōs-*.

NAV *'azéé'* /z é: ?/. Dependent stem noun. YM 1987, p. 4. YM 1992, p. 741.

TUR *ağız* /a ɣ ɯ z/. Räsänen 8b.

name noun, Swadesh 100

ALB *emër* /e/. PAlb *emën*, earlier *emon-*, PIE *H₁n̥Hmen*. Pokorny 321. Mann 89. Janson 23. Huld 61. Walde 132.

ENG *name* /n e/. OE *nama*, PG *namōn-*, PIE *H₁nHmn̥*, root *H₁noH*. Skeat. OED. Pokorny 321. Watkins *nŏ-men-*.

FRE *nom* /n ɔ̃/. LAT.

GER *Name* /n a:/. OHG *namo*, PG *namōn-*, see ENG. Kluge *Name*.

HAW *inoa* /i n o a/. PPN *hingoa*. PE.

LAT *nomen* /n o:/. PIE *H₁noHmn̥*, root *H₁noH*. EM *nomen*. Pokorny 321. Sihler 85, 297. Walde 132.

NAV *'ázhi'* /ʒ i/. Dependent stem noun with high prefix, from glottalized perfective of 'call by name'. PA *zhi*, PPA *shi*. YM 1992, p. 773 *ZHI'*.

TUR *ad* /a d/.

narrow adj, Swadesh 200

ALB *ngushtë* /n g/ LOAN. PAlb *ngustë*. Loan from LAT *angustus*. Huld 98.

ENG *narrow* /n a r/. OE *nearu*, PG *narwaz*, perhaps from root *(s)ner* 'twist'. Pokorny 975. Skeat. OED. Watkins *(s)ner*.

FRE *étroit* /e t r/. Lat. *strictus*, part. of *stringo*, PIE root *streig*, perhaps extension of *ster* 'stiff'. Dauzat. BW. Pokorny 1036, 1022. Watkins *streig-*.

GER *eng* /e ŋ/. OHG *engi*, PG *anguz*, PIE *angʰu*, root *H₂anĝh* 'constrict'.

HAW *lā'iki* /i k i/ DERIVED FROM 'SMALL'. *Lā'ā* 'wide' + *iki* 'small'.

LAT *angustus* /a n g/. PIE *H₂anĝhostos*, root *H₂anĝh*. EM *ango*. Pokorny 43. Sihler 62. Walde 62. Watkins *angh-*.

NAV *'átts'óózí* /ts' ó z í/. 'It is narrow'. YM 1992, p. 642 *TS'ÓZÍ*.

TUR *dar* /d a r/. Räsänen 463a.

near lim, Swadesh 200

ALB *afër* /a f ə r/. There is much speculation that this is a loan, but nothing very compelling. Huld 36. Mann 196.

ENG *near* /n/. OE *nǣr* 'nearer' (possibly influenced semantically by ON *nǣr* 'near'), comp. of *nēah* 'near', PG *nǣhwiz*, see GER. Pokorny 39. Skeat. OED. Barnhart. Watkins *[nēhw-iz]*.

FRE *près de* /p r/ DERIVED FROM 'SQUEEZE'. Low Lat. *pressē*, see *presser* 'squeze'. BW. Dauzat. Watkins *per-5*.

GER *nah* /n/ COGNATE WITH 'AT'. OHG *nāh*, PG *nǣhwiz*, perhaps PIE local particle *nē*, possibly connected to *an* 'at'. Pokorny 39. Kluge *nah(e)*.

HAW *kokoke* /k o k e/. Base is /koke/ 'fast'. PE.

LAT *prope* /p r o/ COGNATE WITH 'FAR'. Assim. form of PIE *prokʷe*, base *pro*, root *per* 'forward'. EM *prope*. Pokorny 810. Walde 47*. Watkins *per-1*.

NAV *'áhání* /ɣ a n/. Apparently from root *ghan* 'to dwell', PA *g̱han*, PPA *x̱an*. Hoijer 15. YM 1992, p. 223.

TUR *yakın* /j a/ COGNATE WITH 'NEW'. From *yak* 'side'; perhaps connected to *yan* 'side'. Räsänen 180a.

neck noun, Swadesh 100

ALB *qafë* /c a f/ LOAN. Most likely a loan from Tur. *kafa*. Huld 106.

ENG *neck* /n/. OE *hnecca*, PG *hnekkōn-*, PIE *knek-*, ext. of root *ken* 'press together'. Pokorny 558. Skeat. OED. Barnhart. Watkins *ken-5*.

FRE *cou* /k u/. LAT *collum*. BW.

GER *Hals* /h a l/. OHG *hals*, PG *h(w)alsaz*, PIE *kʷolso-*, root *kʷel* 'turn'. Watkins *kʷel-1*.

HAW *'ā'ī* /ʔ a: ʔ i:/. PNP *ka(a)kii*. PE.

LAT *collum* /k o l/. PIE *kʷolso-*, root *kʷel* 'turn'. EM *collum*. Pokorny 639. Walde 434. Watkins *kʷel-1*.

NAV *'ak'os* /k' o s/. No obvious connection to *k'os* 'cloud'. YM 1992, p. 355.

TUR *boyun* /b o j u n/. Räsänen 80a.

new adj, Swadesh 100

ALB *ri* /r i/. Huld 108.

ENG *new* /n u/ COGNATE WITH 'NOW'. OE *nīwe*, PG *neujaz*, PIE *newyo-*, base *new*, ext. of *nū* 'now'. Pokorny 769. Skeat. OED. Watkins *newo-*.

FRE *nouveau* /n u v/. LAT *novellus*, from *novus*. BW.

GER *neu* /n oi/ COGNATE WITH 'NOW'. OHG *niuwi*, PG *neujaz*, PIE *newyo-*, see ENG. Pokorny 769. Kluge *neu*.

HAW *hou* /h o u/. PPN *fo'ou*, POc *paqoRu*, PMP *baqeRu*.

LAT *novos* /n o w/ COGNATE WITH 'NOW'. PIE *newos*, base *new*, see ENG. EM *nouus*. Pokorny 769, 770. Sihler 174. Walde 324*. Watkins *newo-*.

NAV *'ániid* /n i: d/. 'It is new', neuter. YM 1992, p. 449 *NIID*.

TUR *yeni* /j e n/ COGNATE WITH 'NEAR'. From *yan* 'side'. Perhaps connected to *yak* 'side' (*yakın* 'near'). Räsänen 185b.

night noun, Swadesh 100

ALB *natë* /n a/. PIE *nokʷt-*, root *nekʷ* 'get dark'. Huld 96. Walde 338*. Pokorny 762.

ENG *night* /n ai/. OE *niht*, PG *naht*, PIE *nokʷt*, root *nekʷ* 'get dark'. OED. Skeat. Pokorny 762. Watkins *nekʷ-t-*.

FRE *nuit* /n ɥ i/. LAT *noctem*, acc. of *nox* q.v. BW.

GER *Nacht* /n a x/. OHG *naht*, PG *naht*, see ENG. Pokorny 762. Kluge *Nacht*.

HAW *pō* /p o:/. PPN *poo*. PE.

LAT *nox* /n o k/. PIE *nokʷt-*, root *nekʷ* 'get dark'. EM *nox*. Pokorny 762. Sihler 113, 282. Walde 338*. Watkins *nekʷ-t-*.

NAV *tl'éé'* /tˡ é: ʔ/. Indepedent stem noun. YM 1987, p. 3. YM 1992, p. 577.

TUR *gece* /g e dʒ/. Orig. locative of *gec* 'late'. Räsänen 245b.

nose noun, Swadesh 100

ALB *hundë* /h u n/. Pre-PAlb *hun-To*. Huld provides a possible etymology from PIE *naH₂s*. Mann 55. Huld 76.

ENG *nose* /n o z/. OE *nosu*, PG *nusō*, prob. from PIE root *naH₂s*. Skeat. OED. Pokorny 755. Lubotsky, p. 60. Watkins *nas-*.

FRE *nez* /n e/. LAT. BW.

GER *Nase* /n a: z/. OHG *nasa*, PG *nasō*, PIE root *naH₂s*. Kluge *Nase*. Pokorny 755. Lubotsky, p. 60.

HAW *ihu* /i h u/. PPN *isu*, POc *isuŋ*, PCEMP, PMP *ijuŋ*. PE. Blust.

LAT *nasus* /n a: s/. PIE *naH₂s-*. EM *nasus*. Pokorny 755. Walde 318*.

NAV *'áchį́h* /tʃ í: h/. Dependent stem noun. PA *chįxy*. YM 1987, p. 4. YM 1992, p. 94.

TUR *burun* /b u r u n/. Räsänen 90a.

not lim, Swadesh 100

ALB *nuk* /n u/. Mann derives from PIE *nekʷe*. Mann 200. Huld 99.

ENG *not* /n/. Shortened from *nought*, OE *nōwiht*, from *ne* 'not' + *ōwiht* 'aught'. *Ne* is a PIE negative. Pokorny 824. Skeat. OED. Watkins *ne*.

FRE *ne...pas* /p a z/. Ringe 1993 has /nə/, but in Modern French, the negative force is in *pas*. For pronunciation, cf. *pas à pas*. Lat. *passus* 'step', from *pat-no-*, perhaps from PIE root *petH* 'spread'. BW. Pokorny 824. Watkins *petə*.

GER *nicht* /n/. OHG *niwiht*, from **ne aiwin wihtes*. The negative particle is PIE *ne*. Pokorny 824. Kluge *nicht*.

HAW *'a'ole* /ʔ o l e/. Base is *'ole* 'not', PEP *kole*. PE.

LAT *non* /n/. Earlier *nōenum* from **ne oinom*. Base is PIE *ne*. Pokorny 824. EM *non*. Sihler 55. Walde 319*. Watkins *ne*.

NAV *doo...da* /d oː/. YM 1992, p. 960. Hoijer 27.

TUR *değil* /d e ɣ i l/. Räsänen 469b.

now lim, Swadesh 200

ALB *tani* /t a n i/. Best guess is *ta-* (PIE *to*), dem. part. prefixed to *ni* 'now' (PIE *nū*). Mann 199. Pokorny 1086. Not in Huld, Janson.

ENG *now* /n au/ COGNATE WITH 'NEW'. OE *nū*, PG *nū*, PIE *nu*. Skeat. Pokorny 769.

FRE *maintenant* /t ə n/ DERIVED FROM 'HAND', COGNATE WITH 'HEAR', 'HOLD'. From *maintenir*, Pop. Lat. **manūtenīre*, from **manū tenēre* 'hold in hand', *ten-ē-* stative extension of PIE root *ten* 'stretch'. BW. Dauzat. Watkins *ten-*.

GER *nun* /n uː/ COGNATE WITH 'NEW'. OHG, PG, IE *nu*. The *-n* was added in 13th c. Kluge *nun*. Pokorny 769.

HAW *ānō* /ʔ aː n oː/. PE.

LAT *nunc* /n u/ COGNATE WITH 'NEW'. From *nun-ce*, PIE *nu* 'now'. Pokorny 770. EM *num*. Walde 340*.

NAV *k'ad* /k' a d/. YM 1992, p. 951.

TUR *şimdi* /i m/. Cf. *imdi* 'now'. Räsänen 41b.

old adj, Swadesh 200

ALB *vjetër* /v j e t/ LOAN, COGNATE WITH 'YEAR'. PAlb *vjeter*, earlier *vetr̥-*, loan from Lat. *veterem*, q.v. Huld 131. Janson 166. Not in Mann, Pokorny.

ENG *old* /o l/. Angl. OE *ald*, PWG

aldaz, *-to* suffix on PIE *al* 'grow, nourish'. Skeat. OED. Pokorny 26. Watkins *al-3*.

FRE *vieil* /v j ɛ/. Pop. LAT *veclus*, from *vetlus*, from *vetulus*, from *vetus*, q.v. BW. Dauzat.

GER *alt* /a l/. OHG *alt*, PWG *aldaz*, see ENG. Pokorny 26. Kluge *alt*.

HAW *o'o* /o ʔ o/. PCP *oko*. PE.

LAT *vetus* /w e t/. PIE *wet-es-*, root *wet* 'year'. EM *uetus*. Pokorny 1175. Sihler 353. Walde 251. Watkins *wet-2*.

NAV *sání* /s á n/. YM 1992, p. 466 *SẠ́*.

TUR *ihtiyar* /h j r/ LOAN. Arabic *ixtiyār* 'choice, old', root *xyr* 'choose'. Wehr 309a.

one lim, Swadesh 100

ALB *një* /ɲ ə/. PAlb *ñi*, *ñë*, earlier *ni-*, *njo*. Huld sees as PIE *smiH2*, root *sem* 'one'. Pokorny 902. Huld 101.

ENG *one* /w ə/ COGNATE WITH 'IF'. OE *ān*, PG *ainaz*, PIE *oinos*, root *i*. Skeat. OED. Pokorny 281. Watkins *oi-no-*.

FRE *un* /y/. LAT. BW.

GER *eins* /ai/ COGNATE WITH 'SOME'. OHG *ein*, PG *ainaz*, see ENG. Kluge *ein*. Pokorny 281. Watkins *oi-no-*.

HAW *'ekahi* /k a h i/ COGNATE WITH 'FEW', 'SOME'. The /ʔe-/ is a counting prefix that is also used for 'two', etc. PPN *tasi*, PMP *tasa*. PE. Blust.

LAT *unus* /uː/ COGNATE WITH 'HE', 'THAT', 'THERE'. OLat, PIE *oinos*, base *oi*, root *i*. Pokorny 281. EM *unus*. Tucker. Sihler 405. Walde 101. Watkins *oi-no-*.

NAV *łáa'ii* /ɬ á ʔ/ COGNATE WITH 'OTHER', 'SOME'. YM 1992, p. 932.

TUR *bir* /b i r/ COGNATE WITH 'WITH'. Räsänen 76b.

other lim, Swadesh 200

ALB *tjetër* /j e/. *Të* 'the' + *jetër* from PAlb *εtVr-*, PIE *etro-*, comparative of *e* 'that'. Mann 109. Huld 177. Pokorny 284. Not in Janson.

ENG *other* /ə/. OE *ōþer*, PG *anþaraz*, PIE *anteros*, root *an* 'there' with a

comparative suffix -teros. Skeat.
OED. Pokorny 37. Watkins an-2.

FRE autre /o/. LAT alter, orig. 'other of
two', replacing alius in Low Latin.
Same root, PIE al 'beyond'.

GER anderer /a n/. OHG ander, PG
anþaraz, see ENG. Pokorny 37.
Kluge ander.

HAW 'ē a'e /ʔ e:/. 'Ē means 'different,
strange'; PEP kee, PPN kese. A'e
is 'up'. PE.

LAT alius /a l/ COGNATE WITH 'SOME'.
PIE H_2alyos 'other', root H_2al
'beyond'. EM alius. Pokorny 25.
Sihler 206. Walde 86. Watkins al-1.

NAV łah /ł a h/ COGNATE WITH 'ONE',
'SOME'. I don't know how to say
other. Łah 'at one time' is the basis
for expressions like łahdi 'at some
other place', łahgo 'in another
way'. YM 1992, p. 951.

TUR başka /b a ʃ/ DERIVED FROM
'HEAD'. Originally dative of baş
'head'. Räsänen 65a.

path noun, Swadesh 100

ALB shteg /ʃ t e g/ COGNATE WITH
'GO'. PIE stoigho 'path', root
steigh 'step'. Mann 50. Huld 114.
Pokorny 1017. Walde 614*.

ENG path /p a θ/ MOTIVATED. OE pœþ,
PWG paþa-. Usu. referred to PIE
pent 'path', but since the
consonants are not Grimm-shifted
(cf. find, as expected), it is
assumed that this is borrowed from
an Iranian dialect. But it seems
much more likely that it is
onomatopoetic for the sound of
footsteps (Sommer): cf.
pitter-patter. Skeat. OED. Pokorny
809. Kluge Pfad. Watkins pent-.

FRE sentier /s ã/. Extension of LAT
semita, perhaps a *sēmitārius.
BW. Dauzat.

GER Pfad /pf a: d/ MOTIVATED. OHG
pfad, PWG paþa-, see ENG.
Pokorny 809. Kluge Pfad.

HAW ala /a l a/. PPN 'ara. PE.

LAT semita /s e: m/. Possibly related to
meo 'go', trames 'side street'. EM
semita.

NAV 'atiin /t i: n/. Dependent stem

noun. YM 1992, p. 994, 516 'road,
path, trail'.

TUR yol /j o l/. May be Old Turkic
borrowing from Arm. uli. Dankoff,
p. 161. Räsänen 205b.

play verb, Swadesh 200

ALB luan /l u a/. Not in Mann, Huld,
Janson, Pokorny.

ENG plays /p l e/. OE plegean, PWG
plegan. Skeat. OED. Barnhart.
Watkins [plegan].

FRE joue /ʒ u/. Lat. jocare (replacing
Class. ludere), from jocus 'word
play', PIE yokos, root yek 'speak'.
Pokorny 503. BW. Dauzat.
Watkins yek-.

GER spielt /ʃ p i: l/. OHG spilōn, orig.
'dance'. Kluge Spiel.

HAW pā'ani /p a: ʔ a n i/. Could this be
qualitative/stative pā-? PE.

LAT ludit /l u: d/. From lūdus 'game',
perhaps PIE loidos, root leid.
Etruscan loans have also been
hypothesized. Pokorny 666. EM
ludus. Sihler 203. Walde 402*.
Watkins leid-.

NAV naashné /n é/. 'I play',
continuative imperf. YM 1992,
p. 423 NE'.

TUR oynayor /o j/. Räsänen 359a.

pull verb, Swadesh 200

ALB tërheq /h e c/. Prefix tër (tra-) +
heq 'pull, withdraw'. Root is
obscure. Huld suggests a H_2olg^{w}ey-.
Huld 73, 151. Not in Janson,
Pokorny.

ENG pulls /p ʊ l/. OE pullian. Skeat.
OED. Barnhart.

FRE tire /t i r/. Dauzat has it as a loan
from Germanic. BW sees it as a
backformation from martirier
'martyr', taken as 'unfortunate
(mar) pulling'. Replaced traire
(from trahere) in 16th cent.
Dauzat. BW.

GER zieht /ts i:/. OHG ziohan, PG
teuhan, PIE root deuk. Kluge
ziehen.

HAW huki /h u k/. PPN futi. Possibly
transitive -i. PE.

LAT trahit /t r a h/. Would reflect a

PIE *tragh*, a theoretically illegal root shape. Pokorny 1089. EM *traho*. Sihler 148. Walde 752. Watkins *tragh-*.

NAV *yisdzį́įs* /dz ĩː z/. 'I pull an object', completive momentaneous. YM 1992, p. 176 *DZĮ́ĮZ*.

TUR *çekiyor* /tʃ e k/. Räsänen 102b.

push verb, Swadesh 200

ALB *shtyn* /ʃ t y/. PIE *studny-, (s)teud,* root *(s)teu.* Pokorny 1033. Mann 65. Not in Janson, Huld.

ENG *pushes* /p u/ LOAN. Borrowed from OF *pousser,* q.v. Skeat. OED.

FRE *pousse* /p u/ COGNATE WITH 'DUST'. Lat. *pulso,* freq. of *pello* 'drive', PIE *pel.* Dauzat. Pokorny 801. BW.

GER *stößt* /ʃ t oː/ COGNATE WITH 'STICK'. OHG *stōzan,* PG *stautan,* PIE *steud,* root *(s)teu.*

HAW *pahu* /p a h u/. PPN *pasu.* PE.

LAT *trudit* /t r/ COGNATE WITH 'WIPE'. PIE *treud* 'squeeze', perhaps ext. of *terH* 'rub'. Pokorny 1095, 1071. EM *trudo.* Walde 755, 728. Watkins *treud-*.

NAV *béshhííł* /ɣ íː ɬ/. 'I push', momentaneous completive. PA *ghił,* PPA *χił.* YM 1992, p. 703 *YIL.*

TUR *itiyor* /i t/. Räsänen 174b.

rain noun, Swadesh 100

ALB *shi* /ʃ i/. Most popular guess is PIE *seuH* 'juice', but consonant is suspicious. Mann 85, Huld 113. Walde 468*. Pokorny 912.

ENG *rain* /r e/. OE *regn,* PG *regnaz,* PIE *reḱ* 'damp' (or possibly *reĝh,* if Alb. *rrjedh* 'flow' is related). Skeat. OED. Pokorny 857. Watkins *reg-2.*

FRE *pluie* /p l/ COGNATE WITH 'FULL'. Pop. LAT **ploia,* from *pluia,* from Class. *pluvia;* influenced by the verb *plovo* (**pluvo*). PIE root *pleu* 'flow', perhaps from a *pel* 'flow, fill'. BW. EM *pluo.*

GER *Regen* /r eː g/. PG *regnaz,* PIE *reḱ* 'damp'. Pokorny 857. Kluge *Regen.*

HAW *ua* /u a/. PPN *'uha,* POc *qusan,* PMP *quZan,* Proto-Austronesian *quZaL.* PE. Ross. Blust.

LAT *pluvia* /p l/ COGNATE WITH 'FULL'. Verb *pluit* 'it rains', PIE root *pleu* 'flow'. Perhaps an extension of *pel* 'flow, fill'. Pokorny 835. EM *pluo.* Sihler 174. Walde 94*, 55*, 56*. Watkins *pleu-.*

NAV *nı́łtsą́* /ts ã́/. From *tsąąd* 'become pregnant', *tsą́* 'belly'. YM 1992, p. 990.

TUR *yağmur* /j a ɣ/. Cf. *yağmak,* verb. Räsänen 177b.

red adj, Swadesh 100

ALB *kuq* /k u c/ LOAN. Loan from Lat. *cocceus* 'scarlet', from *coccum* 'kermes', from Greek *kókkos.* Mann 112. Huld 84. EM.

ENG *red* /r ɛ d/. OE *rēad,* PG *raudaz,* PIE *roudhos,* root *reudh.* Skeat. OED. Pokorny 872. Watkins *reudh-1.*

FRE *rouge* /r u ʒ/. LAT *rubeus,* PIE root *reudh.* BW. Dauzat.

GER *rot* /r oː t/. OHG *rōt,* PG *raudaz,* see ENG. Pokorny 872. Kluge *rot.*

HAW *'ula* /ʔ u l a/. PPN *kula.* PE.

LAT *ruber* /r u b/. PIE *rudhro,* root *reudh* 'red'. Pokorny 872. EM *ruber.* Sihler 148. Walde 358*. Watkins *reudh-1.*

NAV *łichííh* /tʃ íː ʔ/. 'It is red', neuter imperfective. PA *tshigy,* PPA *tshig.* YM 1992, p. 89: root *CHÍÍʼ,* 'red ochre'.

TUR *kızıl* /k ɯ z/. Cf. *kızmak* 'get red'. Räsänen 269a.

right lim (hand), Swadesh 200

ALB *djathtë* /d j a θ/. PIE, perhaps *deḱsyo-,* root *deḱs,* perhaps from *deḱ* 'take'. Huld 53. Janson 135. Walde 784. Pokorny 190.

ENG *right* /r ai/. OE *riht,* PG *rehtaz,* PIE *reḱtos,* root *reĝ* 'straight'. Skeat. OED. Pokorny 855. Barnhart. Watkins *reg-1.*

FRE *droit* /r w a/ COGNATE WITH 'STRAIGHT'. Same word as 'straight'; replaced *destre* by 16th c. PIE root *reĝ.* BW.

GER *recht* /r e x/. OHG *reht,* PG *rehtaz,* see ENG. Pokorny 855. Kluge *recht.*

HAW *'ākau* /k a u/. PCP *katau*, PPN *ma-ta'u*, POc, PMP *ma-taqu*. *'Ā-* is a common prefix in Hawaiian. PE. Blust.

LAT *dexter* /d e k/. PIE *dek̂siteros*, comp. from *dek̂s-*, perhaps root *dek̂* 'take'. EM *dexter*. Pokorny 190. Sihler 68. Walde 784. Watkins *deks-*.

NAV *nish'ná* /n á/. YM 1992, p. 406.

TUR *sağ* /s a ɣ/.

river noun, Swadesh 200

ALB *lumë* /l u/. Guesses include PIE *slubno* 'be slick' and *lei* 'pour'. Huld 88. Pokorny 664. Walde 392*. Mann 97.

ENG *river* /r ɪ/ LOAN. Borrowed from OF *rivere*, from Pop. Lat. *rīpāria*, from *rīpa* '(river) bank', PIE *reip*, ext. of *rei* 'scratch, tear'. Pokorny 858. Skeat. OED. EM *ripa*. Watkins *rei-1*.

FRE *fleuve* /f l/ LOAN, COGNATE WITH 'FLOWER', 'LEAF', 'SWELL'. Borrowed from LAT *fluvius*, replacing *flum* from Lat. *flumen*; both from verb *fluere* 'flow', PIE *bhleu*, root *bhel* 'blow, swell'. Dauzat. BW. Watkins *bhleu-*.

GER *Fluss* /f l/ DERIVED FROM 'FLOW', COGNATE WITH 'BIRD', 'FLY', 'FULL', 'MANY', 'WING'. OHG *fluz*; a nominalization of *fließen* 'flow', q.v. Kluge *Fluß*.

HAW *kahawai* /k a h a/. *Kaha* (PPN *tafa*) 'place' + *wai* 'water'. PE.

LAT *flumen* /f l/ DERIVED FROM 'FLOW', COGNATE WITH 'BLOW', 'FLOWER', 'LEAF'. Derived from *fluo* 'flow', q.v.

NAV *tooh* /t o/ COGNATE WITH 'LAKE', 'SEA', 'WATER'. Independent stem noun. YM 1992, p. 995, 520, 518. Hoijer 17.

TUR *nehir* /n h r/ LOAN. Loan from Arabic *nahr*, root *nhr* 'flow copiously'. Stachowski, v. 3, p. 20. Wehr 1176a.

root noun, Swadesh 100

ALB *rrënjë* /r: ə/. PIE approx. *wrH₂dnyaH₂*, root *wraH₂d*, like

LAT, ENG. Huld 109. Mann 35, 71. Pokorny 1667.

ENG *root* /r u t/ LOAN. Borrowed from ON *rót*, PIE *wraH₂d* 'branch, root'. Pokorny 1167. Skeat. OED. Watkins *wrād-*.

FRE *racine* /r a/. Low LAT *rādīcīna*, from *rādix*, see LAT. Dauzat. BW.

GER *Wurzel* /v u r ts/. OHG *wurzala*, from *wurz-wala*, PG first element *wurtiz*, PIE *wr̥H₂di-*, root *wraH₂d*. Kluge *Wurzel*. Pokorny 1167. Watkins *wrād-*.

HAW *a'a* /a ʔ a/. PPN *(w)aka*, PMP *wakaR*. PE. Blust.

LAT *radix* /r a: d/. PIE *wr̥H₂dīks*, root *wraH₂d*. Pokorny 1167. EM *radix*. Sihler 179. Watkins *wrād-*.

NAV *'akétl'óól* /k e: ʔ/ DERIVED FROM 'FOOT', COGNATE WITH 'CLAW'. *-kétl'óól* 'foot' (*kee'*) + 'cord'. YM 1992, p. 997.

TUR *kök* /k ø k/. Räsänen 287a.

rotten adj, Swadesh 200

ALB *kalbur* /k a l b/. Nominalization of *kalb* 'I rot'. Not in Huld, Janson, Mann, Pokorny.

ENG *rotten* /r ɑ/ LOAN. Borrowed from ON *rotinn*, part. of PG *reut*, PIE *reud* 'tear'; extension of *reu-2* 'tear down'? Pokorny 869. Barnhart. Skeat. OED.

FRE *pourri* /p u/ MOTIVATED. Pop. LAT **putrītus*, from Class. *putrescere*, *putrēre*, same root as *putridus*, see LAT. BW. Watkins *pū̆-*.

GER *faul* /f au/ MOTIVATED. OHG *fūl*, PG *fūlaz*, PIE *pūlos* 'rotten', root *pu*, interjection of disgust. Pokorny 849. Kluge *faul*.

HAW *pilau* /p i l a u/. PPN *pilau*. PE.

LAT *putridus* /p u/ MOTIVATED. *-id-* deverbative of *putreo*, from *puter*, PIE root *pu*, probably interjection of disgust. EM *puteo*. Pokorny 849. Walde 82. Watkins *pū̆-*.

NAV *diłdzííd* /dz í: d/. 'It rots', momentaneous imperfect. PA *gyid*, PPA *gud*. YM 1992, p. 168.

TUR *çürük* /tʃ y r y/. Cf. *çürümek* 'to rot'. Räsänen 121b.

round adj (spherical), Swadesh 100

ALB *rrumbullák* /r: u m b u ɫ/.
Rrumbull itself means 'in a ball'.
Not in Huld, Mann, Pokorny,
Janson, Boretzky.

ENG *round* /r/ LOAN. Borrowed from
OF *rund*, see FRE. PIE root *retH*.
Skeat. OED.

FRE *rond* /r/. Pop. LAT *retundus*,
Class. *rotundus*, q.v. Dauzat. BW.

GER *rund* /r/ LOAN, COGNATE WITH
'STRAIGHT'. Borrowed from FRE
rond, q.v. Kluge.

HAW *poepoe* /p o e/. Base *poe* also
means 'round'. PE.

LAT *rotundus* /r o t/. Participial *-ndo-*
derivative of a **rotor* 'roll', from
rota 'wheel', PIE root *retH*. EM
rota. Pokorny 866. Sihler 625.
Walde 368*. Watkins *ret-*.

NAV *nímaz* /m á: z/. 'It is round',
neuter. PA perhaps *wạts'*. YM
1992, p. 396 *MÁÁZ* 'roll, round'.

TUR *yuvarlak* /j u v a/. From
yuvarlamak 'roll', from *yuva-*
'turn'. Räsänen 212a.

rub verb, Swadesh 200

ALB *fërkon* /f ə r/ LOAN, COGNATE
WITH 'FALL', 'SNOW'. Loan from
LAT *fricat*, PIE *bher* 'apply a
sharp tool to, hit'. Huld 64. Mann
193.

ENG *rubs* /r/ COGNATE WITH 'WORM'.
ME *rubben*, perhaps PIE *wreib*,
root *wer* 'turn'. Skeat. OED.
Barnhart.

FRE *frotte* /f r/. Perhaps connected to
OF *freter*, from Pop. Lat. **frictāre*,
same root as *fricare*, see LAT.
Watkins *bhrēi-*. BW. Dauzat.

GER *reibt* /r/ COGNATE WITH 'THROW',
'WORM'. OHG *rīban*, perhaps PIE
wreib, see ENG. Pokorny 1152.
Kluge *reiben*.

HAW *'ānai* /ʔ a: n a/. Various forms:
'a'anai, *'āna'anai*, *'ānainai*.
Possibly transitive *-i*. PE.

LAT *fricat* /f r/ COGNATE WITH 'HIT'.
Frio 'pulverize', PIE *bhreiH*,
perhaps ext. of *bher* 'apply a sharp
tool to'. Pokorny 166. EM *frico*.

Walde 194*, 159*.

NAV *bídinishhish* /ɣ i: ʒ/. 'I rub
something against something else',
repetitive durative. PA *ghish-gy*.
YM 1992, p. 718.

TUR *sürtüyor* /s y r/. From *sür* 'lead
to; rub'. Räsänen 437b.

salt noun, Swadesh 200

ALB *kripë* /k r/ COGNATE WITH
'SHORT', 'THIN'. Pokorny lists as
krūpā with PIE *kreu* 'break'.
Extension of *(s)ker* 'cut'? That
root may be connected to *(s)kel*
'cut'. Pokorny 622. Huld 82. Mann
92. Walde 482.

ENG *salt* /s ɔ l/. OE *salt*, PG *saltam*,
PIE *sald*, ext. of root *sal*, perhaps
related to *sal* 'dirty grey'.
Lubotsky (p. 60) follows Kortlandt
in reconstructing PIE *saH₂l*. Skeat.
OED. Pokorny 878. Watkins *sal-1*.

FRE *sel* /s ɛ l/. LAT. BW.

GER *Salz* /z a l/. OHG *salz*, PG *saltam*,
see ENG. Pokorny 878. Kluge *Salz*.

HAW *pa'akai* /k a i/ DERIVED FROM
'HOLD', COGNATE WITH 'SEA'. I
assume this is *pa'a* 'solid' + *kai*
'sea'. *Kai* by itself (viz., *tasik*) is
often used for 'salt' in
Central/Eastern-Malayo-
Polynesian languages. PE.
Blust.

LAT *sal* /s a l/. PIE root noun *sal*, see
ENG. EM *sal*. Sihler 44. Walde
452*.

NAV *'áshịịh* /ʒ ī: h/. Stem + prefix
noun. YM 1980, p. 10.

TUR *tuz* /t u z/. Räsänen 502b.

sand noun, Swadesh 100

ALB *rërë* /r/ LOAN. Loan from LAT.
Mann 63. Janson 57.

ENG *sand* /s/. OE *sand*, PG *sandam*,
PIE *samHdho*. Pokorny further
links with Lat. *sabulum* and Greek
psammos by positing underlying
bhs- from root *bhes* 'rub away'.
Skeat. OED. Watkins *bhes-1*.

FRE *sable* /s/. Lat. *sabulum*,
conceivably *(bh)s-abh-lo-m*, PIE
root *bhes* 'rub away'. Dauzat. BW.
Watkins *bhes-1*.

GER *Sand* /z/. OHG *sant*, PG *sandam*, see ENG.

HAW *one* /o n e/. PPN *'one*, POc *qone*, PCEMP *qənay*, PMP *qenay*. PE. Blust.

LAT *arena* /a r/. No known etymology, though *-ena* is apparently an Etruscan suffix.

NAV *séí* /z é í/. *t*-classifier + *zéí*, PA *saxy*, PPA *sax* 'crumble'. Independent noun. YM 1987, p. 3. YM 1992, p. 739.

TUR *kum* /k u m/. Räsänen 299b.

say verb, Swadesh 100
ALB *thotë* /θ/. PIE *k̂eH₁* 'proclaim'. Mann 33. Huld 119. Pokorny 566 *k̂ens*.

ENG *says* /s e/ COGNATE WITH 'SEE'. OE *secgan*, PG *sagjan*, PIE *sokʷ*, root *sekʷ* 'say', perhaps same as *sekʷ* 'see'. Skeat. OED. Pokorny 897. Watkins *sekʷ-3*.

FRE *dit* /d i/. LAT. BW.

GER *sagt* /z a: g/ COGNATE WITH 'SEE'. OHG *sagēn*, PG *sagjan*, see ENG. Skeat. OED. Pokorny 897. Kluge *sagen*.

HAW *'ōlelo* /l e l o/ DERIVED FROM 'TONGUE'. PEP *koolelo*. Stem must be *lelo* 'tongue'. PE.

LAT *dicit* /d i: k/. PIE root *deik̂* 'point'. EM *dix*. Sihler 52. Walde 776, 772.

NAV *ní* /n i/. 'He says', neuter imperfect. PA *ni*. YM 1992, p. 447 *NIID*. Hoijer 27. Pinnow 29.

TUR *diyor* /d e/. Räsänen 467b.

scratch verb, Swadesh 200
ALB *gërvish* /g ə r/. Pokorny lists *gër(r)y(e)j*, 'scratch'; could this be a fusion between that and *vishkoj* 'I scrape'? PIE root is *gred*. Pokorny 405. Not in Mann, Huld, Janson.

ENG *scratches* /s k r a tʃ/ MOTIVATED. Late ME blend of *scrat* and *cratch*, both 'scratch'. *Scrat* may be an extension of PG *krattōjan*, PIE *gred*, perhaps via OF *esgrater*. OED. Barnhart. Skeat. Pokorny 405. Watkins *grat-*.

FRE *gratte* /g r a t/ LOAN, MOTIVATED.

Dauzat: ultim. from PG *krattōjan*, PIE *gred*. BW. Dauzat. Watkins *grat-*.

GER *kratzt* /k r a ts/ MOTIVATED. OHG *krazzōn*, PG *krattōjan*, perhaps PIE *gred*, perhaps via OF *esgrater*. Pokorny 405. Kluge *kratzen*.

HAW *walu* /w a l u/. PPN *waru*. Passive stem is *waluh*. PE.

LAT *scabit* /s k a b/. PIE *skab*. Pokorny 931. EM *scabo*. Walde 562*. Watkins *skep-*.

NAV *'ashch'id* /tʃ' i d/. 'I scratch', repetitive durative. PA *tsh'ᵂit'*, PPA *k'ᵂit'*. YM 1992, p. 106 *CH'ID*. Pinnow 22.

TUR *tırmalıyor* /t ɯ r/ COGNATE WITH 'CLAW'. From *tırma-* 'scratch'. Räsänen 479a.

sea noun, Swadesh 200
ALB *det* /d e/. Perhaps PIE *dheuboto-*, root *dheub*. Huld 50. Not in Janson, Mann, Pokorny.

ENG *sea* /s i/. OE *sǣ*, PG *saiwiz*. PIE root could be *sai*. Skeat. OED. Pokorny 877.

FRE *mer* /m ɛ r/. LAT. BW.

GER *See* /z e:/ COGNATE WITH 'LAKE'. OHG *sēo*, PG *saiwiz*. Same word as *See* 'lake', differentiated by gender since 16th c. Possibly a PIE *sāi* 'pain'. Kluge *See*. Pokorny 877.

HAW *kai* /k a i/ COGNATE WITH 'SALT'. PNP *tai*, PPN *tahi*, PMP *tasik*. PE. Blust.

LAT *mare* /m a r e/. PIE *mori*. Pokorny 748. EM *mare*. Sihler 65. Walde 234*. Watkins *mori-*.

NAV *tónteel* /t o/ COGNATE WITH 'LAKE', 'RIVER', 'WATER'. *Tó* 'water' + *nteel* 'it is broad'. YM 1980, p. 15. YM 1992, p. 518.

TUR *deniz* /d e n i z/. Räsänen 474a.

see verb, Swadesh 100
ALB *sheh* /ʃ/. PIE *sokʷésẑk̂-*, causative of *sekʷ* 'see'? Mann 131, 161, Huld 114. Not in Pokorny.

ENG *sees* /s i/ COGNATE WITH 'SAY'. OE *sēon*, PG *sehwan*, PIE *sekʷ*, may be source of *say*. Skeat. OED. Pokorny 897. Watkins *sekʷ-2*.

FRE *voit* /v w a/. LAT. BW.

GER *sieht* /z e:/ COGNATE WITH 'SAY'. OHG *sehan*, PG *sehwan*, see ENG. Kluge *sehen*. Pokorny 897.

HAW *'ike* /ʔ i k e/ COGNATE WITH 'KNOW'. Same word as 'know'. PPN *kite*, POc, PMP *kita*. Blust.

LAT *videt* /w i d/. PIE stative *wideH₁*, root *weid*. EM *video*. Pokorny 1125. Sihler 178, 531. Walde 237. Watkins *weid-*.

NAV *yish'į́* /ʔ ī: ʔ/. 'I see an object', cursive neuter perfective. PA *'en*. YM 1992, p. 250. Pinnow 30.

TUR *görüyor* /g ø r/. Räsänen 292a.

seed noun, Swadesh 100

ALB *farë* /f a r/. PIE approx. *sporaH₂*, root *sper*. Mann 61. Huld 63. Not in Pokorny.

ENG *seed* /s i/. OE *sǣd*, PG *sǣdiz*, from root 'sow', PIE *seH₁tis*, root *seH₁*. Skeat. OED. Pokorny 390. Watkins *sē-1*.

FRE *graine* /g r ɛ/. Lat. *grāna*, reinterp. from pl. of *grānum*, PIE *ĝṛHnom*, root *ĝerH*. Pokorny 390. BW. Watkins *grə-no-*.

GER *Same* /z a:/. OHG *samo*, to PIE root *seH₁* 'sow'. Pokorny 390. Kluge *Same(n)*.

HAW *'ano'ano* /ʔ a n o/. PE.

LAT *semen* /s e:/. PIE root *seH₁* 'sow'. Pokorny 390. EM *sero*. Sihler 48. Walde 460*. Lubotsky, p. 55. Watkins *sē-1*.

NAV *k'eelyéí* /k' é:/. 'Seeds of cultivated plants' from *k'éé-* 'plant' + *lá* 'things'. YM 1992, p. 1053.

TUR *tohum* /t o h u m/ LOAN. Loan from Persian *tuxm*. Steingass 288b. Räsänen 497a.

sew verb, Swadesh 200

ALB *qep* /c e p/. Mann reconstructs a PIE *kep* as in Lith. *kempù* 'adhere'. Mann 34. Not in Janson, Pokorny, Huld.

ENG *sews* /s o/. OE *siowan*, PG *siwjan*, PIE root *syuH* 'sew'. Skeat. OED. Pokorny 916. Watkins *syū-*.

FRE *coud* /z/. Pop. Lat. *cōsere*, Class. *consuere* 'sew together', *con-* +

suere. PIE root *syuH*. Dauzat. BW.

GER *näht* /n e:/. OHG *najan*, PG *nǣjan*, PIE *(s)nē* 'turn'. Kluge *nähen*.

HAW *humu* /h u m u/. PNP *sumu*. PE.

LAT *suit* /s u/. PIE root *suH*, variant of *syuH*. Pokorny 916. EM *suo*. Sihler 535. Walde 515*. Watkins *syū-*.

NAV *náíłkad* /k a d/. 'He sews it', repetitive durative. PA *Kȧt'*, PPA *Kat'*. YM 1992, p. 285 *KAD*. Hoijer 50.

TUR *dikiyor* /d i k/. From *dik* 'straight'. Räsänen 479b.

sharp adj, Swadesh 200

ALB *mprehtë* /p r/ COGNATE WITH 'DULL'. Adj. of *mpreh* 'sharpen'. *M-* is clearly a prefix *en-*; cf. *preh*, *pref*, same meaning. Mann derives which from *per-ok̂u-*, but maybe a connection to *pres*, etc., 'cut' is to be sought? Mann 82. Not in Huld, Janson.

ENG *sharp* /ʃ a r/ COGNATE WITH 'SHORT'. OE *scearp*, PG *skarpo-*, PIE root *(s)ker* 'cut'. Pokorny 941. OED. Watkins *sker-1*.

FRE *tranchant* /t r/ COGNATE WITH 'THREE'. *Trancher*, OF *trenchier*, prob. from Pop. Lat. **trēnicāre* 'cut in 3', crossing of *trēs* 'three' and **trīnicāre*, base *trīni*, from root *trei*. BW. Dauzat.

GER *scharf* /ʃ a r/ COGNATE WITH 'SHORT'. OHG *scarpf*, PG *skarpo-*, PIE root *(s)ker* 'cut'. Pokorny 941. OED. Kluge *scharf*.

HAW *'oi* /ʔ o i/. PPN *kohi*. PE.

LAT *acutus* /a k/. Stative *aceo* 'be sharp', PIE root *H₂ak̂*. EM *ac-*. Pokorny 18. Sihler 45. Walde 33. Watkins *ak-*.

NAV *deení* /n í/ COGNATE WITH 'DULL'. 'It is sharp'; neuter imperfective, stem *ní*, from PA *yen*. YM 1992, p. 431.

TUR *keskin* /k e s/ DERIVED FROM 'CUT'. From *kes-* 'cut'. Räsänen 257b.

short adj, Swadesh 200

ALB *shkurtër* /ʃ k u r/ LOAN, COGNATE

WITH 'SALT', 'THIN'. Adverb is *shkurt*. Loan from Lat. *curtus*, root *(s)ker* 'cut'. May be a connection to *(s)kel* 'cut'. EM *curtus*. Not in Huld, Janson, Mann.

ENG *short* /ʃ ɔ r/ COGNATE WITH 'SHARP'. OE *scort*, PG *skurto-*, PIE *skr̥do*, root *(s)ker* 'cut'. Skeat. OED. Barnhart. Pokorny 941. Watkins *sker-1*.

FRE *court* /k u r/ COGNATE WITH 'BARK'. Lat. *curtus*, PIE root *(s)ker*. EM. BW.

GER *kurz* /k u r/ LOAN, COGNATE WITH 'SHARP'. Loan from Lat. *curtus*, PIE root *(s)ker*. EM. Kluge *kurz*.

HAW *pōkole* /k o l e/. *Poko* (PNP *poto*) also means 'short'. *Pō-* often expresses state. PE.

LAT *brevis* /b r e w/. PIE *mreĝhw-*, root *mreĝh*. Pokorny 751. EM *brevis*. Sihler 211. Walde 314*. Watkins *mregh-u-*.

NAV *yázhí* /j á ʒ/ COGNATE WITH 'SMALL'. Can function postpositively as an adjective. PA *yazh-*, PPA *yash-*. YM 1992, p. 678.

TUR *kısa* /k ɯ s/. From *kıs* 'compress'. Räsänen 267b.

sing verb, Swadesh 200

ALB *këndon* /k ə n/ LOAN. Loan from LAT *canto*, iterative of *cano*, q.v. Mann 30. Janson 119. Huld 80.

ENG *sings* /s ɪ ŋ/. OE *singan*, PG *singan*, PIE root *sengʷh* 'sing'. Skeat. OED. Barnhart. Pokorny 906. Watkins *sengʷh-*.

FRE *chante* /ʃ ã/. Lat. *canto*, see ALB. BW.

GER *singt* /z i ŋ/. OHG *singan*, PG *singan*, see ENG. Pokorny 906. Kluge *singen*.

HAW *mele* /m e l e/. PNP *umele*. PE.

LAT *canit* /k a n/. PIE root *kan*. EM *cano*. Pokorny 525. Sihler 45. Walde 351. Watkins *kan-*.

NAV *hataał* /t aː l/. 'He sings', durative. Prob. originally 'kick'. Perhaps from *táál* 'kick' (i.e., kick-dance), PA *tatł'*. YM 1992, p. 490 *TÁÁL*. YM 1980, p. 3.

TUR *şarkı söylüyor* /ʃ a r k ɯ/. I.e.,

şarkı 'song' *söylemek* 'speak'.

sit verb (stative), Swadesh 100

ALB *(rri) ndenjur* /d e ɲ/. PIE *en-ten-*, root *ten* 'stretch'. Mann 162. Huld 109. Pokorny 1066.

ENG *sits* /s ɪ t/. OE *sittan*, PG *sitjan*, PIE *sedyo-*, root *sed*. Skeat. OED. Pokorny 885. Watkins *sed-1*.

FRE *(est) assis* /s i z/. *Asseoir* is from Pop. Lat. *assedere*, from *assidere* rebuilt on basis of *sedere* 'sit', q.v.

GER *sitzt* /z i ts/. OHG *sizzan*, PG *sitjan*, see ENG. Pokorny 885. Kluge *sitzen*.

HAW *noho* /n o h o/. PPN *nofo*, POc *nopo*. PE. Blust.

LAT *sedet* /s e d/. PIE *sedeH₁-*, stative from root *sed*. Pokorny 885. EM *sedeo*. Sihler 497. Walde 483*.

NAV *sidá* /d á/. 'He sits', neuter perfective. PA *da*, PPA *daw*. YM 1992, p. 116 *DÁ*.

TUR *oturuyor* /o/. I.e., *remain being*, from *ol* 'be'. Räsänen 361a, 360b.

skin noun, Swadesh 100

ALB *lëkurë* /l ə k/. From PIE *lH₁k-er-*, root *leH₁k* 'tear', like Lat. *lacer*, although there's a chance the Albanian is a loan. Huld 86. Walde 419*. Pokorny 674. EM *lacer*.

ENG *skin* /s k/ LOAN. Borrowed from ON *skinn*, PG *skinþa*, PIE *skend-* 'flay', root *sek* 'cut'. Skeat. OED. Pokorny 929. Barnhart. Watkins *sek-*.

FRE *peau* /p o/. Lat. *pellis* 'hide', replacing *cutis*; PIE *pelnis*, root *pel* 'skin'. Dauzat. BW. Pokorny 803. Watkins *pel-4*.

GER *Haut* /h au/. OHG *hūt*, PG *hūdiz*, dental ext. of PIE *(s)keu(H)* 'cover'. Kluge *Haut*. Pokorny 803.

HAW *'ili* /ʔ i l i/ COGNATE WITH 'BARK'. Same word as 'bark'. PPN *kili*, POc, PMP *kulit*. PE. Blust.

LAT *cutis* /k u/. PIE *ku-ti-*, root *(s)keu* 'cover'. Pokorny 803. EM *cutis*. Walde 549*. Watkins *(s)keu-*.

NAV *'akágí* /k á: ʔ/ COGNATE WITH 'BARK'. From *-káá'* 'surface' + *-í* nominalizer. The word can also

apply to the bark of e.g. aspen.
YM 1992, p. 981.
TUR *deri* /d e r i/. Räsänen 475b.

sky noun, Swadesh 200
ALB *qiell* /c i e/ LOAN. Loan from LAT.
Huld 107. Not in Janson, Mann,
Pokorny.
ENG *sky* /s k ai/ LOAN. Borrowed from
ON *ský* 'cloud', PG *skeujam*, PIE
(s)keu 'cover'. Skeat. Pokorny 803.
FRE *ciel* /s j ɛ/. LAT. BW.
GER *Himmel* /h/. OHG *himil*, PG
himena, PIE root *k̂em* 'cover'.
Watkins derives from PIE *k̂e-men-*,
from root *ak̂* 'sharp'. Kluge
Himmel. Pokorny 556. Watkins *ak-*
'sharp'.
HAW *lani* /l a n i/. PPN *laŋi*, POc *laŋit*,
PMP *láŋit*, Proto-Austronesian
laŋiC. PE. Ross. Blust.
LAT *caelum* /k ae/. PIE *kaid*, perhaps
ext. of *(s)kāi* 'bright'. Pokorny 916.
EM *caelum*. Walde 537*.
NAV *yá* /y á/. Independent stem noun.
PA *ya*. Hoijer 15. YM 1992,
p. 1004.
TUR *gök* /g ø k/. Räsänen 287a.

sleep verb, Swadesh 100
ALB *fle* /l e/. Mann has as a compound,
e.g. *apo-legh-* 'lie down', root PIE
legh 'lie'. Mann 169. Not in Huld,
Pokorny.
ENG *sleeps* /s l i/. OE *slǣpan*, PG
slǣpan, PIE *(s)leHb* 'be weak',
root perhaps *sleH* 'weak'. Skeat.
OED. Barnhart. Pokorny 655.
Watkins *slēb-*.
FRE *dort* /d ɔ r/. LAT. BW.
GER *schläft* /ʃ l a:/. OHG *slāfan*, PG
slǣpan, see ENG. Kluge *schlafen*.
Pokorny 655.
HAW *moe* /m o e/ COGNATE WITH 'LIE'.
Same word as 'lie'. PPN *mohe*. PE.
LAT *dormit* /d o r/. PIE *dr̥myo-*, base
drem, perhaps root *der(H)* 'sleep',
plus present suffix of indeterminate
aspect, *-em-*. EM *dormio*. Pokorny
226. Walde 821. Watkins *drem-*.
NAV *'athosh* /ɣ a: ʒ/ MOTIVATED. 'He is
asleep', repetitive durative. PA
ghạzh, PPA *x̣ʷa'sh*. Basic meaning

appears to be the onomatopoetic
'make a bubbling noise'. YM 1992,
p. 234: root *GHAAZH(1)*. Hoijer
16.
TUR *uyuyor* /u/. Räsänen 508a.

small adj, Swadesh 100
ALB *vogël* /v o g/. Mann 108. Huld 131.
Walde 247. Not in Pokorny.
ENG *small* /s m ɔ l/. OE *smæl*. PIE
(s)mēlo 'small animal'. OED.
Skeat. Pokorny 724. Watkins
mēlo-, mel-5.
FRE *petit* /p ə t/ MOTIVATED. Some
pitt- root from child language?
Dauzat. BW.
GER *klein* /k l/. OHG *klein*, PWG
klainiz, PIE *ĝel* 'bright'. Kluge
klein. Pokorny 366. Watkins *gel-2*.
HAW *iki* /i k i/ MOTIVATED, COGNATE
WITH 'NARROW'. PPN *'iti*, POc
rikiq, PMP *dikiq*. PE. Blust.
LAT *parvos* /p a/ COGNATE WITH
'CHILD', 'FEW'. PIE root *pōu* plus
-ro-, with metathesis. EM *pau-*.
Pokorny 105. Walde 75*. Watkins
pau-.
NAV *yázhí* /j á ʒ/ COGNATE WITH
'SHORT'. PA *yazh-*, PPA *yash-*. Can
function postpositively as an
adjective. YM 1992, p. 678.
TUR *küçük* /k y tʃ y/. Räsänen 269b.

smell verb, Swadesh 200
ALB *marr erë* /e r/ LOAN, DERIVED
FROM 'WIND'. 'Takes' + 'wind', q.v.
ENG *smells* /s m e l/. ME *smellen*,
smüllen. Perhaps PIE *smel*
'smoulder'. Skeat. OED. Pokorny
969.
FRE *sent* /s ã t/. Lat. *sentīre* 'perceive',
replacing OF *oloir* from Lat. *olere*.
PIE *sentyo-*, perhaps from root
sent 'go in a certain direction'.
BW. EM *sentio*. Pokorny 908.
Watkins *sent-*.
GER *riecht* /r i:/ COGNATE WITH
'SMOKE'. OHG *riohhan*, PG
reukan, orig. 'smoke', PIE *reug*,
root perhaps *reu*. Pokorny 872.
Kluge *riechen*. Watkins *reug-*.
HAW *honi* /h o n/. PPN *songi*. Possibly
transitive *-i*. PE.

LAT *olfacit* /o l/. Compound of *ole-* 'smell' and *facio* 'make'. Stative verb *oleo* 'I smell', earlier *olo*, PIE root *H₃od*. Several words have /d/:/l/ alternation in Latin, but may also be influenced by *olea* 'olive', or be a Sabine borrowing. Pokorny 772. EM *oleo*. Sihler 121. Walde 174. Watkins *od-1*.

NAV *yishcháá'* /tʃ i n/. 'He smells something', perfective. PA *tshan*, imperf. *tshin*. YM 1992, p. 82 *CHÁÁ'*. Pinnow 28.

TUR *koklayor* /k o k/. From *kok-* 'smell'. Räsänen 276a.

smoke noun, Swadesh 100

ALB *tym* /t y/ LOAN. Nominalization of *tymos* 'smoke', a loan from Old Church Slavic *timijasati* 'burn incense', a loan from Greek *thumiáza* incense, from *thūō* 'I smoke', PIE *dhuHmos*, root *dheuH* 'swirl in the air'. Lubotsky, p. 59. Huld 117. Pokorny 261.

ENG *smoke* /s m o k/. OE *smoca*, PG *smuk-*, PIE *(s)meug* 'smoke'. OED. Skeat. Pokorny 971. Watkins *smeug-*.

FRE *fumée* /f y/. Extension of OF *fum*, LAT *fumus*, PIE root *dheuH*.

GER *Rauch* /r au/ COGNATE WITH 'SMELL'. OHG *rouh*, PG *raukiz*, PIE *reug*, root perhaps *reu*. Kluge *Rauch*. Pokorny 872.

HAW *uahi* /u/. PEP *au-afi* (literally 'fire smoke'), PPN *'ahu*, PMP *qasu*. PE. Blust.

LAT *fumus* /f u:/. PIE *dhuHmos*, root *dheuH* 'float in air'. Pokorny 261. EM *fumus*. Sihler 47. Walde 835. Watkins *dheu-1*.

NAV *łid* /ł i d/. PPA *łid*. YM 1987, p. 3. YM 1992, p. 370.

TUR *duman* /d u/. From **tum* 'cold', from **tu* 'block the way'. Räsänen 498b.

smooth adj, Swadesh 200

ALB *lëmuet* /l ə m/. Cf. *lëmoj* 'I file, sand, smooth'. Not in Mann, Janson, Huld, Pokorny.

ENG *smooth* /s m u ð/. OE *smōþ*, PG

smanþ-. It is very hard to see the motivation in Pokorny's derivation from *sem* 'one'. Skeat. OED. Pokorny 904. Barnhart.

FRE *lisse* /l i/. From verb *lisser*, Lat. *lixare*, from *lixa*, PIE *wliks-*, root *wleik-*; perhaps crossed with *allisus*, 'used cloth'. BW. EM *lix, liquo*. Pokorny 669. Watkins *wleik-*.

GER *glatt* /g l/ COGNATE WITH 'YELLOW'. OHG *glat*, PG *gladaz*, perhaps PIE root *ĝhel* 'yellow'. Pokorny 429. Kluge *glatt*. Watkins *ghel-2*.

HAW *malino* /l i n o/. POc *ma-lino*; *ma-* is a stative prefix. PE. Ross.

LAT *levis* /l e:/. PIE *leiwis*, root *(s)lei* 'slimy'. Lubotsky treats as u-stem: PIE *leH₁i-u-s* to Pre-Lat. **lēyus*, reformed to i-stem **lējwis* to *lēuis*. Pokorny 664. EM *leuis*. Walde 390*.

NAV *dilkǫǫh* /k ō ?/ COGNATE WITH 'SWIM'. 'It is smooth', neuter imperfective. No obvious connection to *kǫ'* 'fire'. YM 1992, p. 330 *KǪ'*.

TUR *düz* /d y z/. Räsänen 508a.

snake noun, Swadesh 200

ALB *gjarpër* /ɟ a r p/. PIE root *serp* 'creep'. Huld 67. Janson 20. Walde 502*. Pokorny 912.

ENG *snake* /s n e k/. OE *snaca*, PG *snakōn-*, PIE *snogon-*, root *(s)nHg*. Skeat. OED. Barnhart. Watkins *sneg-*.

FRE *serpent* /s ɛ r p/. The usual Lat. word, *serpens* 'the crawler', a euphemism for *anguis*. PIE *serp* 'creep', as in ALB. Pokorny 912. BW. EM. Watkins *serp-2*.

GER *Schlange* /ʃ l a ŋ/. OHG *slango*, from verb *schlingen* 'creep, coil; swallow?', PIE *slongʷh*. Pokorny 961. Kluge *Schlange*.

HAW *naheka* /n a h e k a/ LOAN. Apparently a loan from Hebrew *nahaš* or ENG *snake*, or rather a blend of the two. POc is *mwata*. Blust. PE.

LAT *anguis* /a n g w/. Ringe 1992 has /angwis/. Older, less common word for *serpens*. PIE *ang^w(h)is*, root *ang^w(h)*. Pokorny 43. EM *anguis*. Sihler 163. Walde 63. Watkins *ang^{w}hi-*.

NAV *tl'iish* /t$^{l'}$ i: ʃ/. From 'zigzag'. PA *tl'ighish* 'eel, leech, snake'. YM 1992, p. 582 *TŁIISH*.

TUR *yılan* /j ɯ l/. From *yıl* 'glide'. Räsänen 200a.

snow noun, Swadesh 200

ALB *borë* /b o r/ COGNATE WITH 'FALL', 'RUB'. Because *bie* 'fall' is used for precipitation, may have same origin, in PIE *bher*. Huld 45. Not in Janson, Pokorny.

ENG *snow* /s n o/. OE *snāw*, PG *snaiwaz*, PIE *snoig^{w}hos*, root *sneig^{w}h*. Skeat. Walde 695*. OED. Pokorny 971. Watkins *sneig^{w}h-*.

FRE *neige* /n ɛ/. From the verb *neiger*, from Pop. LAT *nivicare*, built from the stem *nivi-* of *nix*, the noun. PIE root *sneig^{w}h*. BW.

GER *Schnee* /ʃ n e:/. OHG *snēo*, PG *snaiwaz*, see ENG. Walde 695*. Kluge *Schnee*. Pokorny 971.

HAW *hau kea* /h a u/ DERIVED FROM 'ICE'. *Hau* 'ice' (q.v.) + *kea* 'white'. PE.

LAT *nix* /n i k/. PIE *snig^{w}h*, root *sneig^{w}h*. EM *nix*. Pokorny 974. Sihler 163. Walde 695. Watkins *sneig^{w}h-*.

NAV *zas* /z a s/. Alternates with *yas*; verb stem is *dzaas*. PA *ya(x)s*, PPA *yixs*. YM 1992, p. 167, 1006.

TUR *kar* /k a r/. Räsänen 235a.

some lim, Swadesh 200

ALB *disa* /s a/. *Di* 'I know' (q.v.) + *sa* 'how much'. Çabej 182. Not in Janson, Huld, Mann, Pokorny.

ENG *some* /s ə m/. OE *sum*, PG *sumas*, PIE *sm̥mos*, root *sem* 'one'. Skeat. OED. Barnhart. Pokorny 902. Watkins *sem-1*.

FRE *quelques* /k/ COGNATE WITH 'WHAT', 'WHO'. 12th century compound *quel que*, the head from Lat. *qualis*, root *qu*, PIE interrog/rel. stem *k^w*. BW. EM *qualis*.

GER *einige* /ai/ DERIVED FROM 'ONE'. Extension of *ein(s)* 'one'. Kluge *ein*. Pokorny 281.

HAW *kekahi* /k a h i/ DERIVED FROM 'ONE', COGNATE WITH 'FEW'. I assume this is 'the' + 'one'. PE.

LAT *aliqui* /a l/ DERIVED FROM 'OTHER'. Compounded from *alio-* 'other', q.v., and *quis* 'who', q.v. EM *alius*. Pokorny 25. Walde 86. Watkins *al-1*.

NAV *ła'* /ɬ a ʔ/ COGNATE WITH 'ONE', 'OTHER'. YM 1992, p. 934.

TUR *bazı* /b z/ LOAN. Loan from Arabic *ba'ḍ* 'portion, some', root *b'ḍ* 'divide into parts'. Räsänen 66a. Wehr 82a.

spit verb, Swadesh 200

ALB *pështyn* /p ə ʃ/ MOTIVATED. PIE *spyeuH* as in LAT. Çabej 255. Mann 65. Pokorny 999.

ENG *spits* /s p ı t/ MOTIVATED. OE *spittan*, onomat. within Germanic, perhaps connected to PIE *spyeuH*. Skeat. OED. Barnhart. Watkins *spyeu-*.

FRE *crache* /k r a ʃ/ LOAN, MOTIVATED. Earlier *krakka*, either a borrowing from Germanic, or an independent onomatopeia. Dauzat. BW.

GER *spuckt* /ʃ p u k/ LOAN, MOTIVATED. Borrowed from Low German. Kluge *spucken*.

HAW *kuha* /k u h a/ MOTIVATED. PCP *tufa*. Onomatopoeia? Replaces POc *qanusi*. PE. Blust.

LAT *spuit* /s p u/ MOTIVATED. PIE *spu*, variant of *spyeuH* 'spit'. Prob. imitative of the sound and/or action. EM *spuo*. Pokorny 999. Walde 683*. Watkins *spyeu-*.

NAV *dishsheeh* /ʒ é: ʔ/ MOTIVATED. 'I spit' (momentaneous completive). PA *zheK'*, PPA *sheK'*, sounds onomatopoetic. YM 1992, p. 771 *ZHÉÉ'*.

TUR *tükürüyor* /t y k y r/ MOTIVATED. Räsänen 504a. Onomatopoetic.

split verb, Swadesh 200

ALB *çan* /tʃ/. PIE *skedny-*, root *sked* 'split', ext. of *sek* 'cut'. Pokorny

919. Mann 183. Walde 558*. Not in
Janson, Huld.

ENG *splits* /s p l/ LOAN. Borrowed from
Middle Dutch *splitten*, PG *splītan*,
PIE *spleid*, from *(s)plei* 'split', ext.
of *(s)pel* 'split'. Skeat. OED.
Barnhart. Pokorny 985, 1000.
Watkins *splei-*.

FRE *fend* /f a/. LAT. BW.

GER *spaltet* /ʃ p a l/. OHG *spaltan*, PIE
root *(s)pel*. Pokorny 985. Kluge
spalten.

HAW *wāhi* /w a: h/. PCP *waasi*, PNP
faʻasi. Possibly transitive *-i*. PE.

LAT *findit* /f i/. Nasal infix (cf. *fissum*)
of PIE root *bheid*, possibly an ext.
of *bhei* 'hit'. Pokorny 116. EM
findo. Sihler 508. Walde 138*.
Watkins *bheid-*.

NAV *'attániishgéésh* /g i ʒ/ COGNATE
WITH 'CUT'. 'I split it lengthwise
by cutting', momentaneous
imperfective. PA *Gidzh*. YM 1992,
p. 211 *GIZH*.

TUR *yarıyor* /j a r/. Räsänen 189a.

squeeze verb, Swadesh 200

ALB *shtrydh* /ʃ t r y ð/. Çabej simply
says it is native. Not in Mann,
Huld, Janson, Pokorny.

ENG *squeezes* /s k w i z/ MOTIVATED.
Perhaps a blend of various words
like *squash* and *quease*; cf. also
squiss, *squish*, *squize*, *quash*.
Quease is from OE *cwiesan*, of
unknown etym.; *squash* is borrowed
from OF *esquasser* 'crush', from
Pop. Lat. **exquasso*, base *quasso*
'shatter', freq. of *quato* 'shake',
PIE *kʷeHt*. In the end, sibilants
keep popping up, and are prob.
naturally expressive of squashing.
Pokorny 632. OED. Barnhart.
Skeat.

FRE *presse* /p r/ COGNATE WITH
'NEAR'. LAT *presso*, freq. of *premo*,
q.v. PIE root *per* 'hit'. BW.
Watkins *per-5*.

GER *drückt* /d r/ COGNATE WITH
'GUTS'. OHG *thruken*, PWG
þrukkja, PIE *terH* 'rub'. Pokorny
1071. Kluge *drücken*.

HAW *'uwī* /ʔ u w iː/ MOTIVATED. Same

word means 'squeak, squeal';
onomatopoetic? Perhaps connected
to the word for 'thin'. Replaces
POc *poRos*. Blust.

LAT *premit* /p r/. PIE root *per* 'hit'
with durative present suffix *-em-*.
EM *premo*. Pokorny 818. Sihler
583. Walde 43*. Watkins *per-5*.

NAV *yiishnih* /n iː ʔ/. 'I squeeze it',
semelfactive. PA *nigy*, PPA *ni'g*.
YM 1992, p. 454 *NII'*.

TUR *sıkıyor* /s ɯ k/. Possibly connected
to Arm. /jig/ 'stretch tight'.
Dankoff, p. 166. Räsänen 415b.

stab verb, Swadesh 200

ALB *ther* /θ e r/. PIE *k̞r̥*, root *k̞er(H)*
'to wound'. Pokorny 578. Janson
127. Walde 410. Not in Mann,
Huld.

ENG *stabs* /s t a b/. Scottish, else
uncertain. Skeat. Barnhart.

FRE *poignarde* /p w a ɲ/ MOTIVATED.
Poignard 'dagger', rebuilding of
poignal, Pop. Lat. **pugnalis*, from
Class. *pugnus* 'fist'. PIE root
p(e)uĝ. Dauzat. BW.

GER *sticht* /ʃ t e x/. OHG *stehhan*,
PWG *stekan*, PIE root *steig*.
Pokorny 1016. Kluge *stechen*.

HAW *hou* /h o u/. Different from 'new':
PPN *fohu*.

LAT *fodit* /f o d/ COGNATE WITH 'DIG'.
Same word as 'dig'. PIE *bhodhye*
(for expected *bhedhye*), root *bhedh*
'stab'. Pokorny 113. EM *fodio*.
Sihler 121, 536. Watkins *bhedh-*.

NAV *bighá'níshgééd* /g eː d/ COGNATE
WITH 'DIG'. 'I stab him',
momentaneous imperfective. PA
Ged, PPA *Gʷe'd*. YM 1992, p. 202
GEED.

TUR *hançerliyor* /h n tʃ r/ LOAN. Root
is 'dagger', from Arabic *xanjar*.
Stachowski, v. 1, p. 90. Räsänen
155a. Wehr 304b.

stand verb (stative), Swadesh 100

ALB *(rri) më këmbë* /k ə m b/ DERIVED
FROM 'FOOT'. 'On + foot', qv. PIE
root *kamp* 'bend'.

ENG *stands* /s t a/. OE *standan*, PG
standan, PIE *stH₂nt-*, root *steH₂*.

OED. Skeat. Pokorny 1004.
Watkins *stā-*.
FRE *(est) debout* /b u/ LOAN,
MOTIVATED. *Bout*, from *bouter*
'place' from 'shove', from Frankish
*bōtan, PG *bautan*, *-d-* pres. of PIE
root *bh(a)u*. Could the root be
onomatopoetic? Pokorny 112.
Dauzat. BW. Barnhart *beat*.
Watkins *bhau-*.
GER *steht* /ʃ t e:/. OHG *stēn*, PG *stǣ-*,
PIE root *steH₂*. Pokorny 1004.
Kluge *stehen*.
HAW *kū* /k u:/. Passive stem is *kūl-*.
PPN *tuʻu*, POc *tuqur*, PMP *tuqud*.
PE. Blust.
LAT *stat* /s t a:/. PIE *stH₂*, with stative
-eH₁ suffix; root *steH₂*. EM *sto*.
Pokorny 1004. Sihler 50. Walde
603*. Watkins *stā-*.
NAV *sizį́* /z ĩ/. 'He is standing', neuter
perfective. PA *hen, yen*, PPA *hen*.
YM 1992, p. 746.
TUR *duruyor* /d u r/. Räsänen 500a.

star noun, Swadesh 100
ALB *yll* /y ɬ/. Best guess is a PIE
H₁usli, root *H₁eus* 'burn'. But
many try to connect with PIE
sāwel 'sun'. Huld 132. Pokorny 881.
ENG *star* /s t ɑ r/. OE *steorra*, PG
sterrōn-, PIE *H₂sters-*, root
H₂ster. Skeat. OED. Pokorny 1027.
Watkins *ster-3*.
FRE *étoile* /e t w a/. Gallo-Roman
stēla, from LAT *stēlla*. BW.
GER *Stern* /ʃ t e r/. OHG *sterno*, PG
sternōn-, see ENG. Kluge *Stern*.
Pokorny 1027.
HAW *hōkū* /h o: k u:/. PPN *fetuʻu*
(PMP *bituqen*); crossed with *hoku*
(PEP *fotu* or *sotu*) 'night of the
full moon'? PE. Blust.
LAT *stella* /s t e:/. PIE *H₂stērlaH₂-*,
root *H₂ster*. EM *stella*. Pokorny
1027. Walde 635*. Watkins *ster-3*.
NAV *sǫʼ* /s õ ʔ/. Independent stem
noun. PA *siny'*. YM 1987, p. 3.
TUR *yıldız* /j ɯ l d ɯ z/. Räsänen 210a.

stick noun, Swadesh 200
ALB *shkop* /ʃ k o p/. Kiçi & Aliko
(1969). Not in Janson, Huld,

Mann, Çabej, Pokorny.
ENG *stick* /s t ɪ k/. OE *sticca*, PG
stikkōn-, PIE *stig*, root *steig* 'stick'.
OED. Barnhart. Pokorny 1016.
Watkins *steig-*.
FRE *bâton* /b a t/ LOAN. **Bastonem*,
ext. of Low Lat. *bastum*, perhaps
from *basto* 'carry', from Greek
bastazo 'carry a burden'. BW.
Dauzat. EM *bastum*.
GER *Stock* /ʃ t o/ COGNATE WITH
'PUSH'. OHG *stoc*, PG *stukkaz*, PIE
steug-, extension of *steu* 'push'.
Kluge *Stock*. Pokorny 1032.
Watkins *(s)teu-*.
HAW *lāʻau* /l a:/ COGNATE WITH 'LEAF',
'TREE'. Same as 'tree'. *lā-* is the
form *lau* 'leaf' takes in compounds;
ʻau can mean 'stalk, shaft'; perhaps
'leaves with stalks'? PEP *laʻakau*.
LAT *baculum* /b a k/. PIE *baktlom*, root
bak 'cane'. Pokorny 93. EM
baculum. Walde 104*. Watkins
bak-.
NAV *tsin* /ts i n/ COGNATE WITH
'TREE', 'WOODS'. Independent
stem noun. Also 'tree'. YM 1992,
p. 1001, 608.
TUR *değnek* /d e ɣ/. From *değ-* 'touch'.
Räsänen 468b.

stone noun, Swadesh 100
ALB *gur* /g u r/. PIE *gʷerH* 'mountain'.
Huld 66. Mann 82. Walde 682.
Pokorny 477.
ENG *stone* /s t o/. OE *stān*, PG *stainaz*,
PIE *stoinos*, root *stei*. Pokorny
1010. Skeat. OED. Watkins *stei-*.
FRE *pierre* /p j ɛ r/ LOAN. Lat. *petra*,
from Greek. Not in Pokorny. EM
petra. BW. Watkins *petra*.
GER *Stein* /ʃ t ai/. OHG *stein*, PG
stainaz, see ENG. Pokorny 1010.
Kluge *Stein*.
HAW *pōhaku* /h a k u/. PCP *poofatu*,
PPN *fatu*, PMP *batu*. Base *haku*
means 'core, stone'. Blust.
LAT *lapis* /l a p/. PIE root *lep* 'stone'.
Pokorny 678. EM *lapis*. Walde
431*.
NAV *tsé* /ts é/. Independent stem noun.
YM 1987, p. 3. YM 1992, p. 599.
TUR *taş* /t a ʃ/. Räsänen 466a.

straight adj, Swadesh 200

ALB *drejtë* /r e j/ LOAN. Loan from LAT *directus*, compound of *rectus*, qv. Huld 55. Mann 3. Not in Janson.

ENG *straight* /s t r/. Past part. of *stretch*, OE *streccan*, PWG *strakkjan*, from *strakko-* 'straight', perhaps PIE *streg*, ext. of *(s)ter* 'stiff'. Skeat. OED. Pokorny 1023.

FRE *droit* /r w a/ COGNATE WITH 'RIGHT'. LAT *directus*, from *rectus*, q.v. Same word as 'right'. BW.

GER *gerade* /r a: d/ COGNATE WITH 'ROUND'. OHG *gihradi* 'fast', PIE *rotH* 'wheel', root *retH* 'go'.

HAW *pololei* /p o l o l e i/. PE.

LAT *rectus* /r e g/. Past. part. of *rego* 'go in a straight line', PIE root *reĝ*. Pokorny 855. EM *rego*. Sihler 85. Walde 363*. Watkins *reg-1*.

NAV *k'éhózdon* /d o n/. 'It is straight', neuter perfective. PA *dim'* 'become taut'. YM 1992, p. 149 *DQQD*.

TUR *doğru* /d o ɣ r u/ COGNATE WITH 'TRUE'. Also means 'true'. *Straight* appears to be the root idea. Räsänen 454a.

suck verb, Swadesh 200

ALB *thith* /θ i/ MOTIVATED. Note *thith(ë)* 'nipple', *cicë* 'nipple', *sise* 'nipple'. Çabej notes that *c* often changes to *th*. Clearly sounds like a series of nursery words. Not in Huld, Janson, Mann.

ENG *sucks* /s ə/ MOTIVATED. OE *sūcan*, PG *sūkan*, PIE *suHg*, ext. of *seuH* 'juice'. Pokorny 912. Skeat. OED. Barnhart. Watkins *seuə-2*.

FRE *suce* /s y/ MOTIVATED, COGNATE WITH 'WIPE'. Pop. LAT *suctio*, Class. *sugo*.

GER *saugt* /z au/ MOTIVATED. OHG *sūgan*, PG *sūgan*, PIE root *seuH* 'suck'. Pokorny 912. Kluge *saugen*.

HAW *omo* /o m o/. PEP *'omo*. Replaces POc *sosop*. PE. Blust.

LAT *sugit* /s u:/ MOTIVATED. PIE *suH* with a guttural extension; 'juice', 'suck', probably imitating the sound of slurping. Root *seuH*. Pokorny 912. EM *sugo*. Walde 469*. Watkins *seuə-2*.

NAV *yisht'o'* /t' ó: d/ MOTIVATED. 'I suck on it', repetitive durative. PPA *t'ut'* sounds onomatopoetic. YM 1992, p. 560 *T'ÓÓD*. Pinnow 23.

TUR *emiyor* /e m/ MOTIVATED. Sound symbolism? Räsänen 41b.

sun noun, Swadesh 100

ALB *diell* /d i/. Most likely PIE *swelwo-*. Mann 46. Huld 50. Walde 447*.

ENG *sun* /s ə/. OE *sunne*, PG *sunnōn-*, PIE *sun*, *swen*. Skeat. Walde 446*. OED. Pokorny 881. Watkins *sāwel-*.

FRE *soleil* /s ɔ/. Pop. Lat. **soliculus*, ext. of Class. *sol*. BW. Dauzat.

GER *Sonne* /z o/. OHG *sunna*, PG *sunnōn-*, see ENG. Kluge *Sonne*.

HAW *lā* /l a:/ COGNATE WITH 'DAY'. Same as 'day'. PE.

LAT *sol* /s o:/. PIE *suH₂al*; with same slightly mysterious root as in ENG *sun*, GER *Sonne*, ALB *diell*. EM *sol*. Pokorny 881. Sihler 84. Walde 446*. Watkins *sāwel-*.

NAV *shá* /ʃ á/. YM 1987, p. 3: *shá*. YM 1992, p. 468.

TUR *güneş* /g y n/ DERIVED FROM 'DAY'. Räsänen 309a.

swell verb, Swadesh 200

ALB *ënj* /ə ɲ/. Probably PIE *H₂anH₁* 'breathe'. Pokorny 39. Mann 193. Janson 155. Not in Huld.

ENG *swells* /s w ɛ l/. OE *swellan*, PG *swelnan*. Skeat. OED. Barnhart.

FRE *enfle* /f l/ COGNATE WITH 'FLOWER', 'LEAF', 'RIVER'. Lat. *inflo*, *in* 'in' (q.v.) and *flo* 'blow' (q.v.). PIE root *bhel-*. BW.

GER *schwillt* /ʃ v e l/. OHG *swellan*, PG *swelnan*. Kluge *schwellen*.

HAW *ho'opehu* /p e h u/. The root is *pehu* 'swollen'. PCP *pefu* or *pesu*. PE.

LAT *tumet* /t u/. Stative stem *tumē-*, from *-m-* ext. of PIE root *teuH* 'swell'. EM *tumeo*. Pokorny 1082. Walde 708. Watkins *teuə-*.

NAV *niishchaad* /tʃ a: d/. 'I swell up', momentaneous imperfective. YM

1992, p. 73 *CHAAD.*
TUR *şişiyor* /ʃ i ʃ/. Räsänen 424a.

swim verb, Swadesh 100
ALB *noton* /n o/. PIE *snaH₂* as in
LAT. Huld 148.
ENG *swims* /s w ɪ m/. OE *swimman*,
PG *swimjan*, perhaps PIE *swem*
'move'. OED. Skeat. Barnhart.
Pokorny 1046. Watkins *swem-*.
FRE *nage* /n a/. Lat. *navigare* 'sail',
compound, 1st part from *navis*
'ship', PIE *naH₂us*; no connection
to Lat. *no.* Dauzat. BW. Pokorny
755.
GER *schwimmt* /ʃ v i m/. PG *swimjan*,
see ENG. Pokorny 1046. Kluge
schwimmen. Watkins *swem-*.
HAW *'au* /ʔ a u/. PPN *kau*, POc
perhaps *kakaRu.* Blust.
LAT *nat* /n aː/ COGNATE WITH 'CLOUD'.
PIE root *snaH₂*, conceivably
connected to *nubes* 'cloud'. EM *no.*
Pokorny 972. Sihler 483. Walde
692*.
NAV *'ashkǫ́ǫ́h* /k õː ʔ/ COGNATE WITH
'SMOOTH'. 'I swim', momentaneous.
Possibly connected to *kǫ'* 'smooth'.
YM 1992, p. 331. Pinnow 31.
TUR *yüzüyor* /j y z/. Räsänen 214a.

tail noun, Swadesh 100
ALB *bisht* /b i ʃ t/. Huld 45. Not in
Pokorny.
ENG *tail* /t e/. OE *tægel*, PG *taglaz*,
PIE *doklos*, root *dek̂* 'fringe'. OED.
Skeat. Pokorny 191. Watkins *dek-2*.
FRE *queue* /k ø/. LAT *cōda*, by-form of
cauda. BW.
GER *Schwanz* /ʃ v a n ts/. From
schwanken 'wag', PG *swankan*,
variant of PIE root *sweng.* Kluge
Schwanz. Watkins *sweng(w)-*.
HAW *huelo* /h u e l o/.
LAT *cauda* /k au d/. EM *cauda.*
NAV *'atsee'* /ts eː ʔ/. Dependent stem
noun. YM 1987, p. 4: *-tsee'.* YM
1992, p. 603.
TUR *kuyruk* /k u j r u k/. Räsänen 296b.

that lim (neuter), Swadesh 100
ALB *ai* /a/ MOTIVATED, COGNATE WITH
'HE', 'THERE', 'THEY'. Also *ay.* A-
'that' + PIE *so* 'that'. The *a-* for

designating remote objects is
unexplained; possibly expressive?
Same word as 'he'. Huld 37.
Pokorny 979.
ENG *that* /ð/ COGNATE WITH 'THERE',
'THEY', 'THIS'. OE *þæt*, PG *þat*,
PIE *tod-*, root *t* 'that'; ending *-at*
cognate with 'what'. Skeat. OED.
Barnhart. Pokorny 1086. Watkins
to-.
FRE *cela* /s/ COGNATE WITH 'THIS'. *Ce-*
(also in *ceci* 'this', *ici* 'here') is
from Pop. Lat. *ecce + hoc*; plus *là*,
Lat. *illāc. Ecce* has the *-ce* clitic as
in *hic*, etc., but it's not clear what
it is attached to; Pokorny guessed
ed from *e-* 'that'. Pokorny 282,
609. BW. EM *ecce.* BW.
GER *das* /d/ COGNATE WITH 'THERE',
'THIS', 'THOU'. PG *þat*, see ENG.
Kluge *der.* Pokorny 1086.
HAW *kēlā* /k eː/ COGNATE WITH 'THIS'.
Initial morpheme cognate with
'this, some'. PNP *teelaa.* PE.
LAT *illud* /i/ COGNATE WITH 'HE',
'ONE', 'THERE'. OLat *ol-*, initial *i*
under influence of *is* 'he'; *ol* is
usually connected to *alius* 'other'.
EM *ille.* Pokorny 25. Sihler 393.
Walde 84.
NAV *'éi* /ʔ á/ MOTIVATED, COGNATE
WITH 'THERE'. 'He, that one': *'á*
'there' + *ii*, relative enclitic. YM
1992, p. 931.
TUR *o* /o/ COGNATE WITH 'HE',
'THERE', 'THEY'. Räsänen 356a.

there lim, Swadesh 200
ALB *aty* /a/ MOTIVATED, COGNATE
WITH 'HE', 'THAT', 'THEY'. *A* is
morpheme of distance, as in 'that,
they'. Mann 197. Not in Huld,
Janson, Pokorny.
ENG *there* /ð/ COGNATE WITH 'THAT',
'THEY', 'THIS'. OE *þēr*, PG *þēr*,
r-locative on PIE demonstrative
root *t.* Skeat. Watkins *to-*. Pokorny
1086.
FRE *là* /l/ COGNATE WITH 'HE', 'THEY'.
Same as the final morpheme in *cela*
'that'. Lat. *illāc*, adverbial
extension of *ille*, see *il* 'he'. PIE
root *al* crossed with *i.* EM *ille.*

BW. Pokorny 24.

GER *da* /d/ COGNATE WITH 'THAT', 'THIS', 'THOU'. OHG *dā(r)*, PWG *þǣr*, see ENG. Kluge *da*.

HAW *laila* /l a i l a/. PEP *laila* or *leila*. PE.

LAT *ibi* /i/ COGNATE WITH 'HE', 'ONE', 'THAT'. Locative (cf. *ubi* 'where') extension of *i-* as in *is* 'he', *unus* 'one'. Pokorny 281. EM *ibi*. Sihler 79. Walde 100.

NAV *'áadi* /ʔ á: / MOTIVATED, COGNATE WITH 'THAT'. Dem. *'áa-* plus enclitic *-di* 'at'. YM 1992, p. 935.

TUR *orada* /o/ DERIVED FROM 'THAT', COGNATE WITH 'HE', 'THEY'. Locative of *ora* 'that place', from *o-* demonstrative morpheme.

they link, Swadesh 200

ALB *ata* /a/ MOTIVATED, COGNATE WITH 'HE', 'THAT', 'THERE'. Same initial morpheme as in 'that', 'there'. Mann 115. Not in Huld, Janson, Pokorny.

ENG *they* /ð/ LOAN, COGNATE WITH 'THERE', 'THAT', 'THIS'. Mostly reflects ON *þei-*, PG *þai*, PIE demonstrative root *t-*. OED. Skeat. Barnhart. Watkins *to-*.

FRE *ils* /i l/ DERIVED FROM 'HE', COGNATE WITH 'THERE'. Plural of *il* 'he', q.v. Cross of PIE *al* 'beyond' and *i*.

GER *sie* /z/. PIE demonstratives *so*, *sā*, root *s*. Pokorny 979. Kluge *sie*.

HAW *lākou* /l a:/. *-kou* is plural suffix. PNP *kilaatou*. PE.

LAT *ei* /e/. In Lat. *e* suppletively alternates with *i* (cf. singular *is*), but apparently they are unrelated. Pokorny 281. EM *is*.

NAV *daabí* /b í/ DERIVED FROM 'HE'. Plural of *bí* 'he'. YM 1992, p. 930.

TUR *onlar* /o/ DERIVED FROM 'THAT', COGNATE WITH 'HE', 'THERE'. Plural of word for 'that'. Räsänen 356a.

thick adj, Swadesh 200

ALB *trashë* /t r a ʃ/. *Trash* 'I thicken'. Not in Janson, Huld, Mann, Pokorny.

ENG *thick* /θ ɪ k/. OE *þicce*, PG *þikuz*, PIE *tegu-*. Skeat. OED. Barnhart. Pokorny 1057. Watkins *tegu-*.

FRE *épais* /e p ɛ/. Lat. *spissus*, *-to-* form of **spid*, perhaps ext. of PIE root *speH₁* 'thrive'. Dauzat. BW. Pokorny 984.

GER *dick* /d i k/. OHG *dicki*, PG *þikuz*, see ENG. Pokorny 1057. Kluge *dick*. Pokorny 1057.

HAW *mānoa* /n o a/. *Mā-* is a common stative prefix, so maybe the root is *noa*. PE.

LAT *crassus* /k r/. Perhaps PIE base *kert* 'turn', ext. of *(s)ker*. Pokorny 584, 935. EM *crassus*. Walde 421.

NAV *ditą́* /t ą́/. 'It is thick', neuter absolute imperfective. PA *tany*. YM 1992, p. 482 *TĄ́*.

TUR *kalın* /k a l ɯ n/. Räsänen 226b.

thin adj, Swadesh 200

ALB *hollë* /h o ɬ/ COGNATE WITH 'SALT', 'SHORT'. PIE *skēlo*, root *(s)kel* 'cut', like Lat. *culter* 'knife'. May be a connection to *(s)ker* 'cut'. Pokorny 924. Huld 75. Mann 95. Walde 591*. Not in Janson.

ENG *thin* /θ ɪ n/ COGNATE WITH 'THINK'. OE *þynne*, PG *þunn-*, PIE root *ten* 'stretch'. Skeat. OED. Pokorny 1069. Barnhart. Watkins *ten-*.

FRE *mince* /m ɛ̃/. From verb *mincier* 'mince', Pop. Lat. *minūtio*, from *minūtus* 'small', from *minuo* 'decrease', PIE *minu*, root *mei*. Dauzat. BW. Pokorny 711. EM *minuo*. Watkins *mei-2*.

GER *dünn* /d y n/ COGNATE WITH 'THINK'. OHG *dunni*, PG *þunn-*, see ENG.

HAW *wīwī* /w i:/. PPN *iwiiwi*. Connected to *'uwī* 'squeeze'?

LAT *tenuis* /t e n/ COGNATE WITH 'HOLD'. PIE *tenú-* 'thin' from root *ten* 'stretch'. Lubotsky, p. 59: root *tenH*. Pokorny 1069. EM *tenuis*. Sihler 97. Walde 724. Watkins *ten-*.

NAV *'átt'ą́'í* /t' ą́ ʔ í/ COGNATE WITH 'LEAF'. 'It is thin', neuter imperfective. PA *t'an'* 'leaf'. YM 1992, p. 534, *T'ĄH*.

TUR *ince* /i n dʒ e/. Räsänen 203b.

think verb, Swadesh 200

ALB *mendon* /m e n/ LOAN. From *mend* 'mind', loan from Lat. *mentem*, PIE root *men* 'think'. Pokorny 726. Huld 91. Mann 72. EM *mens*. Not in Janson.

ENG *thinks* /θ ɪ n/ COGNATE WITH 'THIN'. OE *þencan*, factitive of PG *þinkan*, PIE *teng*, possibly extension of *ten* 'stretch'. Skeat. OED. Pokorny 1088. Kluge *denken*. Watkins *tong-*.

FRE *pense* /p ɑ̃/ LOAN. Pop. Lat. *pensare* 'weigh, judge' from *pendere* 'weigh', PIE root *(s)pen* 'draw'. Nasal must be influenced by written language, i.e., a partial loan from Latin. Dauzat. BW. Pokorny 988, 982. Watkins *(s)pen-*.

GER *denkt* /d e n/ COGNATE WITH 'THIN'. PG *þinkan*, PIE *teng*, possibly extension of *ten* 'stretch'. Pokorny 1088. Kluge *denken*.

HAW *manaʻo* /n a ʔ o/. PPN *manako*. Passive stem is *manaʻol-*. Reduplicates as *mananaʻo*. PE.

LAT *cogitat* /g/. Prefix *co-* 'with' plus *ago* 'drive', PIE *aĝ*. Pokorny 4. EM *ago*. Walde 35.

NAV *nízin* /z i n/ COGNATE WITH 'KNOW'. 'He thinks'. YM 1992, p. 756 *ZÍÍʼ*. YM 1980, p. 391.

TUR *düşünüyor* /d y ʃ/ DERIVED FROM 'FALL'. From *düş-* 'fall'. Räsänen 507b.

this lim (neuter), Swadesh 100

ALB *ky* /k/ COGNATE WITH 'HERE', 'WHAT', 'WHO'. *k-* is a proximate demonstrative, perhaps from PIE indefinites in *kʷ*. (Or the enclitic *-ke*: would that be same as FRE?) *-y* is from PIE *so* 'that', as in *ay* (*ai*) 'that'. The former may therefore have same origin as *çʼ* 'what'. Mann 114. Huld 84. Walde 509*. Pokorny 929.

ENG *this* /ð/ DERIVED FROM 'THAT', COGNATE WITH 'THERE', 'THEY'. OE *þis*, PG *þasi-*, demonstrative *þa-* with proximative *-se/-si* ending. PIE dem. root *t-*. OED. Skeat. Pokorny 1086. Watkins *to-*.

FRE *ceci* /s/ COGNATE WITH 'THAT'. Same root as 'that' (Lat. *-ce* as in *hoc*), with the *-ci* prox. suffix as in *ici* 'here'.

GER *dieses* /d/ DERIVED FROM 'THAT', COGNATE WITH 'THERE', 'THOU'. Same source as *das* 'that' but with proximative *-se/-si* ending. PIE dem. root *t*. Pokorny 1086.

HAW *kēia* /k eː/ COGNATE WITH 'THAT'. PCP *teeia*. Same initial morpheme as 'that'. PE.

LAT *hoc* /h/ COGNATE WITH 'HERE'. PIE *ĝho* plus suffixal *-c(e)*. Pokorny 609. EM *hic*. Sihler 393. Walde 542.

NAV *díí* /d íː/. YM 1992, p. 931.

TUR *bu* /b/ COGNATE WITH 'HERE', 'I', 'WE'. Räsänen 85a.

thou link (familiar), Swadesh 100

ALB *ti* /t/. PIE *tū* 'thou'. Root may be demonstrative *t*. Mann 114. Huld 116. Walde 745. Pokorny 1097.

ENG *you* /j u/ COGNATE WITH 'YOU'. Same word as 'you' pl. OE *ēow* (obl.), PG *iuwiz*, PIE root *yu*. Skeat. OED. Pokorny 514. Skeat. Watkins *yu-1*.

FRE *tu* /t/. Nominative. Ringe 1993 has /tua/, a disjunctive. LAT *tu*. PIE root may be demonstrative *t*.

GER *du* /d/ COGNATE WITH 'THAT', 'THERE', 'THIS'. OHG *du*, PG *þū*, PIE *tū*. Root *t-*, which may have something to do with pronominal stem *to-* 'that'. Pokorny 1097. Kluge *du*.

HAW *ʻoe* /ʔ o e/ COGNATE WITH 'YOU'. PPN *koe*. Cf. PMP *i-kahu*. PE. Blust.

LAT *tu* /t/. PIE *tū*, root *t*. Pokorny 1097. Walde 745. Watkins *tu-*.

NAV *ni* /n i/ COGNATE WITH 'WE', 'YOU'. Cf. *nihi* 'we', also 'you' duoplural. Emphatic independent, subjective case. YM 1992, p. 930.

TUR *sen* /s/ COGNATE WITH 'YOU'. Räsänen 409b.

three lim, Swadesh 200

ALB *tre* /t r e/. PIE *treyes*, root *trey*.
Huld 117. Mann. Walde 753.
Pokorny 1090.

ENG *three* /θ r i/. OE *þrēo*, PG *þrijiz*,
PIE *treyes*, root *trey*. Skeat. OED.
Pokorny 1090. Watkins *trei-*.

FRE *trois* /t r w a/ COGNATE WITH
'SHARP'. LAT *trēs*, PIE root *trey*.

GER *drei* /d r ai/. OHG *drī*, PG *þrijiz*,
see ENG. Pokorny 1090. Kluge
drei. Watkins *trei-*.

HAW *kolu* /k o l u/. PPN *tolu*, PMP
telu. PE. Blust.

LAT *tres* /t r e:/. PIE *treyes*, root *trey*.
Pokorny 1090. EM *tres*. Sihler 410.
Walde 753. Watkins *trei-*.

NAV *táá'* /t á: ʔ/. YM 1992, p. 932,
485. Hoijer 10.

TUR *üç* /y tʃ/. Räsänen 518a.

throw verb, Swadesh 200

ALB *hedh* /h e/. PIE root *(s)keud-*
'throw', perhaps from a *(s)keu*.
Mann 155. Huld 72. Walde 554*.
Pokorny 956.

ENG *throws* /θ r/. OE *þrāwan* 'turn,
twist', PG *þrǣw-*, PIE *treH₁*,
variant of root *terH₁* 'rub'. Skeat.
OED. Barnhart. Pokorny 1071.
Watkins *terə-1*.

FRE *lance* /l ā s/ LOAN. Low Lat. *lanceo*
'handle a lance', from *lancea*
'lance', from Celtic, PIE root *plāk*
'hit'. Dauzat. BW. Pokorny 832.
EM *lancea*.

GER *wirft* /v e r/ COGNATE WITH 'RUB',
'WORM'. OHG *werfan*, PG *werpan*,
PIE *wer* 'turn'. Kluge *werfen*.
Pokorny 1152.

HAW *nou* /n o u/. PCP *no(o)u*. PE.

LAT *iacit* /j a/ COGNATE WITH 'LIE'.
PIE root *yē*. EM *iacio*. Pokorny
502. Sihler 531. Walde 199.
Watkins *yē-*.

NAV *hishhan* /ɣ a n/. 'I throw (single
object)', momentaneous
completive. PA *ghan* 'tie together',
PPA *xan*. YM 1992, p. 224 *GHAN*.

TUR *atıyor* /a t/. Räsänen 31a.

tie verb, Swadesh 200

ALB *lidh* /l i ð/. PIE *leiĝ*, like LAT *ligo*.
Pokorny 668. Mann 174. Huld 86.

Tucker *ligo*, Walde 400*.

ENG *ties* /t ai/. OE *tīegan*, PIE
doukeyo-, PIE root *deuk* 'pull'.
Skeat. OED. Pokorny 220. Watkins
deuk-.

FRE *attache* /t a ʃ/ LOAN. OF *estachier*,
replacing the initial with the prefix
a-. The noun is *estache* 'pole',
Frankish *stakka* 'stake', PG
stikkon-, PIE *stig*, root *steig* 'stick'.
BW. Pokorny 1017.

GER *bindet* /b i n d/. OHG *bintan*, PG
bindan, PIE root *bhendh*. Pokorny
127. Kluge *binden*.

HAW *hīki'i* /k i ʔ/. PE: root *ki'i* is found
with half a dozen different prefixes.
PCP *-tiki*. Possibly transitive *-i*.
PE.

LAT *ligat* /l i g/. PIE *leig*. EM *ligo*.
Pokorny 668. Walde 400*. Watkins
leig-1.

NAV *be'eshtł'ó* /tˡʼ ó/. 1st person.
Conclusive imperfect. PA *tł'u-ny*,
PPA *tł'u*. YM 1992, p. 587 *TŁ'Ǫ*.

TUR *bağlıyor* /b a ɣ/. Räsänen 53a.

tongue noun, Swadesh 100

ALB *gjuhë* /ɟ u h/. PAlb *gʎuhë*, earlier
gluxo; pre-Alb. *ĝhundsḱe*. Could be
related to LAT, ENG, GER if there
was a spontaneous metathesis from
dnĝhu. Mann 91. Huld 71. Walde
792.

ENG *tongue* /t ə ŋ/. OE *tunge*, PG
tungōn-, PIE root *dnĝhū*. Skeat.
OED. Pokorny 223. Watkins *dnĝhū*.

FRE *langue* /l ā g/. LAT.

GER *Zunge* /ts u ŋ/. OHG *zunga*, PG
tungōn-, see ENG. Kluge *Zunge*.

HAW *alelo* /l e l o/ COGNATE WITH
'SAY'. Also *elelo*, *lelo*. Suspiciously
close to *'ōlelo* 'language, speak'.
Replaces POc. *maya*. PE. Blust.

LAT *lingua* /l i n gw/. OLat *dingua*.
PIE *dnĝhwaH₂*, root *dnĝhū*. The
initial *l* is either a matter of
possibly dialectal *d:l* alternation
(cf. 'smell'), and/or a crossing with
lingo 'lick'. EM *lingua*. Pokorny
223. Sihler 39. Walde 792. Watkins
dnĝhū.

NAV *'atsoo'* /ts o: ʔ/. Dependent stem
noun. YM 1987, p. 4. YM 1992,

p. 617.
TUR *dil* /d i l/. Räsänen 478a.

tooth noun, Swadesh 100
ALB *dhëmb* /ð ə m b/. PIE *ĝombhos* 'tooth', root *ĝembh* 'bite'. Mann 30. Huld 58. Pokorny 369. Walde 575.
ENG *tooth* /t/ COGNATE WITH 'EAT'. OE *tōþ*, PG *tanþuz*, PIE *dent-*, pres. part. of *H₁ed*. Skeat. OED. Pokorny 287. Watkins *dent-*.
FRE *dent* /d/. LAT *dentem*, acc. of *dens*, q.v. BW.
GER *Zahn* /ts/ COGNATE WITH 'EAT'. OHG *zan*, PG *tanþuz*, see ENG. Pokorny 287. Kluge *Zahn*.
HAW *niho* /n i h o/. PPN *nifo*, POc *nipon*, PCEMP *nipən*, PMP *nipen*. PE. Blust.
LAT *dens* /d/ COGNATE WITH 'EAT'. PIE *dent-*, a participle (i.e., *H₁dn̥t*) of root *H₁ed* 'eat'. Pokorny 287. EM *dens*. Sihler. Walde 120. Watkins *dent-*.
NAV *'awoo'* /ɣ o: ʔ/. Dependent stem noun. YM 1987, p. 4. YM 1992, p. 660.
TUR *diş* /d i ʃ/. Räsänen 481b.

tree noun, Swadesh 100
ALB *dru* /d r u/. PAlb *drun-*, root *deru*. Pokorny 215. Mann 33, 69. Huld 56. Walde 805.
ENG *tree* /t r i/ COGNATE WITH 'TRUE'. OE *trēow*, PG *trewam*, PIE *drewom*, root *deru*. Skeat. OED. Watkins *deru*.
FRE *arbre* /a r b/. LAT *arbor, arborem*. BW.
GER *Baum* /b au/. OHG *boum*, PWG *baumaz*, possibly PIE root *bheuH* 'grow'. Pokorny 146. Kluge *Baum*. Watkins *bheuə-*.
HAW *lā'au* /l a:/ COGNATE WITH 'LEAF', 'STICK'. *Lā-* is the form *lau* 'leaf' takes in compounds; *'au* can mean 'stalk, shaft': perhaps 'leaves with stalks'? PEP *la'akau*. PE.
LAT *arbor* /a r b/. Possibly *erdh* 'tall'. EM *arbos*. Pokorny 339. Walde 149.
NAV *tsin* /ts i n/ COGNATE WITH 'STICK', 'WOODS'. YM 1987, p. 3:

tsin 'wood, tree', independent stem noun.
TUR *ağaç* /a ɣ a tʃ/. Räsänen 7b.

true adj, Swadesh 200
ALB *vërtetë* /v ə r/ LOAN. Clearly connected to LAT *verus*; loan from *veritatem*? PIE *wēr*. Not in Huld, Mann, Janson, Pokorny.
ENG *true* /t r u/ COGNATE WITH 'TREE'. OE *trēowe*, PG *trewwjaz*, PIE *drew* 'firm', perh. metaph. extension of meaning 'tree'. Skeat. OED.
FRE *vrai* /v r/. From *verai*, Pop. LAT *vērācus*, replacing Class. *vērāx*, from *vērus*, q.v. BW.
GER *wahr* /v a: r/. OHG *wār*, PG *wǣraz*, PIE root *wēr*. Pokorny 1166. Kluge *wahr*.
HAW *'oia'i'o* /ʔ o i a/. Seems to be a compound of *'oia* and *'i'o*, both of which mean 'true'. PE.
LAT *verus* /w e: r/. PIE *wēr*. EM *uerus*. Pokorny 1166. Walde 286. Watkins *wēro-*.
NAV *'aanii* /n í:/. YM 1992, p. 451 *NÍÍD*.
TUR *doğru* /d o ɣ r u/ COGNATE WITH 'STRAIGHT'. 'Straight' seems to be the original idea. Räsänen 454a.

two lim, Swadesh 100
ALB *dy* /d y/. PIE, perhaps the fem. *duwai*; root is *dw*. Mann 31, Huld 56. Walde 817. Pokorny 228.
ENG *two* /t u/. OE *twā*, PG *twai*, PIE *duwai*, root *dw-*. Skeat. OED. Pokorny 228. Barnhart. Watkins *dwo-*.
FRE *deux* /d ø/. LAT *duos*, acc. of *duo*, q.v.
GER *zwei* /ts v/. OHG *zwei* neut., PIE root *dw-*. Pokorny 228. Kluge *zwei*.
HAW *lua* /l u a/. POc *rua*, PMP *duha*. PE. Blust.
LAT *duo* /d u/. The /o/ is a dual suffix, as in *ambo*. PIE *dw*. Pokorny 228. EM *duo*. Walde 817. Watkins *dwo-*.
NAV *naaki* /n a:/. PA *nà-tV*. YM 1992, p. 932. Hoijer 11.
TUR *iki* /i k i/. Räsänen 39a.

vomit verb, Swadesh 200

ALB *vjell* /v j e ɫ/. PIE *wel* 'turn'.
Mann 157, Huld 130. Pokorny 1142.

ENG *vomits* /v ɑ m/ LOAN. Borrowed
from Lat. *vomit-*, where the *t* is
either past part. or frequentative;
PIE root *wemH*. Skeat. OED.
Watkins *wemə-*.

FRE *vomit* /v ɔ m/ LOAN. Pop. LAT
*vomīre, Class. *vomere*. The /o/
must be under the infl. of the
written language (partial loan).
PIE root *wemH*. BW.

GER *erbricht* /b r/. Prefix *er-* on
brechen 'break', OHG *brehhan*, PG
brekan, PIE *bhreg*, perhaps from
bher 'apply a sharp tool to'.
Pokorny 165. Kluge *brechen*.

HAW *luaʻi* /l u a/. PPN *luaʻaki*. PPN
lua (POc *luaq*), plus a
transitivizing suffix *-aʻi*. Cf. also
luea 'nausea'. PE. EP, p. 87.

LAT *vomit* /w o m/. Assim. from
*vemo, PIE root *wemH* 'spit'.
Pokorny 1146. EM *uomo*. Walde
262. Watkins *wemə-*.

NAV *náshkwi* /kw i/. 'I vomit',
conclusive. PA *Кнy*, PPA *Kʷiy*.
YM 1992, p. 334 *KWI*.

TUR *kusuyor* /k u s/. Räsänen 304b.

wash verb, Swadesh 200

ALB *lau* /l a/. Connected to LAT *lavo*,
but posibly via loan. PIE root is
law. Huld 85. Janson 177. Mann
146.

ENG *washes* /w ɑ/ COGNATE WITH
'WATER', 'WET'. OE *wæscan*, PWG
waskan, PIE *H₂wed-n/r*. Skeat.
OED. Pokorny 80. Barnhart.
Watkins *wed-1*.

FRE *lave* /l a v/. LAT. BW.

GER *wäscht* /v a ʃ/ COGNATE WITH
'WATER'. OHG *wascan*, PWG
waskan, see ENG. Kluge *waschen*.
Pokorny 80. Watkins *wed-1*.

HAW *holoi* /h o l o/. PPN *soloʻi*, with
transitivizing *-i*, cf. Samoan *solo*
'wipe'. PE. EP, p. 88.

LAT *lavat* /l a w/. PIE root *lau* 'wash'.
Pokorny 692. EM *lavo*. Walde 441.
Watkins *leu(ə)-*.

NAV *yiisgis* /g i z/. 'I wash (a
permeable object such as

clothing)', semelfactive imperfect.
Orig. 'wring'. PA *Gĭts'*, PPA *Gits'*.
YM 1992, p. 207 *GIZ*.

TUR *yıkıyor* /j/. From *jajka*, from *jaj*.
Räsänen 179a.

water noun, Swadesh 100

ALB *ujë* /u j/. PIE, perhaps *udryom*;
root is perhaps *H₂wed* 'moisten'.
Man 92. Huld 121. Pokorny 79.

ENG *water* /w ɑ t/ COGNATE WITH
'WASH', 'WET'. OE *wæter*, PG
watar, PIE *wodō*, base *H₂wed*.
Skeat. OED. Pokorny 79. Watkins
wed-1.

FRE *eau* /o/. LAT. BW.

GER *Wasser* /v a s/ COGNATE WITH
'WASH'. OHG *wazzar*, PG *watar*,
see ENG. Pokorny 79. Kluge
Wasser.

HAW *wai* /w a i/. PPN *wai*, PCEMP
waiR, PMP *wahiR*. PE. Blust.

LAT *aqua* /a kw/. PIE *akʷaH₂-*, root
akʷ. Pokorny 23. EM *aqua*. Walde
34. Watkins *akʷā-*.

NAV *tó* /t o/ COGNATE WITH 'LAKE',
'RIVER', 'SEA'. Independent stem
noun. YM 1987, p. 3. Hoijer 10.
YM 1992, p. 518.

TUR *su* /s u j/. Räsänen 431a.

we link (exclusive), Swadesh 100

ALB *né* /n/. PIE *nos*, oblique of root
ne. Mann 28. Huld 96. Walde 320.
Pokorny 758.

ENG *we* /w i/. OE *wē*, PG *wīz*, PIE
weies, root *we*. Skeat. OED *I*.
Pokorny 1114. Watkins *we-*.

FRE *nous* /n/. LAT *nos*, unstressed.
BW.

GER *wir* /v i:/. OHG *wir*, PG *wīz*, see
ENG. Kluge *wir*.

HAW *mākou* /m a:/. *-kou* is suffix for
more than 2 people. PEP *maatou*,
PNP *kimaatou*, PPN *kimautolu*,
POc *k-ami*. PE. Blust.

LAT *nos* /n/. PIE root *ne* (orig.
non-nominative). Pokorny 758. EM
nos. Walde 320*. Watkins *nes-2*.

NAV *nihí* /n i/ DERIVED FROM 'THOU',
COGNATE WITH 'YOU'. This is also
2nd duoplural, and resembles
singular *ni* 'thou'. YM 1992,

p. 930: emphatic independent, subjective case.

TUR *biz* /b/ DERIVED FROM 'I', COGNATE WITH 'HERE', 'THIS'. Räsänen 66a, 333b.

wet adj, Swadesh 200

ALB *lagët* /l a g/. *Lag* 'moisten', PIE *w̧lg*, root *welg* 'wet'. Mann 36. Pokorny 1145. Not in Huld, Pokorny.

ENG *wet* /w ɛ t/ COGNATE WITH 'WASH', 'WATER'. OE *wǣt*, PIE *wēdo-*, root *wed* 'water'. Skeat. OED. Pokorny 80. Watkins *wed-1*.

FRE *mouillé* /m u j/. Verb *mouiller*, Pop. Lat. **mollio* 'soften bread by moistening it', from *mollis* 'soft', PIE *m̧ldwis*, root *mel* 'soft'. BW. Pokorny 716. Dauzat. Watkins *mel-1*.

GER *nass* /n a s/. OHG *naz*, PG *nataz*. Kluge *naß*.

HAW *pulu* /p u l u/. PCP *pulu*. PE.

LAT *umidus* /u:/ COGNATE WITH 'WIFE'. *-id-* deverbative of stative *ūmeo* 'be wet', PIE *ugʷsmo-*, root *wegʷ* 'wet' (some have connected this to *uxor* 'wife'). Pokorny 1118. EM *umeo*. Walde 248. Watkins *wegʷ-*.

NAV *dittéé'* /t̞ e: ʔ/. 'It is wet', neuter. PA *t̞eK'*, PPA *t̞eK'ʷ*. YM 1992, p. 568 *T̞ÉÉ'*. Pinnow 22.

TUR *ıslak* /ɯ s l a/. Cf. *isla-* 'to wet'.

what link, Swadesh 100

ALB *ç'* /tʃ/ COGNATE WITH 'HERE', 'WHO', 'THIS'. It is tempting to see this as coming from PIE *kʷid*, but the phonetics is obscure. Mann 38. Huld 47. Pokorny 644.

ENG *what* /w/ COGNATE WITH 'WHO'. OE *hwæt*, PG *hwat*, PIE *kʷod*, root *kʷ* interrog. pron. Skeat. OED. Pokorny 647. Watkins *kʷo-*.

FRE *quoi* /k/ COGNATE WITH 'SOME', 'WHO'. LAT. BW.

GER *was* /v/ COGNATE WITH 'IF', 'WHO'. OHG *hwaz*, PG *hwat*, see ENG. Kluge *was*. Pokorny 647.

HAW *aha* /a h a/. PEP *afa*, PPN *haa* and probably *hafa*, PCEMP *(s)apa*,

PMP *apa*. PE. Blust.

LAT *quid* /kw/ COGNATE WITH 'BECAUSE', 'WHO'. PIE interrog. pronoun root *kʷ*. Pokorny 644. EM *quis*. Walde 519. Watkins *kʷo-*.

NAV *ha'át'íísh* /h a:/ COGNATE WITH 'BECAUSE', 'WHO'. Root is *ha(a)-*, interrog. pron. YM 1992, p. 935.

TUR *ne* /n e/. Räsänen 352a.

white adj, Swadesh 100

ALB *bardhë* /b a r/. PIE *bhr̥Hĝ*, root *bherHĝ* 'bright', ext. of *bher*. Huld 40. Walde 170*. Pokorny 139.

ENG *white* /w ai t/. OE *hwīt*, PG *hwītaz*, PIE *k̂weitos*, root PIE *k̂weit* 'shine'. Skeat. OED. Pokorny 628. Watkins *weit-*.

FRE *blanc* /b l/ LOAN. Borrowed from a Germanic language, PG *blank-* 'shining, pale', PIE root *bhlag* 'burn, shine', extension of *bhel* 'shine'. Dauzat. BW. Barnhart *blank*. Pokorny 125. Watkins *bhel-1*.

GER *weiß* /v ai s/. OHG *hwīz*, PG *hwītaz*, see ENG. Kluge *weiß*.

HAW *ke'oke'o* /k e ʔ o/. PCP *tekoteko*. Base *ke'o* means 'white, clear'. PE.

LAT *albus* /a l/. PIE *albhos*, perhaps root *al* with *-bho* suffix. Pokorny 30. EM *albus*. Walde 92. Watkins *albho-*.

NAV *łigai* /g a i:/. 'It is white', neuter. PA *G̱aghy*, PPA *G̱ay*. YM 1992, p. 195 *GAII*. Pinnow 36.

TUR *ak* /a k/. Räsänen 12b.

who link, Swadesh 100

ALB *kush* /k u/ COGNATE WITH 'HERE', 'THIS', 'WHAT'. PIE *kʷu*, root interrog. *kʷ*. This could be same root as in 'what'. The *-sh* is secondary. Pokorny 647. Mann 122. Huld 84.

ENG *who* /h/ COGNATE WITH 'WHAT'. OE *hwā*, PG *hwaz*, PIE *kʷos*, root *kʷ*, interr. pron. Skeat. OED. Pokorny 647. Watkins *kʷo-*.

FRE *qui* /k/ COGNATE WITH 'SOME', 'WHAT'. Pop. Lat. *quī*, same base as LAT *quis*, q.v. BW.

GER *wer* /v/ COGNATE WITH 'IF', 'WHAT'. PG *hwiz*, PIE *kʷis*, root *kʷ*, interr. pron. Pokorny 647. Kluge *wer.*

HAW *wai* /w a i/. PPN *hai*, PMP *sai*. PE. Blust.

LAT *quis* /kw/ COGNATE WITH 'BECAUSE', 'WHAT'. PIE *kʷis*, root *kʷ*, interr. pron.

NAV *háí* /h aː/ COGNATE WITH 'BECAUSE', 'WHAT'. YM 1992, p. 935. Root is *ha(a)-*, interrog. pron.

TUR *kim* /k i m/. Räsänen 271a.

wide adj, Swadesh 200

ALB *gjerë* /ɟ e r/. Best guess is a PIE *sHn-* 'send away'. Pokorny 889. Janson 28. Huld 68. Walde 463*.

ENG *wide* /w ai/ COGNATE WITH 'WITH'. OE *wīd*, PG *wīdaz*, PIE *wītos*, from *wi-* 'apart' + *itos* 'gone' (root *i* as in *year*). Skeat. OED. Barnhart. Pokorny 294. Watkins *wi-*.

FRE *large* /l a/. Rebuilt from fem. of Lat. *largus* 'generous', perhaps **laiesagos*, from PIE root *lai* 'fat'. Dauzat. BW. Pokorny 652. EM *largus*. Barnhart.

GER *weit* /v ai/. OHG *wīt*, PG *wīdaz*, see ENG. Kluge *weit*. Pokorny 294.

HAW *ākea* /aː k e a/. PE.

LAT *latus* /t l aː/. PIE *stļH₂tos*, root *stelH₂* 'spread'. Pokorny 652. EM *latus*. Walde 643*. Watkins *stelə-*.

NAV *niteel* /t eː l/. Or *ntel* with syllabic *n*. PA *tèl*, PPA *te'l*. YM 1992, p. 498 *TEEL*. Hoijer 11.

TUR *geniş* /g e n/. Also *gen*. Räsänen 253a.

wife noun, Swadesh 200

ALB *grua* /g r/ COGNATE WITH 'WOMAN'. Same word as 'woman'. PIE *ĝraH₂u-*, like Greek *graûs* 'old woman', root *ĝerH₂* 'to age'. Mann 28, 69, 85, 99. Huld 66. Pokorny 473, 390.

ENG *wife* /w ai f/ COGNATE WITH 'WOMAN'. OE *wīf*, PG *wīban*. Skeat. OED. Barnhart.

FRE *femme* /f a/ COGNATE WITH 'WOMAN'. Same word as 'woman'. Lat. *fēmina* 'female', from 'nursing', *-meno-* participle of PIE

root *dheH₁* 'suck'. BW. Pokorny 652. EM *fecundus*. Watkins *dhē(i)-*.

GER *Gattin* /g a t/ DERIVED FROM 'HUSBAND', COGNATE WITH 'GOOD'. *Gatte* 'husband' + suffix *-in*, feminine. PIE root *ghedh* 'unite'. Pokorny 423. Kluge.

HAW *wahine* /h i n e/ COGNATE WITH 'WOMAN'. Same word as 'woman'. PCP *wafine*, PPN *fafine* 'woman'; replaces POc *qasawa* 'spouse'. PE. Blust.

LAT *uxor* /u k/ COGNATE WITH 'WET'. Suffixal *-sor* as in *soror* 'sister'; root may be *euk* 'be accustomed' or *wegʷ* 'wet' as in *umidus*. Pokorny 1118. EM *uxor*. Walde 250. Watkins *euk-*.

NAV *'a'áád* /ʔ áː d/. Dependent stem noun. Hoijer 10. YM 1992, p. 968, 27.

TUR *karı* /k a r ɯ/

wind noun, Swadesh 200

ALB *erë* /e r/ LOAN, COGNATE WITH 'SMELL'. Loan from Lat. *aer*, from Greek *aér*, PIE *aH₂wer-*, perhaps connected to root *H₂wer* 'raise'. EM *aer*. Pokorny 82, 1150. Mann 91. Huld 62. Watkins *wer-2*.

ENG *wind* /w ɪ/ MOTIVATED, COGNATE WITH 'WING'. OE *wind*, PG *windaz*, PIE *wēntos*, pres. part. of root *H₂weH₁* 'blow'. Skeat. OED. Pokorny 82. Watkins *wē-*.

FRE *vent* /v ã/ MOTIVATED, COGNATE WITH 'BLOW'. LAT *ventus*. BW.

GER *Wind* /v i/ MOTIVATED. OHG *wint*, PG *windaz*, see ENG. Pokorny 82. Kluge *Wind*.

HAW *makani* /k a n i/. PPN *ma-tangi*, POc *jaŋi*. PE. Ross.

LAT *ventus* /w e/ MOTIVATED. PIE *H₂weH₁ntos*, see ENG. EM *uentus*. Walde 220. Watkins *wē-*.

NAV *nílch'i* /tʃ i/ COGNATE WITH 'BLOW'. From verb *ch'i* 'wind blows'. YM 1992, p. 104, *CH'I*. Hoijer 14.

TUR *rüzgar* /r y z/ LOAN. Loan from Persian *rūzgār*, 'era, wind, day'; perhaps from *rūz* 'day'. Räsänen 390b. Steingass 593, 592b.

wing noun, Swadesh 200

ALB *fletë* /f l/ MOTIVATED, COGNATE
WITH 'FLY'. Appears related to the
word for 'fly'. Çabej considers this
a motivated, expressive formation.
Çabej 196. Not in Janson, Mann,
Huld.

ENG *wing* /w/ LOAN, COGNATE WITH
'WIND'. ME *wenge*, borrowed from
ON *vængr*, pl. of *vængr*, perhaps
PG *wē-ingjaz*, from PIE root
H₂weH₁ 'raise'. Skeat. OED.
Barnhart. Watkins *wē-*.

FRE *aile* /ɛ/. LAT *ala*. BW.

GER *Flügel* /f l/ DERIVED FROM 'FLY',
COGNATE WITH 'BIRD', 'FLOW',
'FULL', 'MANY', 'RIVER'. Derived
from *fliegen*, 'fly', OHG *fliogan*,
PG *fleugan*, PIE *pleuk-*, PIE *pleu-*
'flow', possibly from root *pel(H)*
'fill, pour'. Pokorny 835. Kluge
fliegen, *Flügel*.

HAW *'ēheu* /ʔ e: h e u/. PCP *ke(e)feu* or
ke(e)seu. PE.

LAT *ala* /aː/. PIE *ak̂slaH₂-*, root *ak̂s*
'axis'. Some have connected to *ag̑*
'drive', as in 'think'. Pokorny 4.
EM *ala*. Walde 37. Watkins *aks-*.

NAV *'at'a'* /t' a ʔ/ DERIVED FROM
'FLY', COGNATE WITH 'FEATHER'.
'Wing, feather', from 'fly'. PA
t'aG. YM 1992, p. 532, *T'A'*.

TUR *kanat* /k a n a d/. Räsänen 230b.

wipe verb, Swadesh 200

ALB *fshin* /ʃ/. Mann derives from an
apo-ksūny- (root *kes* 'scratch'),
Pokorny from root *bhes* 'rub away'.
Mann 147. Pokorny 146. Not in
Huld, Janson.

ENG *wipes* /w ai p/. OE *wīpian*, PG
wīpjan, PIE root *weib*. Skeat. OED.
Pokorny 1132. Watkins *weip-*.

FRE *essuie* /s ɥ/ MOTIVATED, COGNATE
WITH 'SUCK'. Lat. *exsūco* 'express
juice', *ex-* 'out from' + *sūcus*
'juice', PIE root *seuH* 'take liquid'.
Dauzat. BW. Pokorny 913.
Watkins *seuə-2*.

GER *wischt* /v i ʃ/. PG *wiska-*, PIE *weis*
'turn'. Kluge *Wisch*. Pokorny 1133.
OED *whisk*.

HAW *kāwele* /k aː w/ LOAN. From Eng.

towel, OF *toaille*, PWG *þwahljō*,
from *þwahan* 'wash'. PE. OED.

LAT *terget* /t e r/ COGNATE WITH
'PUSH'. May be an extension of *tero*
'rub', PIE root *terH₁*. Pokorny
1071. EM *tergeo*. Walde 732.

NAV *nánisht'od* /t' óː d/ MOTIVATED,
COGNATE WITH 'SUCK'. 'I give it a
series of wipes (with a cloth)',
repetitive seriative imperfective.
Root meaning is supposed to be
shred, apparently unconnected with
homophonous root 'suck'. But the
FRE parallel is hard to ignore. YM
1992, p. 559 *T'ÓÓD*.

TUR *siliyor* /s i l/. Räsänen 421b.

with link, Swadesh 200

ALB *me* /m e/. From a PIE *mōt*,
approximately source of GER *mit*.
Huld 91. Mann 207.

ENG *with* /w i/ COGNATE WITH 'WIDE'.
OE *wið* 'against', PG *wiþer*, PIE
wi- 'apart' + comp. suffix *-tero-*.
Skeat. OED. Pokorny 1177.
Watkins *wi-*.

FRE *avec* /a v/. OF *avuec*, Pop. Lat. *ab
hoc* 'thence', Lat. *ab* 'from', PIE
apo. BW. Pokorny 53. Watkins
apo-.

GER *mit* /m i/. OHG *mit*, PG *medi*,
PIE root *medh* 'middle', from *me*.
Kluge *mit*. Pokorny 702.

HAW *me* /m e/. PCP *me*. PE.

LAT *cum* /k o/. Earlier *com*; prefix is
com and *co*. PIE *kom, ko*. Pokorny
612. EM *cum*. Walde 458. Watkins
kom.

NAV *bił* /i ɬ/. Postposition. YM 1980,
p. 83. YM 1992, p. 241.

TUR *ile* /i/ DERIVED FROM 'ONE'. From
birle 'unite'. Räsänen 171a.

woman noun, Swadesh 100

ALB *grua* /g r/ COGNATE WITH 'WIFE'.
Also 'wife'. PIE *g̑raH₂u-*, like
Greek *graûs* 'old woman', root
g̑erH₂ 'to age'. Mann 28, 69, 85,
99. Huld 66. Pokorny 473, 390.

ENG *woman* /w u/ DERIVED FROM
'WIFE'. OE *wīfman*, from *wīf*
'woman' + *man* 'human'. Skeat.
OED.

FRE *femme* /f a/ COGNATE WITH
'WIFE'. Also used for 'wife'. Lat.
fēmina 'female', from 'nursing',
-meno- participle of PIE root
dheH₁ 'suck'. BW. Pokorny 652.
EM *fecundus*. Watkins *dhē(i)-*.

GER *Frau* /f r/ COGNATE WITH 'FAR'.
OHG *frouwa* 'lady', PG *frōwō-*,
PIE *prōwo-*, root *per*. Pokorny 810.
Kluge *Frau*. Watkins *per-1*.

HAW *wahine* /h i n e/ COGNATE WITH
'WIFE'. Same word as 'wife.' PCP
wafine, PPN *fafine*, POc *(pa)pine*,
PCEMP *bai, (ba)b-in-ay*, PMP
bahi, (ba)b-in-ahi. Commonly *-hine*
in kinship terms, so the *wa-* is
reduplicative. PE. Blust.

LAT *mulier* /m u l/. Perhaps PIE *mel*
'grind'. Pokorny 716. EM *mulier*.
Walde 285*.

NAV *'asdzání* /dz á n/. '(Young)
woman', perhaps from *dzą́* 'having
a hole' (p. 166). YM 1992, p. 1015,
166.

TUR *kadın* /k a d/ LOAN. From 'lady',
from 'wife of the khan', from Sgd.
xwātūn 'queen', from *xwātāw*
'king'. Räsänen 157a.

woods noun, Swadesh 200

ALB *pyll* /p y ł/ LOAN, COGNATE WITH
'FULL'. Loan from Pop. Lat.
*padūlem 'swamp', Class. *palūdem*,
nom. *palūs*; PIE *plHoud*, from
pelH-eu. The root *pelH* may be the
'fill' word or the 'flow' word, and
indeed the two roots may be the
same. Mann 31. Huld 105. EM
palūs. Pokorny 799, 835.

ENG *woods* /w ʊ d/. OE *wudu*, from
widu, PG *widuz*, PIE *widhus* 'tree',
perhaps from *weidh* 'divide'. Skeat.
OED. Pokorny 1177. Watkins
widhu-.

FRE *bois* /b w a z/ LOAN. Borrowed
from Germanic, cf. Eng. *bush*. For
root, cf. *boisage*. Dauzat. BW.
Barnhart.

GER *Wald* /v a l/. OHG *wald*, PG
walþuz, PIE *welt* 'wild', root
perhaps *wel* 'hair'. Pokorny 1139.
Kluge *Wald*. Watkins *welt-*.

HAW *ulu lā'au* /u l u/. 'Group' (PPN

'ulu) + 'tree'. PE.

LAT *silva* /s i l/. Walde 507*.

NAV *tsintah* /ts i n/ DERIVED FROM
'TREE', COGNATE WITH 'STICK'.
Tsin 'tree' + *tah* 'among'.

TUR *orman* /o r m a n/. Räsänen 365b.

worm noun, Swadesh 200

ALB *krimb* /k r i m/. PIE *kʷr̥mi-*
'worm'. The initial consonant does
not match with the Germanic and
Lat. forms, but the parallel is so
close there must be some kind of
(analogical?) connection. Huld 82.
Mann 36. Pokorny 649. EM
uermis. Walde 523. Not in Janson.

ENG *worm* /w ə r/ COGNATE WITH
'RUB'. OE *wyrm*, PG *wurmiz*, PIE
wr̥mis, root *wer* 'turn'. Skeat.
OED. Pokorny 1152. Watkins
wer-3.

FRE *ver* /v ɛ r/ COGNATE WITH
'LAUGH'. LAT *vermis*. BW.

GER *Wurm* /v u r/ COGNATE WITH
'RUB', 'THROW'. OHG *wurm*, PG
wurmiz, PIE *wr̥mis*, root *wer*
'turn'. Kluge *Wurm*. Pokorny 1152.

HAW *koʻe* /k o ʔ e/. PPN *toke*. PE.

LAT *vermis* /w e r/ COGNATE WITH
'LAUGH'. PIE *wr̥mis*, root *wer*
'turn'. Pokorny 1152. EM *uermis*.
Walde 270. Watkins *wer-3*.

NAV *ch'osh* /tʃ' o ʃ/. 'Worm, bug'. PA
tsh'ish, PPA *tsh'ish*. YM 1992,
p. 115 *CH'OSH*.

TUR *solucan* /s o l u/ LOAN. Probably a
native word, but influenced by
Arm. /sołun/ 'creeping thing'.
Räsänen 425b. Dankoff, p. 168.

year noun, Swadesh 200

ALB *vit* /v i t/ COGNATE WITH 'OLD'.
PIE *wetes*, root *wet* 'year'. Huld
129. Pokorny 1175. Mann 96.

ENG *year* /j/. Angl. OE *gēr*, PG *jǣram*,
PIE *yērom*, base *yēr*, possibly root
ei 'go'. Skeat. OED. Pokorny 297.
Watkins *yēr-*.

FRE *année* /a n/. Extension (perhaps
already Pop. Lat. *annata*) on LAT
annus. PIE root *at*. BW.

GER *Jahr* /j/. OHG *jār*, PG *jǣram*, see
ENG. Kluge *Jahr*. Pokorny 297.

HAW *makahiki* /m a k a/ DERIVED FROM 'EYE'. PEP *matafiti* or *matasiti*. Could this be a compound of *maka* 'eye, beginning' + *hiki* 'celestial direction'? PE.

LAT *annus* /a n/. PIE *atnos*, root *at* 'go, year'. Pokorny 69. EM *annus*. Walde 41. Watkins *at-*.

NAV *hai* /h ai/. 'Winter, year'. PA *ẍay*. YM 1992, p. 237.

TUR *sene* /s e n/ LOAN. Loan from Arabic. Stachowski, v. 3, p. 70. Wehr 505a.

yellow adj, Swadesh 100

ALB *verdhë* /v e r/ LOAN. Loan from Lat. *viridis*, q.v. under 'green'. PIE root possibly *weis*. Mann 110. Huld 84.

ENG *yellow* /j ɛ l/. OE *geolo*, PG *gelwaz*, PIE *ĝhelwo-*, root *ĝhel*. Skeat. OED. Pokorny 429. Watkins *ghel-2*.

FRE *jaune* /ʒ o/. Lat. *galbinus* 'pale green', from *galbus*. That itself appears late, may be from PIE *ĝhel-*. Pokorny 429. EM *galbus*. Huld 84. Dauzat.

GER *gelb* /g e l/ COGNATE WITH 'SMOOTH'. OHG *gelo*, PG *gelwaz*, PIE *ĝhelwo*, root *ĝhel*. Pokorny 429.

HAW *melemele* /m e l e/. Base *mele* also means 'yellow'. PE.

LAT *flavos* /f l/ COGNATE WITH 'FLOWER', 'LEAF'. PIE *bhlēwos* 'blue, yellow', ext. of *bhel* 'shining'. Or possibly connected with *flos* 'flower'. Pokorny 160. EM *flauus*. Walde 212*. Watkins *bhel-1*.

NAV *litso* /ts o iː/. 'It is yellow', neuter. PA *tsugh*, PPA *tsuẍ*. YM 1992, p. 613 *TSOII* 'yellow, tan'. Pinnow 37.

TUR *sarı* /s a r ɯ/. Räsänen 403b.

you link (plural), Swadesh 200

ALB *ju* /j u/. PIE *wes*, same root as LAT, but not as ENG. Mann 36. Huld 78. Walde 209. Pokorny 514.

ENG *you* /j u/ COGNATE WITH 'THOU'. Same as 'thou'. OE *ēow* (obl.), PG *iuwiz*, PIE oblique *yu*. Skeat. OED.

Pokorny 514. Watkins *yu-1*.

FRE *vous* /v/. LAT *vos*, unstressed, see LAT. BW.

GER *ihr* /iː/. OHG *ir*, PWG *jiz*, PIE oblique *yu*, but with the vowel crossed with that of *wir* 'we'. Kluge *ihr*. Pokorny 514.

HAW *'oukou* /ʔ o u/ DERIVED FROM 'THOU'. *-kou* is plural suffix. Probably connected to singular *'oe*. PE.

LAT *vos* /w/. PIE *wŏs*, root *we*. Pokorny 514. EM *uos*. Walde 209. Watkins *wŏs*.

NAV *nihí* /n i/ DERIVED FROM 'THOU', COGNATE WITH 'WE'. Emphatic independent, subjective case. YM 1992, p. 930.

TUR *siz* /s/ DERIVED FROM 'THOU'. Plural inflexion of *sen* 'thou'; cf. *ben:biz* 'I:we'. Räsänen 424b.

References

Agresti, Alan. 1990. *Categorical Data Analysis*. New York: Wiley.

Agresti, Alan. 1992. A Survey of Exact Inference for Contigency Tables. *Statistical Science* 7:131–177.

Anttila, Raimo. 1972. *An Introduction to Historical and Comparative Linguistics*. New York: Macmillan.

Babitch, Rose Mary, and Eric Lebrun. 1989. Dialectometry as Computerized Agglomerative Hierarchical Classification Analysis. *Journal of English Linguistics* 22:83–90.

Barnhart, Robert K. (ed.). 1988. *The Barnhart Dictionary of Etymology*. Bronx: Wilson.

Baxter, William H., and Alexis Manaster Ramer. 1996. Review of Ringe 1992. *Diachronica* 13:371–384.

Beekes, Robert S. P. 1995. *Comparative Indo-European Linguistics: An Introduction*. Amsterdam: John Benjamins.

Bender, Marvin L. 1969. Chance CVC Correspondences in Unrelated Languages. *Language* 45:519–531.

Bengtson, John D., and Merritt Ruhlen. 1994. Global Etymologies. In *On the Origin of Languages*, ed. Merritt Ruhlen. 277–336. Stanford: Stanford Univ. Press.

Benveniste, Émile. 1935. *Origines de la formation des noms en indo-européen*. Paris: Adrien-Maisonneuve.

Berlin, Brent. 1994. Evidence for Pervasive Synesthetic Sound Symbolism in Ethnozoological Nomenclature. In *Sound Symbolism*, ed. Leanne Hinton, Johanna Nichols, and John J. Ohala. 76–93. Cambridge, Eng.: Cambridge University Press.

Bloch, Oscar, and W. von Wartburg. 1964. *Dictionnaire étymologique de la langue française*. Paris: Presses Universitaires de France. 4. éd., revue et augm. par W. von Wartburg.

Bloomfield, Leonard. 1984. *Language*. Chicago: Univ. of Chicago Press. Reprint of New York: Holt, Rinehart and Winston, 1933.

Blust, Robert. 1993. Central and Central-Eastern Malayo-Polynesian. *Oceanic Linguistics* 32:241–293.

Bomhard, Allan R. 1990. A Survey of the Comparative Phonology of the So-Called "Nostratic" Languages. In *Linguistic Change and Reconstruction Methodology*, ed. Philip Baldi. 331–358. Berlin: Mouton de Gruyter.

Boretzky, Norbert. 1975–1976. *Der türkische Einfluss auf das Albanische*. Wiesbaden: Harrassowitz.

Çabej, Eqrem. 1982. *Studime etimologjike në fushë të shqipes*. Tiranë: Akademia e Shkencave e RPS të Shqipërisë, Instituti i Gjuhësisë dhe i Letërsisë.

Campbell, Lyle. 1973. Distant Genetic Relationships and the Maya-Chipaya hypothesis. *Anthropological Linguistics* 15:113–135.

Campbell, Lyle. 1988. Review of Greenberg (1987). *Language* 64:591–615.

Campbell, Lyle. 1996. On Sound Change and Challenges to Regularity. In *The Comparative Method Reviewed*, ed. Mark Durie and Malcolm Ross. 72–89. New York: Oxford University Press.

Campbell, Lyle. 1999. *Historical Linguistics: An Introduction*. Cambridge, Mass.: MIT Press.

Christiansen, Tom, and Nathan Torkington. 1998. *Perl Cookbook*. Sebastopol, CA: O'Reilly.

Collinder, Björn. 1947. *La parenté linguistique et le calcul de probabilités*. Språkvetenskapliga Sällskapets i Uppsala Förhandlingar. Uppsala: Almqvist & Wiksells Boktruckeri.

Collinder, Björn. 1965. *An Introduction to the Uralic Languages*. Berkeley: Univ. of California Press.

Covington, Michael A. 1996. An Algorithm to Align Words for Historical Comparison. *Computational Linguistics* 22:481–496.

Cowan, H. K. J. 1962. Statistical Determination of Linguistic Relationship. *Studia Linguistica* 16:57–96.

Czekanowski, Jan. 1928. Na marginesie recenzji P. K. Moszyńskiego o książce: Wstęp do historji słowian. *Lud* ser. 2, vol. 7. Cited by Kroeber & Chrétien, 1937.

Dankoff, Robert. 1995. *Armenian Loanwords in Turkish*. Wiesbaden: Harrassowitz.

Dauzat, Albert. 1938. *Dictionnaire étymologique de la langue française*. Paris: Larousse.

Diffloth, Gérard. 1994. i: Big, a: Small. In *Sound Symbolism*, ed. Leanne Hinton, Johanna Nichols, and John J. Ohala. 107–114. Cambridge, Eng.: Cambridge University Press.

Dolgopolsky, A. B. 1986. A Probabilistic Hypothesis Concerning the Oldest Relationships among the Language Families in Northern Eurasia. In *Typology, Relationship and Time: A Collection of Papers on Language Change and Relationship by Soviet Linguists*, ed. Vitalij V. Shevoroshkin and T.L. Markey. Ann Arbor, MI: Karoma.

Drizari, Nelo. 1957. *Albanian-English and English-Albanian Dictionary.* New York: Frederick Ungar. Enlarged edition.

Durie, Mark, and Malcolm Ross (ed.). 1996. *The Comparative Method Reviewed.* New York: Oxford University Press.

Dyen, Isidore, A. T. James, and J. W. L. Cole. 1967. Language Divergence and Estimated Word Retention Rate. *Language* 43:150–171.

Elbert, Samuel H., and Mary Kawena Pukui. 1979. *Hawaiian Grammar.* Honolulu: University of Hawaii Press.

Embleton, Sheila M. 1986. *Statistics in Historical Linguistics.* Bochum: N. Brockmeyer.

Ernout, A., and A. Meillet. 1979. *Dictionnaire étymologique de la langue latine.* Paris: Klincksieck. 4. éd., 3. tirage, augm. d'additions et de corrections nouvelles par Jacques André.

Fox, Anthony. 1995. *Linguistic Reconstruction: An Introduction to Theory and Method.* Oxford: Oxford University Press.

Good, Phillip. 1995. *Permutation Tests.* New York: Springer. 2nd corr. printing.

Greenberg, Joseph H. 1966. Some Universals of Grammar with Particular Reference to the Order of Meaningful Elements. In *Universals of Language,* ed. Joseph H. Greenberg. 73–113. Cambridge, MA: MIT Press, second edition.

Greenberg, Joseph H. 1971. The Indo-Pacific Hypothesis. In *Linguistics in Oceania,* ed. Thomas A. Sebeok. Current Trends in Linguistics, 8, Vol. 1, 807–871. The Hague: Mouton.

Greenberg, Joseph H. 1987. *Language in the Americas.* Stanford: Stanford Univ. Press.

Greenberg, Joseph H. 1993. Observations Concerning Ringe's *Calculating the Factor of Chance in Language Comparison. Proceedings of the American Philosophical Society* 137:79–89.

Greenberg, Joseph H., and Merritt Ruhlen. 1992. Linguistic Origins of Native Americans. *Scientific American* 267:94–99.

Grimes, Joseph E., and Frederick B. Agard. 1959. Linguistic Divergence in Romance. *Language* 35:598–604.

Guy, J. B. M. 1980. *Glottochronology Without Cognate Recognition.* Canberra: Australian National Univ., Dept. of Linguistics.

Haile, Berard. 1926. *A Manual of Navaho Grammar.* Arizona: St. Michael's.

Hays, William L. 1981. *Statistics.* New York: Holt, Rinehart and Winston. Third edition.

Hinton, Leanne, Johanna Nichols, and John J. Ohala (ed.). 1994. *Sound symbolism.* Cambridge, Eng.: Cambridge University Press.

Hock, Hans Henrich. 1991. *Principles of Historical Linguistics.* Berlin: Mouton de Gruyer. Second edition.

Hock, Hans Henrich. 1993. Swallow Tales: Chance and the "World Etymology" MALIQ'A 'swallow, throat'. In *CLS 29: Papers from the 29th Regional*

Meeting of the Chicago Linguistic Society. Vol. 1, The main session, ed. Katharine Beals et al. Chicago. Chicago Linguistic Society.

Hogben, Lancelot. 1965. *The Mother Tongue*. New York: Norton.

Hoijer, Harry. 1945. *Navaho Phonology*. Albuquerque: Univ. of New Mexico Press.

Hony, H.C., and Fahir İz. 1984. *The Oxford Turkish-English Dictionary*. Oxford: Oxford Univ. Press. 3rd ed., A. D. Alderson and Fahir Iz.

Hopper, Paul J., and Elizabeth Closs Traugott. 1993. *Grammaticalization*. Cambridge, Eng.: Cambridge University Press.

Huld, Martin E. 1984. *Basic Albanian Etymologies*. Columbus, OH: Slavica.

Hymes, Dell H. 1956. Na-Dene and Positional Analysis of Categories. *American Anthropologist* 58:624–638.

Illich-Svitych, Vladislav M. 1971–1984. *Opyt sravnenija nostratičeskix jazykov*. Moskva: Nauka.

International Phonetic Association. 1996. Reproduction of The International Phonetic Alphabet. Revised to 1993, Updated 1996. Available from http://www2.arts.gla.ac.uk/IPA/ipachart.html.

International Phonetic Association. 1999. *Handbook of the International Phonetic Association: A Guide to the Use of the International Phonetic Alphabet*. Cambridge, Eng.: Cambridge University Press.

İz, Fahir, and H. C. Hony. 1955. *An English-Turkish Dictionary*. Oxford: Clarendon Press.

Jakobson, Roman. 1971. Why "Mama" and "Papa"? In *Selected writings*. 538–545. The Hague: Mouton, 2nd edition.

Janson, Bernd. 1986. *Etymologische und chronologische Untersuchungen zu den Bedingungen des Rhotazismus im Albanischen*. Frankfurt am Main: Peter Lang.

Jones, Alex I. 1989. Australian and the *Mana* Languages. *Oceanic Linguistics* 28:181–196.

Justeson, John S., and Laurence D. Stephens. 1980. Chance Cognation: A Probabilistic Model and Decision Procedure for Historical Inference. In *Papers from the 4th International Conference on Historical Linguistics*, ed. Elizabeth Closs Traugott, Rebecca Labrum, and Susan Shepherd. Amsterdam. Benjamins.

Kacori, Thoma. 1979. *A Handbook of Albanian*. Sofia: Sofia University.

Kari, James M. 1976. *Navajo Verb Prefix Phonology*. New York: Garland.

Katzner, Kenneth. 1986. *The Languages of the World*. London: Routledge.

Kessler, Brett. 1995. Computational Dialectology in Irish Gaelic. In *Seventh Conference of the European Chapter of the Association for Computational Linguistics*, 60–66.

Kessler, Brett, and Rebecca Treiman. 1997. Syllable Structure and the Distribution of Phonemes in English Syllables. *Journal of Memory and Language* 37:295–311.

Kiçi, Gasper. 1976. *Albanian-English Dictionary.* Tivoli: Tip. A. Picchi.

Kiçi, Gasper, and Hysni Aliko. 1969. *English-Albanian Dictionary.* Rome: Tipografica Editrice Romana.

Kluge, Friedrich. 1995. *Etymologisches Wörterbuch der deutschen Sprache.* Berlin, Walter de Gruyter. 23., erweit. Aufl., bearb. von Elmar Seebold.

Knuth, Donald E. 1973. *The Art of Computer Programming.* Reading, MA: Addison-Wesley. Second edition. Vol. 1, Fundamental Algorithms.

Koch, Harold. 1996. Reconstruction in Morphology. In *The Comparative Method Reviewed,* ed. Mark Durie and Malcolm Ross. 218–263. New York: Oxford University Press.

Kroeber, A. L., and C. D. Chrétien. 1937. Quantitative Classification of Indo-European Languages. *Language* 13:83–103.

Kruskal, Joseph B., Isidore Dyen, and Paul Black. 1973. Some Results from the Vocabulary Method of Reconstructing Language Trees. In *Lexicostatistics in Genetic Linguistics: Proceedings of the Yale Conference, Yale University, April 3–4, 1971,* ed. Isidore Dyen, 30–55. The Hague. Mouton.

Labov, William. 1994. *Principles of Linguistic Change.* Oxford: Blackwell.

Lass, Roger. 1997. *Historical Linguistics and Language Change.* Cambridge, Eng.: Cambridge University Press.

Lubotsky, Alexander. 1989. Against a Proto-Indo-European Phoneme *"a". In *The New Sound of Indo-European,* ed. Theo Vennemann. 54–66. Berlin: Mouton de Gruyter.

Manaster Ramer, Alexis. 1993. On Illič-Svityč's Nostratic Theory. *Studies in Language* 17:205–250.

Manaster Ramer, Alexis, and Christopher Hitchcock. 1996. Glass Houses: Greenberg, Ringe, and the Mathematics of Comparative Linguistics. *Anthropological Linguistics* 38:601–619.

Mann, Stuart E. 1977. *An Albanian Historical Grammar.* Hamburg: Helmut Buske.

Marck, Jeff. 1996. Kin Terms in the Polynesian Protolanguages. *Oceanic Linguistics* 35:195–257.

Matisoff, James A. 1990. On Megalocomparison. *Language* 66:106–120.

Meillet, Antoine. 1922/1950. *Les dialectes indo-européens.* Paris: É. Champion. Reproduction of the 1922 edition published by H. Champion.

Meillet, Antoine. 1925. *La méthode comparative en linguistique historique.* Oslo: H. Aschehoug.

Meillet, Antoine. 1926. *Linguistique historique et linguistique générale.* Paris: Champion. 2d éd.

Murdock, George Peter. 1959. Cross-Language Parallels in Parental Kin Terms. *Anthropological Linguistics* 1:1–5.

Nerbonne, John, and Wilbert Heeringa. 1998. Computationele vergelijking en classificatie van dialecten. *Taal en Tongval* 50:164–193.

Nerbonne, John, Wilbert Heeringa, and Peter Kleiweg. 1999. Edit Distance and Dialect Proximity. In *Time Warps, String Edits, and Macromolecules*, ed. David Sankoff and Joseph Kruskal. v–xv. Stanford, CA: CSLI.

Newport, Frank. 2000. Women's Most Pressing Concerns Today are Money, Family, Health and Stress. Available from http://www.gallup.com/poll/-releases/pr000317.asp.

Nichols, Johanna. 1996. The Comparative Method as Heuristic. In *The Comparative Method Reviewed*, ed. Mark Durie and Malcolm Ross. 39–71. New York: Oxford University Press.

Ohala, John J. 1994. The Frequency Code Underlies the Sound-Symbolic Use of Voice Pitch. In *Sound Symbolism*, ed. Leanne Hinton, Johanna Nichols, and John J. Ohala. 325–347. Cambridge, Eng.: Cambridge University Press.

Oswalt, Robert L. 1970. The Detection of Remote Linguistic Relationships. *Computer Studies in the Humanities and Verbal Behavior* 3:117–129.

Oswalt, Robert L. 1975. The Relative Stability of Some Syntactic and Semantic Categories. *Working Papers on Language Universals* 19:1–19.

Pedersen, Holger. 1903. Türkische Lautgesetze. *Zeitschrift der Deutschen morgenländischen Gesellschaft* 57:535–561.

Pinnow, Heinz-Jürgen. 1974. *Studie zur Verbstammvariation im Navaho*. Berlin: Mann.

Pokorny, Julius. 1959. *Indogermanisches etymologisches Wörterbuch*. Bern: Framcke.

Poser, William J., and Lyle Campbell. 1992. Indo-European Practice and Historical Methodology. In *Proceedings of the Eighteenth Annual Meeting of the Berkeley Linguistics Society, Feb. 14–17*, ed. Laura A. Buszard-Welcher, Lionel Wee, and William Weigel, 214–236. Berkeley, Calif. Berkeley Linguistics Society.

Price, Glanville. 1971. *The French Language: Present and Past*. London: Edward Arnold.

Pukui, Mary Kawena, and Samuel H. Elbert. 1986. *Hawaiian Dictionary*. Honolulu: Univ. of Hawaii Press. Rev. edition.

Räsänen, Martti. 1969. *Versuch eines etymologischen Wörterbuches der Türksprachen*. Lexica Societatis Fenno-Ugricae, 17, 1. Helsinki: Suomoloais-Ugrilainen Seura.

Ringe, Donald A., Jr. 1992. *On Calculating the Factor of Chance in Language Comparison*. Transactions of the American Philosophical Society, 82, pt. 1. Philadelphia: American Philosophical Society.

Ringe, Donald A., Jr. 1993. A Reply to Professor Greenberg. *Proceedings of the American Philosophical Society* 137:91–109.

Ringe, Donald A., Jr. 1995. The "Mana" Languages and the Three-Language Problem. *Oceanic Linguistics* 34:99–122.

Ringe, Donald A., Jr. 1996. The Mathematics of 'Amerind'. *Diachronica* 13:135–154.

Rosenfelder, Mark. 1998a. Deriving Proto-World with Tools you Probably Have at Home. Available from http://zompist.com/proto.html.

Rosenfelder, Mark. 1998b. How Likely are Chance Resemblances Between Languages? Available from http://zompist.com/chance.htm.

Ross, Alan S. C. 1950. Philological Probability Problems. *Journal of the Royal Statistical Society, Series B (Methodological)* 12:1.

Ross, Malcolm D. 1995. Proto-Oceanic Terms for Meteorological Phenomena. *Oceanic Linguistics* 34(2):261–304.

Ross, Malcolm D. 1996. Contact-Induced Change and the Comparative Method: Cases from Papua New Guinea. In *The Comparative Method Reviewed*, ed. Mark Durie and Malcolm Ross. 180–217. New York: Oxford University Press.

Roy, Dietrich. 1970. *Langenscheidt's Standard Dictionary of the English and German Languages*. Berlin. 6th edition.

Ruhlen, Merritt. 1994. *The Origin of Language: Tracing the Evolution of the Mother Tongue*. New York: John Wiley & Sons.

Salmons, Joe. 1992. A Look at the Data for a Global Etymology: *Tik* 'Finger'. In *Explanation in Historical Linguistics*, ed. Garry W. Davis and Gregory K. Iverson. 207–228. Amsterdam: Benjamins.

Sapir, Edward. 1911. Review of *The Chimariko Indians and Language* by R. B. Dixon. *American Anthropologist* 13:141–143.

Saussure, Ferdinand de. 1989. *Cours de linguistique générale*. Wiesbaden: Harrassowitz, vol. 1. Éd. critique par Rudolf Engler.

Séguy, Jean. 1973. La dialectométrie dans l'Atlas linguistique de la Gascogne. *Revue de linguistique romane* 37:1–24.

Sihler, Andrew L. 1994. *New Comparative Grammar of Greek and Latin*. New York: Oxford University Press.

Simpson, J. A., and Edmund S. Weiner (ed.). 1989. *The Oxford English Dictionary*. Oxford: Oxford University Press. 2nd edition.

Skeat, Walter W. 1980. *A Concise Etymological Dictionary of the English Language*. New York: Putnam's.

Stachowski, Stanisław. 1986. *Studien über die arabischen Lehnwörter im Osmanisch-Türkischen*. Worcław: Zakład narodowy imienia ossolin'skich wydawnictwo Polskiej akademii nauk.

Steingass, F. 1930. *A Comprehensive Persian-English Dictionary*. London: Allen.

Swadesh, Morris. 1950. Salish Internal Relationships. *International Journal of American Linguistics* 16:157–167.

Swadesh, Morris. 1952. Lexico-statistic Dating of Prehistoric Ethnic Contacts. *Proceedings of the American Philosophical Society* 96:452–463.

Swadesh, Morris. 1955. Towards Greater Accuracy in Lexicostatistic Dating. *International Journal of American Linguistics* 21:121–137.

Swadesh, Morris. 1956. Problems of Long-Range Comparison in Penutian. *Language* 32:17–41.

Sweetser, Eve. 1990. *From Etymology to Pragmatics*. Cambridge: Cambridge University Press.

Thomason, Sarah Grey, and Terrence Kaufman. 1991. *Language Contact, Creolization, and Genetic Linguistics*. Berkeley: Univ. of California Press.

Traugott, Elizabeth Closs. 1989. On the Rise of Epistemic Meanings in English: An Example of Subjectification in Semantic Change. *Language* 65:31–55.

Tucker, T. G. 1931. *A Concise Etymological Dictionary of Latin*. Halle (Saale): Niemeyer.

Villemin, F. 1983. Un essai de détection des origines du japonais à partir de deux méthodes statistiques. In *Historical Linguistics*, ed. B. Brainerd. 116–135. Bochum: N. Brockmeyer.

Walde, Alois. 1930. *Vergleichendes Wörterbuch der indogermanischen Sprachen*. Berlin: W. de Gruyter. Hrsg. und bearb. von Jukius Pokorny.

Warnow, Tandy. 1997. Mathematical Approaches to Comparative Linguistics. *Proceedings of the National Academy of Sciences of the United States of America* 94:6585–6590.

Warnow, Tandy, Donald Ringe, and Ann Taylor. 1996. Reconstructing the Evolutionary History of Natural Languages. In *Proceedings of the Seventh Annual ACM-SIAM Symposium on Discrete Algorithms*, 314–322. New York. ACM.

Watkins, Calvert. 1985. *The American Heritage Dictionary of Indo-European Roots*. Boston: Houghton Mifflin.

Wehr, Hans. 1979. *A Dictionary of Modern Written Arabic*. Wiesbaden: Harrassowitz. 4th ed. by J. Milton Cowan.

Weinreich, Uriel. 1966. *Languages in Contact*. The Hague: Mouton.

Wells, John. 1999. SAMPA: Computer Readable Phonetic Alphabet. Available from http://www.phon.ucl.ac.uk/home/sampa/home.htm.

Wickens, Thomas D. 1989. *Multiway Contingency Tables Analysis for the Social Sciences*. Hillsdale, NJ: Lawrence Erlbaum.

Wilkins, David P. 1996. Natural Tendencies of Semantic Change and the Search for Cognates. In *The Comparative Method Reviewed*, ed. Mark Durie and Malcolm Ross. 264–304. New York: Oxford University Press.

Winkler, Robert L., and William L. Hays. 1975. *Statistics: Probability, Inference, and Decision*. New York: Holt, Rinehart and Winston.

Woods, Anthony, Paul Fletcher, and Arthur Hughes. 1986. *Statistics in Language Studies*. Cambridge, Eng.: Cambridge University Press.

Young, Robert W., and William Morgan. 1980. *The Navajo Language*. Albuquerque: Univ. of New Mexico Press.

Young, Robert W., and William Morgan. 1987. *The Navajo Language*. Albuquerque: Univ. of New Mexico Press. Rev. edition.

Young, Robert W., and William Morgan. 1992. *Analytical Lexicon of Navajo*. Albuquerque: Univ. of New Mexico Press.

Index